COASTAL WETLANDS

This volume is part of a series of volumes on Coastlines of the World. The papers included in the volume are to be presented at Coastal Zone '91.

Volume Editor H. Suzanne Bolton
Series Editor Orville T. Magoon

Published by the
American Society of Civil Engineers
345 East 47th Street
New York, New York 10017-2398

ABSTRACT

This proceedings, *Coastal Wetlands*, contains papers presented at Coastal Zone '91, the Seventh Symposium on Coastal and Ocean Management held in Long Beach, California, July 8-12, 1991. This volume is part of a continuing series of volumes of *Coastlines of the World*. Some of the topics covered include environmental considerations, engineering and science; data gathering, and monitoring; legal, regulatory, and political aspects of coastal management; planning, conservation, and development; and public information and citizen participation. This volume provides the professionals, decision-makers, and the general public with a broad understanding of these subjects as they relate to wetlands.

Library of Congress Cataloging-in-Publication Data

Symposium on Coastal and Ocean Management (7th: 1991: Long Beach, Calif.)
    Coastal wetlands/volume edited by H. Suzanne Bolton.
      p.   cm.—(Coastlines of the world)
    "These papers were presented at Coastal Zone '91, the Seventh Symposium on Coastal and Ocean Management held in Long Beach, California, July 8-12, 1991"—T.p. verso.
    Includes indexes.
    ISBN 0-87262-840-X
    1. Coastal zone management—United States—Congresses. 2. Wetland conservation—United States—Congresses. 3. Estuarine area conservation—United States—Congresses. I. Bolton, H. Suzanne. II. Title. III. Series.
HT392.S94   1991
333.91'8'0973—dc20   91-4240
                                        CIP

FOREWORD

Coastal Zone '91, is the seventh in a series of multidisciplinary biennial symposia on comprehensive coastal and ocean management. Professionals, citizens and decision makers met for five days in Long Beach, California, to exchange information and views on matters ranging from regional to international scope and interest. This year's theme was entitled "A SPOTLIGHT ON SOLUTIONS, GLOBAL CONCERNS: MULTI-LEVEL RESPONSIBILITIES", emphasized a recurrent focus on practical coastal problem solving.

Sponsors and affiliates included the American Shore and Beach Preservation Association, American Society of Civil Engineers (ASCE), Coastal Zone Foundation, Department of Commerce, National Oceanic and Atmospheric Administration, as well as many other organizations (see title page). The range of sponsorship hints at the diversity of those attending the Coastal Zone '91 Symposium. The presence of these diverse viewpoints will surely stimulate improved coastal and ocean management through the best of current knowledge and cooperation.

This volume of the "Coastlines of the World" series is included as part of the Coastal Zone '91 Conference. The purpose of this special regional volume is to focus on the coastline and coastal zone managment of Wetlands.

Each volume of the "Coastlines of the World" series has one or more guest volume editors representing the particular geographical or topical area of interest.

All papers have been accepted for publication by the Volume Editors. All papers are eligible for discussion in the Journal of Waterway, Port, Coastal, and Ocean Engineering, ASCE and all papers are eligible for ASCE awards.

An eighth conference is now being planned to maintain this dialogue and information exchange. Information is available by contacting the Coastal Zone Foundation, P.O. Box 279, Middletown, California 95461, U.S.A.

Orville T. Magoon
Coastlines of the World
Series Editor

PREFACE

H. Suzanne Bolton[1]

This decade began with a Presidential mandate of no net loss of wetlands. The national attention prompting a Presidential mandate was stimulated by the efforts and recommendations of the National Wetlands Policy Forum. This forum was convened by the Conservation Foundation with funding provided by the Administrator of EPA, Lee Thomas, and by other supporters of the Foundation's Environmental Dispute Resolution program. Former Governor Thomas H. Kean of New Jersey chaired a 20 member forum, including "3 governors, a state legislator, and heads of state agencies; a town supervisor; chief executive officers of environmental groups and businesses; farmers and ranchers; and academic experts." (The National Wetlands Forum Final Report, 1988, p. vii).

The recommendations in the consensus report recognized the benefits of wetlands and acknowledged that inadequate protection of wetlands should not continue. The recommendations are extensive but can be grouped into three categories. These categories are: (1) protecting the resource; (2) improving the management and protection process; and (3) implementing the recommendations outlined by the report. The most often cited recommendation is the Forum's interim protection goal "to achieve no overall net loss of the nation's remaining wetlands base." (NWPF, p. 3).

The Forum report intensified public interest and pressures on the legislative and executive arms of Federal government for and against changing the status of wetlands protection.

The Federal resource protection and management agencies quietly reinforced and supported, within the Administration, many of the protective elements of the Forum recommendations. Federal and academic wetlands research, developed over the past twenty years, have awakened the scientific community to the ecological and economic value of the Nation's wetlands real estate. This research provides a foundation for the next generation of basic and applied work needed to refine the priority of wetlands ecosystems in the context of an ever developing coastal zone.

The rate of wetlands loss has decreased nationwide since the mid-seventies. This national trend can be traced to the implementation of state and Federal regulations and the increased public awareness and scientific understanding resulting from the research mentioned above. The increasing loss of coastal wetlands is not, however, accurately described by national statistics. Increased national economic pressures and demand for valuable coastal acreage as we enter the twenty first century will place those regulations and our limited coastal wetlands under siege.

The National Fish and Wildlife Service (FWS) assumes that existing coastal wetlands comprise

---

[1]Chief, Ocean and Coastal Services Branch, Constituent Affairs Division, Office of Legislation, Education and Outreach, National Oceanic and Atmospheric Administration.

about 10% of the national wetlands inventory; the addition of coastal margins (including submerged aquatic vegetation), not included in the National Wetland Inventory (NWI), increases this percentage only slightly. The salt marshes, the bogs, mud flats, rock intertidal zones, tidal and nontidal flats and marshes, and aquatic vegetated zones compose the nutrient base for most of the living marine resources supporting a large commercial and sports fishing industry. In addition to nurturing various life stages of marine and anadromous species, coastal wetlands also slow the rush of toxic pesticides, nutrients, and valuable sediment into estuaries and the coastal margins; provide temporary reservoir for flood waters; support wading birds and migrant flocks; moderate local climate; enhance recreational activities; and contribute aesthetics to the coastal margin.

While direct conversion of wetlands for farm usage is the greatest threat nationally to the existing wetlands base, wetlands are threatened by other direct and indirect activities of human populations, including water diversions, over nutrification, toxic contamination, and development.

The values and functions of wetlands vary amongst wetland types. The quantification of the values is a scientifically unsophisticated process and, economic assessments often reflect political biases against environmental attributes. While the scientific community recognizes the multiple values and functions of various wetland types, few researchers have conducted sufficient studies to significantly advance the understanding of the interrelationships within wetlands and between wetlands and adjacent habitats. Understanding these relationships is critical to improving our ability to place economic value on ecological products and services. The science of economics has not been creatively applied to the assessment of ecological values. Instances of the application of economics assessing ecological values have been biased by political and institutional forces prejudicial to a balanced assessment of the worth of wetlands. (Scodari, pp. 3–4, 1990)

The papers presented in the thirteen sessions focusing on the wetlands element of the 1991 Coastal Zone meeting cross the areas of wetlands research, monitoring, regulatory activities, and policy development. Eight papers explore the varied efforts now in progress to map, categorize, and assess temporal changes in wetlands acreage and type. Another series of papers review the multifaceted efforts of the Federal FWS to address wetlands problems in Louisiana. Aspects of functional value and global climate change impacts are also addressed, as well as case studies of wetlands restoration or enhancement efforts.

Despite the great interest demonstrated by the large response to the call for coastal wetlands papers, critical subject areas are not well represented in this forum, a characteristic of most wetlands conferences. The subject areas of concern either cross scientific disciplines or require that established methodology be applied in non-traditional manners. For instance, in the coastal regime, wetland life is dependent upon the physical factors of tides, currents, and ocean generated weather and climate events, yet these often controlling factors are frequently omitted or only cursorily integrated with the biological and chemical elements.

Only one paper addresses the economic assessment of wetlands values. Economists have not yet crossed disciplines to the extent that the terminology suited for tangible commodities can be applied satisfactorily to environmental attributes. Similarly, a coastal zone conference is not a traditional forum for a scholarly thesis on economics and, despite the applicability of this session for the subject, most economists prefer traditional economic conferences for introducing new approaches and methodologies. As these new methodologies gain acceptance, the need to cross disciplines and communicate directly with the coastal zone regulatory and management community will become not only apparent, but also critical to the economist.

Though far from well understood, most of the current work and the following papers review the species interaction and aspects of productivity within specific wetlands. Other functional values of wetlands are harder to study directly. They are even more difficult to assess by acknowledged scientific method. The ability of wetlands to serve as a natural sponge to absorb excess flood waters or contaminated runoff is often assessed in retrospect, after wetlands are lost. The study of the impact of global climatic change on wetlands requires the use of indirect or surrogate phenomena, such as subsidence or compaction, to simulate potential effects of sea level rise. The value of wetlands in moderating local climate has been acknowledged often indirectly by noting climate variation after the conversion of a wetland to other purposes. This functional value, though little studied, may be of great significance in areas of large wetland expanses such as the Louisiana or South Carolina coasts or the Everglades. Wetland loss, if properly assessed, may play a greater role in global change events than has been considered by scientists who have focused their efforts globally, to the exclusion or sublimation of local contributions.

The papers in this volume provide a glimpse into that ecosystem once considered only an eyesore or breeding place for mosquitoes and other undesirable vermin. As coastal real estate continues to increase in value, the threat of loss of the remaining coastal wetlands will accelerate. The existence of a knowledgeable and aggressive cadre of coastal zone managers, regulators, and concerned public will be critical to the assessment of wetland real estate for its inherent functional values. Only responsible coastal zone development can ensure "no net loss" of these highly productive and aesthetically appealing coastal ecosystems.

APPENDIX

The Conservation Foundation, *Protecting America's Wetlands: An Action Agenda,* Final Report of the National Wetlands Policy Forum, 1988. 89 pp.

Scodari, Paul F., *Wetlands Protection: The Role of Economics,* Environmental Law Institute, 1990. 69 pp.

CONTENTS

ASSESSMENT OF A WETLAND FUNCTIONAL VALUE

Alaska Coastal Wetlands Survey
Jonathan V. Hall 1
Relative Fisheries Values of Natural Versus Transplanted Eelgrass Beds, "Zostera
marina," in Southern California
Robert S. Hoffman 16
WET—Its Efficacy in Wetland Functional Assessment
M. Frances Eargle 31

WETLANDS: PRIORITY THREATS

Habitat Degradation and Fishery Declines in the U.S.
James R. Chambers 46
Eutrophication and its Effects on Coastal Habitats
R. Eugene Turner and Nancy N. Rabalais 61
Coastal Systems—On the Margin
Howard Levenson 75
Contaminants and Associated Biological Effects in Coastal and Estuarine
Ecosystems Near Certain Urban Areas
Bruce B. McCain, Sin-Lam Chan, Donald W. Brown, Margaret M. Krahn,
Robert C. Clark, Jr., O. Paul Olson, John T. Landahl, and Usha Varanasi 84
Effects of Fresh Water Development and Water Pollution Policies on the
World's River Delta/Estuary/Coastal Zone Ecosystems
Michael Rozengurt and Irwin Haydock 85
Recent Changes in Estuarine Wetlands of the Coterminous United States
Ralph W. Tiner 100

WETLANDS: SPECIAL AREAS OF FOCUS

Are Mature Wetland and Terrestrial Ecosystems in Carbon Steady State?
James Morris .... *
Effects of Accelerated Sea-Level Rise on Coastal Secondary Production
Robert J. Zimmerman, Thomas J. Minello, Edward F. Klima, and James M. Nance ... 110
Sea Level Rise—Global Processes, Local Decisions
Margaret A. Davidson .... *

---

*Manuscript not available at time of printing.

POLICY IMPLICATIONS FROM ESTUARINE COMPARISONS

Applying Lessons from NEP Pilot Projects to New NEP Designees—The Case of
Massachusetts Bays
  Enid Kumin and Patty Maraj-Whittemore .... *
Trends in Eutrophication in Five Major U.S. Estuaries—Management Implications
  Donald W. Stanley, Virginia Lee, and Alan Desbonnet 125
Compliance and Enforcement in Pretreatment Programs—A Comparison of
Three Programs
  Jon G. Sutinen 127
Approaches to Environmental Management Planning—the CZMA and the National
Estuary Program
  Mark T. Imperial, Donald Robadue, and Tim Hennessey 142

INNOVATIVE APPROACHES TO WETLANDS CREATION

Sediment Diversion as a Form of Wetland Creation in the Mississippi Delta
  Jack B. Moger and Kenneth J. Faust 157
Wetland-Creation Projects on the Toronto Waterfront
  S. Donald Speller .... *
An Innovative Approach to Wetlands Restoration in the San Francisco Bay
Estuary—The Montezuma Wetlands Project
  Stuart W. Siegel and James D. Levin 164
Maintaining Resources Through Re-Establishing Physical Process at Pescadero Marsh
  Thomas L. Taylor 176
New Wetlands in Southern California
  Andrea Bertolotti and James R. Crumpley 181

GUIDELINES FOR WETLAND ECOSYSTEMS

The Effects of Shoreline Erosion in Galveston Bay, Texas
  Robert W. Nailon and Edward L. Seidensticker 193
The Geologic Review Process—An Evolution of a Successful Interagency Program
  Brian J. Harder, James D. Rives, III, and Lynn H. Wellman 207
Guidelines for Tidal Restoration of Impounded Wetlands
  Joseph K. Shisler .... *

LOUISIANA WETLANDS STUDY

Louisiana Coastal Wetland Loss Study—An Overview
  James B. Johnson, Mary C. Watzin, and S. Jeffress Williams .... *
Results of Geologic Processes Studies of Barrier Island Erosion and Wetlands Loss
in Coastal Louisiana
  S. Jeffress Williams, Shea Penland, and Asbury H. Sallenger, Jr. 215
Landscape Simulation of Coastal Wetlands
  Thomas W. Doyle 226
Managing Louisiana Marshes—An Experimental Approach
  Edward C. Pendleton, A.L. Foote, and G.R. Guntenspergen 235

---

*Manuscript not available at time of printing.

WETLANDS MAPPING: NATIONAL PROGRAMS

The U.S. Fish and Wildlife Service's National Wetlands Inventory
Bill O. Wilen .... *
Overview of the Federal Coastal Wetlands Mapping Effort
Sari J. Kiraly 247
NOAA's CoastWatch—Change Analysis Program
James P. Thomas, Randolph L. Ferguson, Jerome E. Dobson, and Ford A. Cross 259
National Mapping Division Programs, Products, and Services That Can Support
Wetlands Mapping
Franklin S. Baxter 268
Monitoring Seagrass Distribution and Abundance Patterns
Robert J. Orth, Randolph L. Ferguson, and Kenneth D. Haddad 281

POLICY, POLITICS, AND LEGAL IMPLICATIONS
OF WETLAND MANAGEMENT

Defining Wetlands—A Balance of Policy, Practicality, and Science
David M. Ivester 301
Coastal Louisiana—Abundant Renewable Natural Resources—In Peril
William S. Perret and Mark F. Chatry 317
State Freshwater Wetlands Protection Legislation Leading to Section 404
Assumption—The New Jersey Experience
Marjorie Ernst and Ernest Hahn .... *
Survey of Coastal States Wetlands Mitigation Policy Procedure and Practices
William B. O'Beirne .... *
Impounding Our Nation's Coastal Wetlands—Win, Lose, or Draw?
J. Scott Feierabend .... *

WETLANDS MAPPING: LOCAL, STATE,
AND REGIONAL APPROACHES

Incorporating Global Positioning System Technology Into Coastal Mapping and
Research Efforts
William K. Michener, William H. Jefferson, David A. Karinshak, and
Charles Gilbert 332
Marine Resource Mapping and Monitoring in Florida
Kenneth D. Haddad and Gail A. McGarry 347
Thurston Regional Wetland Mapping Methodology
Steven W. Morrison 362
Mapping at the National Wetlands Research Center
James B. Johnston and Lawrence R. Handley 376

WETLAND MANAGEMENT: CASE STUDIES

Requiem for a Watermeadow—The Kaiser Site, Grays Harbor, WA
David E. Ortman 384

---

*Manuscript not available at time of printing.

Comparison of Two Marsh Management Plans in Louisiana
L. Phil Pittman and Ricardo W. Serpas 405
Restoring Prospect Island Wetlands—A Proposal for the Sacramento-San Joaquin
Delta of California
Fred Kindel 418
The Muzzi Marsh, Corte Madera, California, Long-Term Observations of a
Restored Marsh in San Francisco Bay
Phyllis M. Faber 424
Designing a Critical Area Program for Narragansett Bay—How Much Should We
Hope For?
Jennie C. Myers .... *
A Community-Based Approach to Mangrove Management in Ecuador
A. Bodero and R. Twilley .... *

RESTORATION/MITIGATION ISSUES IN ESTUARIES

Seagrass Decline—Problems and Solutions
Frederick T. Short, Galen E. Jones, and David M. Burdick 439
The Role of Scientific Research in Improving the Tijuana Estuary Tidal Restoration Plan
J. Zedler, C. Nordby, and T. Griswold .... *
Exploitation of Ecological Growth Models for Management of Seagrass Resources
Randall S. Alberte and Richard C. Zimmerman 454
Successful Tidal Wetland Mitigation in Norfolk, Virginia
Carvel Blair 463
Establishing Eelgrass Beds in California
Michael D. Curtis 477

THE WETLAND PLANNING PROCESS

Sea Management Patterns—Taxonomical Frameworks
Adalberto Vallega 590
Coastal Wetlands Management in Venezuela
Mirady Sebastiani, Alicia Villamizar, and Marie de Lourdes Olivo 503
Iterative Analysis for Planning of Mitigation Alternatives—Batiquitos Lagoon
Enhancement Project
Richard J. Mishaga and Steven L. Da Costa .... *

Subject Index 513

Author Index 516

*Manuscript not available at time of printing.

Alaska Coastal Wetlands Survey

Jonathan V. Hall[1]

Abstract

A determination of the current area of coastal wetland and deepwater habitat types in Alaska was made by the U.S. Fish and Wildlife Service (Service). The sampling study was designed to develop data on three categories of estuarine habitats: (1) Estuarine Intertidal Non-vegetated, (2) Estuarine Intertidal Vegetated, and (3) Estuarine Subtidal. These were the same categories previously used by the Service in a similar study completed in 1983 for the contiguous United States. Statistics on estuarine habitats that were truly national in scope would not be available until an accurate measurement of Alaska's estuarine types was made. Aerial photography was the data source used for the coastal habitat survey. A random sample of the coastal zone consisting of 1000 10.35 square kilometer plots was used to develop statistics with the desired level of precision.

Introduction

In 1984, the Service initiated a study designed to determine the current area of wetland and deepwater habitat types in Alaska. Statistical information on the extent and distribution of wetlands is important for the development or alteration of federal, state, and local wetland management programs and policies. Accurate wetland area data reduce conflicts by permitting more efficient planning and resource allocation. Resource managers are making decisions affecting coastal and other wetland types in Alaska without information on the extent of the resource.

----

[1]Regional Wetlands Coordinator, U.S. Fish and Wildlife Service, 1011 E. Tudor Rd., Anchorage, AK 99503

1

Data on the current status and trends of coastal and inland wetlands in the contiguous United States were developed by the Service in 1983 (Frayer et al. 1983). Following completion of this analysis, the Service initiated the Alaska survey in order to fill a sizable data gap. Wetland area statistics that were truly national in scope would not be available until an accurate measurement of Alaska's wetland resources was made.

The State of Alaska was not included in the earlier study for two reasons. First, suitable aerial photography needed to conduct the statistical sampling study was not available for large areas in Alaska. Secondly, wetland losses relative to the state's total wetland coverage have not been great. It would not have been practical to include Alaska in a study that was initiated primarily in response to the need for accurate information on wetland losses that were known to have occurred in the lower 48 states over the past 20 years. The Alaska study was designed to assess only current wetland area. Coastal and inland wetland area data for Hawaii, the only other state not sampled in the original statistical analysis, were developed by the Service and the University of Massachusetts in 1987 (Griffin et al. 1987).

The Alaska wetlands survey is designed to be completed in two phases: (1) the analysis of coastal wetlands; and (2) the analysis of inland wetlands. This report presents the results of phase 1 of the statewide study. The coastal survey is restricted to marine and estuarine wetlands as defined by the classification of Cowardin et al. (1979). Ocean derived salinity in these wetlands must be greater than 0.5 ppt at some time during the year in all years. Phase 2 of the statewide analysis will measure the extent of Alaska's freshwater wetlands. This includes coastal wetlands that are influenced by tidal water with salinities less than 0.5 ppt. Completion of this phase will depend on the availability of cooperative funding. The coastal wetlands survey is funded jointly by the Service and the National Oceanic and Atmospheric Administrations's National Marine Pollution Program Office.

Survey Area Description

Alaska's 55,000 kilometer coastline crosses an extensive range of both latitude and longitude. The latitudinal variation extends from 50°16'N on Amatignak Island in the Aleutian Island chain to 71°23'N at Point Barrow. Longitude ranges from 130°00'W at Camp Point in southeast Alaska to 172°28'E on Attu Island at the eastern end of the Aleutian chain (Orth 1967). With these distances, there is tremendous variability of climatic, geologic, and tidal conditions along the coast of Alaska. The dynamic aspects of the state's coastal environment have a significant effect on the distribution, structure, and function of coastal wetland habitats.

Marine waters bordering Alaska include the Beaufort, Chukchi, and Bering Seas, and the Gulf of Alaska (Pacific Ocean). Mean tidal amplitudes vary from 9.23 meters at Turnagain Arm in southcentral Alaska to 0.09 meters at Point Barrow on the Beaufort Sea coast (U.S. National Oceanic and Atmospheric Administration 1988). The mean range in Turnagain Arm is among the highest along the entire west coast of North and South America.

Variations of solar radiation (e.g., angle of incidence, intensity, diurnal and seasonal duration) across Alaska's 20° latitudinal span result in markedly different climates. Temperature extremes range from approximately - 60°F on the coastal plain of the North Slope to greater than 90°F in the southcentral and southeast regions. Mean annual precipitation in coastal areas of Alaska vary from less than 12.7 centimeters on the Arctic coast to more than 508 centimeters in several locations in southeast Alaska.

Sea ice in the Beaufort, Chukchi and Bering Seas is conspicuous evidence of the extremely low temperature regimes in the northern latitudes. The seasonal presence of ice is an important variable that influences coastal habitats in Alaska (Batten and Murray 1982). The scouring effects of ice on substrates in intertidal and shallow subtidal areas limits the establishment of vegetation. Ice cover also reduces phytoplankton production due to reduced light transmission and low temperatures (Redburn 1976).

Coastal morphology is another element of the physical environment in Alaska that is extremely diverse. Major coastal landforms include glacier-formed fjords, large deltas, barrier islands, rocky headlands, lagoons, bays, and inlets. Intricate shorelines in some regions, particularly the Alexander Archipelago (southeast Alaska), account for the extraordinarily high coastline length for the state. Approximately 63 percent of the tidal shoreline is in the southeast region where the coast is a labyrinth of fjords, islands, and bays (U.S. Army Corps of Engineers 1971).

Survey Procedure

The objective of the study is to develop area estimates for categories of coastal wetlands and deepwater habitats in Alaska. The coastal survey is a component of

a larger effort to produce statistics for all wetlands in
Alaska. The statewide survey is designed to develop
statistics that will, on the average, have a probability
of 90 percent that estimated totals are within 10 percent
of the true totals, by category. The statistical
methodologies used in the Alaska study are the same as
those used by the Service in the study on the status and
trends of wetlands and deepwater habitats of the 48
conterminous states (Frayer et al. 1983).

Aerial photography was the data source used for the
Alaska coastal wetlands survey. The mean year of the
photography is 1980, with over 90 percent of the photo
coverage acquired within three years of the mean year.
Most of the photography was 1:60,000 scale, color-infrared
imagery flown by the National Aeronautical and Space
Administration for the Alaska High-Altitude Aerial
Photography Program (AHAP). Black-and-white and true-color
imagery of scales between 1:16,800 and 1:76,000 was used
for coastal areas not covered by the AHAP program.

A random sample within Alaska's estuarine and marine
intertidal areas was used to develop the area statistics.
Estuarine areas were delimited and measured on all coastal
1:250,000 scale U.S. Geological Survey (USGS) maps. The
estuarine boundaries were drawn using the estuarine system
definition in the U.S. Fish and Wildlife Service's
"Classification of Wetlands and Deepwater Habitats of the
United States" (Cowardin et al. 1979). In general, the
estuarine system consists of deepwater tidal habitats and
adjacent tidal wetlands that are usually semi-enclosed by
land but have open, partly obstructed, or sporadic access
to the open ocean, and in which ocean water is at least
occasionally diluted by freshwater runoff from the land.

In order to determine the total number of plots
needed to develop statistics with the desired level of
precision, 101 coastal zone plots were sampled in a pilot
study. After analysis of the pilot study plots, it was
determined that 899 more plots would have to be randomly
selected. The total sample size for the coastal zone
stratum is 1,000 plots.

Each sample unit in the coastal wetlands survey
consists of a 10.35 square kilometer plot, 3.22 kilometers
on each side. The photointerpretation of the wetland types
was performed stereoscopically using a 6 power
magnification stereoscope. The smallest wetlands
delineated and classified were approximately .4 hectares
in size. The interpreted information was transferred from
the photos to overlays on the 1:63,360 scale base maps.

Measurements of the wetlands on the overlays were made using a video area measurement system.

In addition to the development of statewide coastal wetland statistics, the sample plot data were analyzed on a regional basis. This analysis was conducted to identify variations in the distribution of the wetland classes along Alaska's extensive coastline. Four coastal regions (Figure 1) were identified based on morphological and climatic factors. A listing of the number of sample plots by coastal zone region is shown in Table 1.

Table 1. Number of samples taken within each coastal zone region.

Region	Number of Plots
Southeast	342
Southcentral	314
Western	170
Northern	174

Four categories of wetland and deepwater habitat were analyzed in the Alaska coastal survey: (1) Marine Intertidal, (2) Estuarine Subtidal, (3) Estuarine Intertidal Non-vegetated and (4) Estuarine Intertidal Vegetated. These categories are the same as the coastal classes used in the study on the status and trends of wetlands and deepwater habitats of the 48 conterminous states. Identical types were selected to facilitate direct comparison of results and to determine national totals. The following section describes the four classes used in the coastal stratum. The classes are listed under definitions of the two ecological systems represented in the study.

Marine System

The Marine System consists of the open ocean overlying the continental shelf and its associated high - energy coastline. Marine habitats are exposed to the waves and currents of the open ocean and the hydrologic characteristics are determined primarily by the ebb and flow of oceanic tides. Salinities generally exceed 30 ppt, with little or no dilution except outside the mouths of estuaries. Shallow coastal indentations or bays without appreciable freshwater inflow, and coasts with exposed

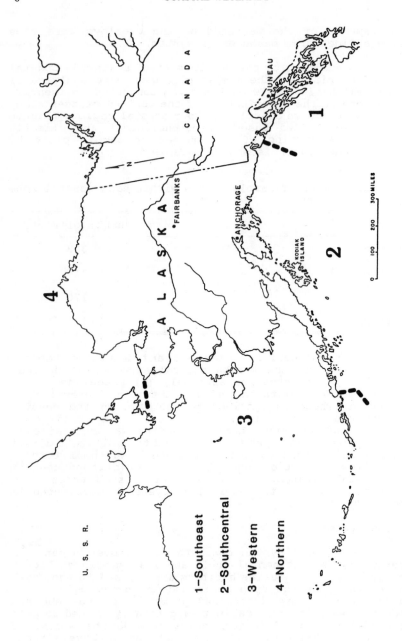

Figure 1. Coastal regions of Alaska

rocky islands that provide the mainland with little or no shelter from wind and waves, are also considered part of the Marine System because they generally support typical marine biota.

Marine Intertidal. This category includes all areas in which the substrate is exposed and flooded by tides, including the associated splash zone. Marine Intertidal habitats consist of rocky shores and unconsolidated shores (beaches, bars and flats). The substrates may be unvegetated or vegetated with aquatic beds such as Fucus and Halosaccion.

Estuarine System

The Estuarine System consists of deepwater tidal habitats and adjacent tidal wetlands that are usually semi-enclosed by land but have open, partly obstructed, or sporadic access to the open ocean. The ocean water is at least occasionally diluted by freshwater runoff from the land. The salinity may be periodically increased above that of the open ocean by evaporation. Along some low - energy coastlines there is appreciable dilution of sea water.

Estuarine Subtidal. All estuarine areas in which the substrate is continuously submerged.

Estuarine Intertidal Non-vegetated. All estuarine areas where: (1) the substrate is exposed and flooded by tides; and (2) vegetation in the form of emergents, shrubs, or trees is nonexistent or negligible. This category is used to estimate the extent of Alaska's estuarine mudflats. Estuarine rocky shores, beaches, and bars are also included, but the areal extent of these habitat types is small in comparison to mudflats. The substrates may be vegetated with algal aquatic beds such as Fucus or vascular aquatic beds such as Zostera.

Estuarine Intertidal Vegetated. This category includes all estuarine areas where: (1) the substrate is exposed and flooded by tides, and (2) vegetation in the form of emergents, shrubs, or trees is evident. This type is used to estimate the extent of areas commonly called "salt marshes" and "brackish tidal marshes." Forested estuarine wetlands are uncommon in Alaska and usually consist of a cover of dead trees over an emergent marsh layer. Estuarine wetlands dominated by shrubs are also a very small component of this category.

Results and Discussion

    Statewide coverage for the four wetland categories
analyzed in the coastal zone stratum are shown in Table 2.
These data represent the extent of the wetland classes in
1980, the mean year of the aerial photography used in the
study.

    The standard error for each entry is expressed as
a percentage of the entry (SE%). In general, the smaller
the SE%, the higher the reliability of the estimate.
Reliability can be stated at the 68 percent confidence
level as the estimate plus or minus the SE%/100 times the
estimate. For example, if an entry is one million hectares
and the SE% is 20, then there is 68 percent confidence that
the true value is between eight hundred thousand and 1.2
million hectares. An equivalent statement for 95 percent
confidence can be made by adding and subtracting twice the
amount to and from the entry.

Table 2.   Areal extent of coastal wetland and deepwater
           habitat classes in Alaska, 1980.

Class	Area (hectares)	SE%	% of Total
Marine Intertidal	18,574	29.2	0.2
Estuarine Subtidal	7,546,729	0.7	90.0
Estuarine Intertidal Non-vegetated	681,167	7.6	8.1
Estuarine Intertidal Vegetated	138,156	14.1	1.7
Total	8,384,626		100.0

    By excluding the area for the Marine Intertidal
class shown in Table 2, the total extent of estuarine
habitat in Alaska was determined to be 8,366,052 hectares.
This figure is slightly larger than the 8,087,179 hectare
estuarine total for the 48 conterminous United States and
Hawaii (Frayer et al. 1983; Griffin et al. 1987).
Combining the data from all areas of the United States
reveals a national total for estuaries of 16,453,231
hectares (Table 3).

Table 3. Area of estuarine habitats for the United States, including Alaska

Wetland Class	Area (hectares)	% of Total
Estuarine Subtidal	13,536,352	82.3
Estuarine Intertidal Non-vegetated	979,854	6.0
Estuarine Intertidal Vegetated	1,937,024	11.7
Total	16,453,230	100.0

The relative abundance and extent of estuarine habitats in Alaska and the combined data for all other coastal regions of the United States are compared in Figure 2. One of the most significant differences between the two data sets is the extent of Estuarine Intertidal Vegetated habitat. In Alaska, this wetland class comprises 138,156 hectares, or 1.7 percent of the state's estuarine total. The same class totals 1,798,868 hectares in the coastal areas of the conterminous United States and Hawaii, or 22.2 percent of the estuarine system. The area of Estuarine Intertidal Vegetated wetland in Alaska is 7.1 percent of the national total for this class.

Interestingly, Estuarine Intertidal Non-vegetated wetlands (flats, beaches, and bars) are more extensive in Alaska than in the remainder of the United States. Alaska's 681,167 hectares of this wetland class is more than twice the amount (298,687 hectares) found along the shoreline of the 23 other coastal states. The area in Alaska accounts for nearly 70 per cent of the national total.

Tables 4,5,6 and 7 present the area data for the southeast, southcentral, western, and northern coastal regions of Alaska. The Marine Intertidal class was excluded from the regional tables due to high standard errors for this limited category. Figure 3 shows the area contributions from each coastal zone toward the statewide totals for the three estuarine habitat types.

In all regions of Alaska, the Estuarine Subtidal class comprises most of the estuarine system. The coverage of this deepwater habitat category ranges from 70.9 percent of the total area in western Alaska to 96.9 percent in the southeast coastal region. The areal extent of Estuarine Subtidal habitat was greatest in the southeast where about

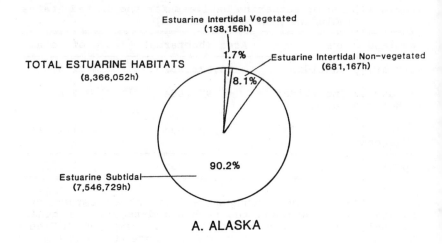

A. ALASKA

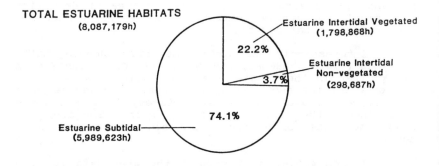

B. CONTERMINOUS U.S. AND HAWAII

Figure 2. Relative abundance and area (hectares) of estuarine habitat classes in (A) Alaska, and (B) the conterminous United States and Hawaii.

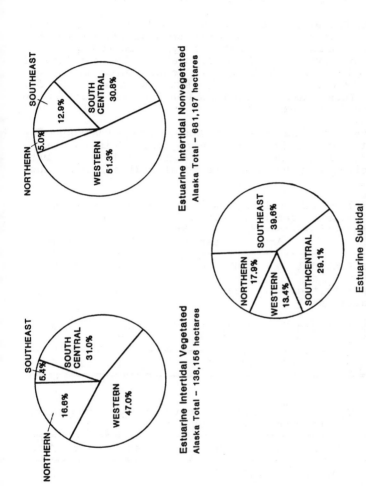

Estuarine Intertidal Nonvegetated
Alaska Total – 681,167 hectares

Estuarine Intertidal Vegetated
Alaska Total – 138,156 hectares

Estuarine Subtidal
Alaska Total – 7,546,729 hectares

Figure 3. Regional distribution of the three estuarine habitat classes in Alaska.

3 million hectares were measured. This accounts for about
40 percent of the total area of this type in Alaska.

The Estuarine Intertidal Non-vegetated wetland class
was most abundant in the western coastal region where
348,981 hectares of this habitat type occurred. The
presence of vast areas of mudflats along the shoreline of
the Yukon/Kuskokwim Delta greatly contributed to this
amount. Over 51 percent of the state's total area of
Estuarine Intertidal Non-vegetated wetland was found in the
western coastal area.

The Estuarine Intertidal Non-vegetated category was
least extensive in the northern region. This coastal
stretch had 339,38 hectares of flats, beaches, and bars
representing 2.4 percent of the total regional estuarine

Table 4.   Area of estuarine habitats in the southeast
coastal region of Alaska.

Class	Area (hectares)	SE%	% of Total
Estuarine Subtidal	2,987,024	0.6	96.9
Estuarine Intertidal Non-vegetated	88,071	18.8	2.9
Estuarine Intertidal Vegetated	7,505	26.6	0.2
Total	3,082,601		100.0

Table 5.   Area of estuarine habitats in the southcentral
coastal region of Alaska.

Class	Area (hectares)	SE%	% of Total
Estuarine Subtidal	2,195,224	1.6	89.7
Estuarine Intertidal Non-vegetated	210,176	14.2	8.6
Estuarine Intertidal Vegetated	42,876	26.7	1.7
Total	2,448,276		100.0

Table 6. Area of estuarine habitats in the western
coastal region of Alaska.

Class	Area (hectares)	SE%	% of Total
Estuarine Subtidal	1,010,770	4.0	70.9
Estuarine Intertidal Non-vegetated	348,981	10.8	24.5
Estuarine Intertidal Vegetated	64,864	22.7	4.6
Total	1,424,615		100.0

Table 7. Area of estuarine habitats in the northern
coastal region of Alaska.

Class	Area (hectares)	SE%	% of Total
Estuarine Subtidal	1,353,711	0.8	96.0
Estuarine Intertidal Non-vegetated	33,939	21.6	2.4
Estuarine Intertidal Vegetated	22,911	23.6	1.6
Total	1,410,560		100.0

area. Only about five percent of the statewide area of
this wetland class is located in the northern zone. Area
determinations for the southeast and southcentral coastal
regions were 88,071 hectares and 210,176 hectares,
respectively.

The Estuarine Intertidal Vegetated class,
representing brackish or salt marsh habitat, was unevenly
distributed among the four coastal areas. The greatest
extent of this type occurred in the western region. The
coastal marsh type totaled 64,864 hectares in this zone,
or about 47 percent of the total for the entire state. The
second highest areal extent for the vegetated category was
found in the southcentral region, where 42,876 hectares
were recorded. This represents 31 percent of the state's
total. The extensive salt marshes in Cook Inlet account
for a significant portion of this area.

The southeast region ranked last in coverage of Estuarine Intertidal Vegetated wetland. Although this area has the largest estuarine acreage and a shoreline that comprises about 63 percent of the state's total, the coastal marsh class covered only 7,505 hectares. This amount comprised 5.4 percent of Alaska's total for the vegetated category. The steep, rocky shorelines that are typical throughout the southeast region preclude development of large coastal marsh areas.

Even in the northern coastal region, the extent of Estuarine Intertidal Vegetated wetland is significantly greater than in southeast Alaska. The northern zone's 22,911 hectares is nearly three times the amount found in the southeast region. Approximately 17 percent of the total estuarine marsh area in the state occurs in the northernmost sample region.

Summary

The results reported are based on a statistically designed study of the current status of wetlands and deepwater habitats of Alaska. The coastal wetlands survey represents the first phase of the investigation of wetland extent for the entire state. The second phase will produce area estimates for the inland wetland types.

Data on the status of coastal and inland wetlands in the conterminous United States and Hawaii have been available for several years. Statistics that are truly national in scope required completion of an analysis of the extent of wetlands in Alaska. The Alaska coastal wetlands survey filled the data gap that existed for the marine and estuarine categories.

The analysis of 1000 sample plots shows that approximately 8.4 million hectares of estuarine habitat occur along the coastline of Alaska. This comprises about 51 percent of the nation's estuarine area. Alaska's 681,000 hectares of Estuarine Intertidal Non-vegetated wetlands (flats, beaches and bars) account for 70 percent of the national total for this category. The Estuarine Intertidal Vegetated wetland class (brackish marsh and salt marsh) has much less coverage. Approximately 138,000 hectares, or 7.1 percent, of the United States total for vegetated estuarine wetlands was measured by the Alaska coastal wetlands survey.

The data on regional wetland distribution show wide variation in the amount of coastal habitat types along Alaska's extensive coastline. Although the southeast region contains 63 percent of the state's shoreline, only

5.4 percent of the Estuarine Intertidal Vegetated wetland class occurs in that coastal area. The western region has the greatest coverage for the two estuarine intertidal categories. The Estuarine Subtidal class is most extensive in the southeast area.

References

Batten, A.R. and D.F. Murray. 1982. A literature survey on the wetland vegetation of Alaska. U.S. Army Corps of Engineers, Washington, D.C. Technical Report Y-82-2. 223p.

Cowardin, L.M., V. Carter, F.C. Golet, and E.T. La Roe. 1979. Classification of wetlands and deepwater habitats of the United States. U.S. Fish and Wildlife Service, Washington, D.C. FWS/OBS-79/31. 103p.

Frayer, W.E., T.J. Monahan, D.C. Bowden, and F.A. Graybill. 1983. Status and trends of wetlands and deepwater habitats in the conterminous United States. 1950s to 1970s. Colorado State University, Department of Forest and Wood Sciences, Ft. Collins, Colorado. 32p.

Griffin, C.R., L.H. Fredrickson, and K. Neithammer. 1987. Semi-annual report (January-July 1986). Management and habitat selection of endangered waterbirds in Hawaii. University of Massachusetts, Department of Forestry and Wildlife Management, Amherst, Massachusetts. 6p.

Orth, D.J. 1967. Dictionary of Alaska place names. U.S. Geological Survey Paper 567. U.S. Government Printing Office, Washington D.C. 1984p.

Redburn, D.R. 1976. Coastal ecosystems of Alaska: a preliminary review of the distribution and abundance of primary producers and consumers in the marine environment. Alaska Department of Environmental Conservation, Juneau, Alaska. 179p.

U.S. Army Corps of Engineers. 1971. National shoreline study, inventory report - Alaska region. U.S. Army Corps of Engineers, North Pacific Division, Portland, Oregon. 39p.

U.S. National Oceanic and Atmospheric Administration. 1988. Tide tables: high and low water predictions, west coast of North and South America. U.S. Department of Commerce, Washington, D.C. 232p.

Relative Fisheries Values of Natural
Versus Transplanted Eelgrass Beds (<u>Zostera</u> <u>marina</u>)
In Southern California

Robert S. Hoffman[1]

Abstract

Eelgrass (<u>Zostera</u> <u>marina</u>) vegetated aquatic areas in Southern California are relatively rare, with much of the eelgrass resource concentrated in the San Diego region. Mitigation projects which include eelgrass transplanting are becoming more numerous as increased pressure for development and/or shoreline modification projects occur in these areas. The functional value of these transplanted beds for fishery resources has been assumed to be equivalent to natural beds. However, little documentation exists on whether the same functional value will in fact eventually develop, and if it does, what the required length of time is for this process to occur.

A recent mitigation project, which included eelgrass transplanting in Mission Bay, was utilized as an opportunity to determine the fishery component of a transplanted bed versus that of an adjacent natural bed. While fishery utilization of transplant area was high beginning with the first sampling period (three months after transplant), approximately one year was necessary before the two areas reached habitat equivalency for fishery resources. Parameters evaluated for this determination included, number of fish captured, biomass of fish captured, species diversity, and percent similarity for the control site as compared to the transplanted site.

INTRODUCTION

The southern California geographic area (Point Conception to the Mexican border) has approximately 25 coastal bay/lagoon or harbor systems located along the coastline (Figure 1.) Of these 25 systems, only ten have
----
[1]Fishery Biologist, National Marine Fisheries Service, 300 S. Ferry St., Terminal Island, CA 90731

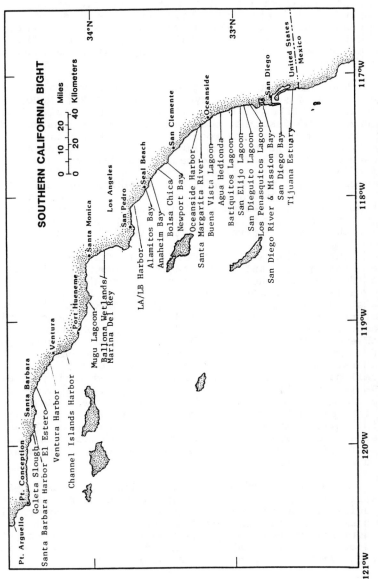

Figure 1. Coastal bay/lagoon and harbor systems in southern California.

eelgrass (Zostera marina) resources associated with them
(Table 1.). Most of these eelgrass resources are
concentrated in the San Diego area. Approximately 98
percent of the estimated total vegetated area is found in
three systems, Agua Hedionda Lagoon, Mission Bay, and San
Diego Bay.

The importance of seagrass as a marine resource has
been well described (Thayer et al., 1975; McRoy and
McMillian, 1977; Phillips, 1984; Kenworthy et al., 1988).
Ziemen (1982) described the various roles seagrass provides
to the surrounding environment. Those functional values
that are of particular importance to marine fishery
resources include, the formation of detrital food webs
associated with decaying leaves, the availability of a
diverse assemblage of epiphytic organisms on the leaves
which serve as potential forage sites, and the physical
structure of the leaves which provide a multi-dimensional
habitat that functions as a shelter or refuge area,
particularly for larval and juvenile life stages.

During the last decade, a considerable number of
eelgrass transplant projects have been initiated in
California and other areas of the United States. Many of
these have been conducted as mitigation for the loss of
existing eelgrass resources from dredging projects.
Unfortunately, few studies have been conducted which
demonstrate the time necessary for transplanted beds to
achieve habitat equivalent to natural beds. One such study
of eelgrass meadows in North Carolina indicated that a
natural bed was more suitable habitat for adult bay
scallops than a recently transplanted bed (Smith et al.,
1989). This functional difference appeared to be related
to the more dense vegetation found in the natural bed,
which reduced predation on the scallops by seabirds.
Similarly, Homziak et al. (1982) found that 60 to 70
percent of expected number of infauna species became
reestablished within 250 days after planting of an eelgrass
bed was completed.

A recent beach expansion project in Mission Bay,
California, and related eelgrass transplant activity,
provided an opportunity to investigate the recovery time
necessary for a transplanted bed to achieve the same
functional value as an adjacent natural bed. However, the
objective of this study was limited to only the comparison
of the fishery component of the two areas.

METHODS

The study site was located in the Sail Bay area of
Mission Bay, California (Figure 2.). This area had been

Table 1. Estimated Area of Eelgrass Coverage for
Coastal Systems in Southern California

System	Acres of Eelgrass	Percent of Total
Mugu Lagoon	< 1	0.07
Los Angeles Harbor	< 1	0.07
Alamitos Bay	2-6	0.27
Anaheim Bay	2-6	0.27
Newport Bay	1-3	0.13
Oceanside Harbor	< 1	0.07
Agua Hedionda Lagoon	70	4.75
Mission Bay	800	53.58
San Diego River Marsh	3-15	0.67
San Diego Bay	600	40.19
TOTAL	**1493**	**100.00**

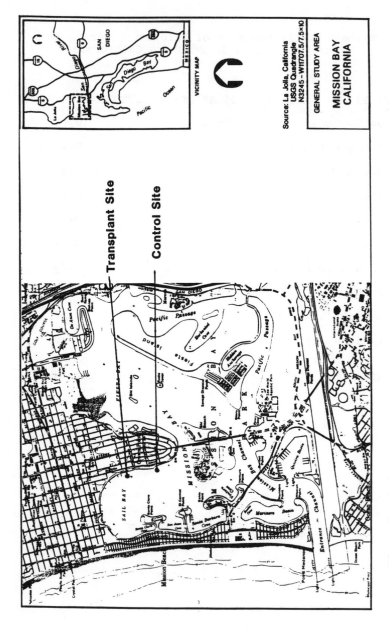

Figure 2.   Study area.

subjected to a beach expansion project in which dredged offshore sand deposits were deposited along an existing beach area. The nearshore area was heavily vegetated by eelgrass (Zostera marina) prior to the deposition of sand material. In order to mitigate for the expected loss of this resource, a transplant project was implemented within the impacted area following completion of dredging activities.

Two areas were selected to be sampled for a comparison of their respective fishery values. An area within the transplant zone where restoration work had been completed in March 1987, and an adjacent unaffected natural bed area (control) were the sites chosen for the sampling program. The study began in July 1987, and continued for a period of two years with sampling occurring on a quarterly basis.

A beach seine of 15.2 meters in length by 1.2 meters in depth with a mesh size of 5 mm. was used to capture juvenile and adult fish in the transplant and control areas. Sampling procedures consisted of two individuals wading out with the rolled up seine until a depth of approximately 1.5 meters was reached. The seine then was unrolled and hauled to shore. Two replicate seine hauls were completed at approximately the +1 foot (MLLW) tide height during the day and also again at night.

With some exceptions, all fish captured during the study were identified, measured (standard length) and weighed in the field. In the case of numerous small individuals of a particular species, groups of individuals were weighed and recorded. The treatment of members of the Gobiidae family deviated from this procedure since field identification of small individuals was not possible. Thus, all individuals from this family, with the exception of the yellowfin goby (Acanthogobius flavimanus), were placed in one group designated as "gobies". The majority of those individual designated as "gobies" were either arrow gobies (Clevelandia ios) or shadow gobies (Quietula y-cauda).

Analysis of the data included the comparison of the two stations on the basis of number of species, number of individuals, total biomass, and percent composition of sample by topsmelt, (Atherinops affinis). The Shannon-Wiener diversity index (H') in which $H' = -\sum_i P_i \log P_i$ where P_i is the proportion of individuals in the ith species was also calculated.

While the Shannon-Wiener index provides a measure of species richness it does not differentiate when changes in species composition occur. Horn (1980) suggested the use

of a percentage similarity index (PS) developed by
Whittaker and Fairbanks (1958) to overcome this potential
deficiency. Therefore, that index also was used to
evaluate the two sites sampled in this study.

PS values range from 0, when two samples contain no
species in common, to 100, when the two samples are
identical in both species composition and relative
abundance. The calculation of the PS index is as follows:
PS = $100(1.0-0.5\sum|P_{ia} - P_{ib}|)$ where P_{ia} is the proportion of
individuals in the $_i$th species of sample a and P_{ib} the same
for sample b.

Finally, given the usual non-normal distribution of
fish catch data, nonparametric statistical methods were
employed to determine whether pooled samples from each
sampling period for the control and transplant site were
significantly different. The Wilcoxon-Mann-Whitney test
was utilized for this comparison.

RESULTS

A total of 10,630 fish representing 25 species and
weighing 48,556 grams were collected in 32 seine hauls from
the transplant site (Table 2.). Individuals collected from
the control site consisted of a total of 8,516 fish
representing 22 species and weighing 32,022 grams (Table
3.). The dominant species for each of the stations were
very similar. The only obvious differences were larger
numbers of shiner surfperch (Cymatogaster aggregata)
present at the transplant station and a similar greater
number of individuals of California killifish (Fundulus
parvipinnis) at the control area.

While it appears that these results indicate that the
transplant site served as more valuable habitat to fishery
resources given the greater total number of individuals
caught, the results from the two stations are not directly
comparable. Because the beach at the transplant site was
widened at a 15:1 slope, as compared to the adjacent
control site slope of approximately 10:1, a much greater
area was sampled at the transplant station. A more
accurate interpretation of the results is depicted in
Figures 3A and 3B which represent the results on a unit
area sampled basis.

Numbers of fish and biomass of fish caught generally
were higher at the control site for the first four sampling
periods or first year of the study. However, Wilcoxon-
Mann-Whitney tests for each of the sampling periods only
indicated that the number of individuals caught during the
January 5, 1988, sampling period were significantly
different between the two sites (P<0.05).

Table 2.

MISSION BAY BEACH SEINE STUDY
Cumulative Total
Transplant Station
(8 sampling periods 7/87-4/89)

Common Name	Species	Individuals number	%	Biomass grams	%
topsmelt	Atherinops affinis	9324	87.7	29345	60.0
dwarf surfperch	Micrometrus minimus	299	2.8	3936	8.1
bay pipefish	Syngnathus leptorhynchus	270	2.5	180	0.4
gobies	Gobiidae	226	2.1	139	0.3
giant kelpfish	Heterostichus rostratus	150	1.4	2183	4.5
shiner surfperch	Cymatogaster aggregata	73	0.7	1784	3.7
spotted sand bass	Paralabrax maculatofasciatus	52	0.5	3453	7.1
california halibut	Paralichthys californicus	49	0.5	916	1.9
staghorn sculpin	Leptocottus armatus	47	0.4	264	0.5
black surfperch	Embiotoca jacksoni	37	0.4	3690	7.6
bay blenny	Hypsoblennius gentilis	35	0.3	128	0.3
slough anchovy	Anchoa delicatissima	17	0.2	50	0.1
barred sand bass	Paralabrax nebulifer	14	0.1	896	1.9
diamond turbot	Hypsopsetta guttulata	8	0.1	123	0.3
round stingray	Urolophus halleri	6	0.1	1268	2.6
queenfish	Seriphus politus	6	0.1	107	0.2
northern anchovy	Engraulis mordax	6	0.1	8	-
black croaker	Cheilotrema saturnum	3	-	5	-
spotted kelpfish	Gibbonsia elegans	2	-	18	-
yellowfin goby	Acanthogobius flavimanus	1	-	39	0.1
deepbody anchovy	Anchoa compressa	1	-	15	-
california killifish	Fundulus parvipinnis	1	-	4	-
kelp bass	Paralabrax clathratus	1	-	3	-
unidentifed rockfish	Sebastes sp.	1	-	1	-
unidentifed turbot	Pleuronichthys sp.	1	-	1	-
	TOTAL	10630		48556	

Table 3.

MISSION BAY BEACH SEINE STUDY
Cumulative Total
Control Station
(8 sampling periods 7/87-4/89)

Common Name	Species	Individuals number	%	Biomass grams	%
topsmelt	Atherinops affinis	6797	80.0	20407	63.7
dwarf surfperch	Micrometrus minimus	560	6.6	3479	10.9
gobies	Gobiidae	412	4.8	225	0.7
bay pipefish	Syngnathus leptorhynchus	307	3.6	177	0.6
giant kelpfish	Heterostichus rostratus	118	1.4	1596	5.0
staghorn sculpin	Leptocottus armatus	113	1.3	733	2.3
California halibut	Paralichthys californicus	46	0.5	732	2.3
bay blenny	Hypsoblennius gentilis	45	0.5	197	0.6
California killifish	Fundulus parvipinnis	34	0.4	182	0.6
spotted sand bass	Paralabrax maculatofasciatus	20	0.2	1678	5.2
black surfperch	Embiotoca jacksoni	15	0.2	1643	5.2
slough anchovy	Anchoa delicatissima	14	0.2	22	0.1
barred sand bass	Paralabrax nebulifer	8	0.1	413	1.3
yellowfin goby	Acanthogobius flavimanus	8	0.1	213	0.7
kelp bass	Paralabrax clathratus	5	0.1	42	0.1
shiner surfperch	Cymatogaster aggregata	5	0.1	20	0.1
black croaker	Cheilotrema saturnum	3	—	3	—
queenfish	Seriphus politus	2	—	10	—
round stingray	Urolophus halleri	1	—	236	0.7
deepbody anchovy	Anchoa compressa	1	—	3	—
diamond turbot	Hypsopsetta guttulata	1	—	10	—
salema	Xenistius californiensis	1	—	1	—
	TOTAL	8516		32022	

NUMBER OF FISH CAPTURED

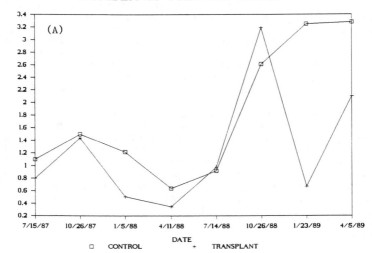

BIOMASS OF FISH CAPTURED

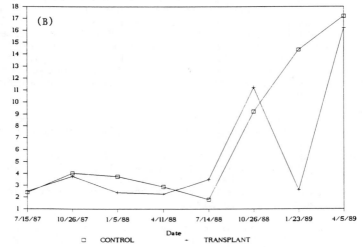

Figure 3. Time series analysis of number of fish (A) and biomass of fish (B) captured.

While little difference was noted between the stations in terms of number of species captured during each sampling period (Figure 4A.), species diversity (H') showed the greatest differences during the first three periods (Figure 4B.). This same tendency was noted for the percent similarity (PS) calculation. PS values were lowest during the first three sampling periods ranging from 43.6 to 58.6 (Figure 5A.). Those values varied from a low of 74.3 to a high of 84.1 during the last five sampling periods.

The comparison of A. affinis percent composition of the total catch (Figure 5B.) revealed extremely high values (95-99 percent) during the first three sampling periods for the transplant station as compared to those calculated for the control station (71-89 percent). Again, similar to the other preceding parameters analyzed, that percentage dropped during the fourth sampling period and more closely approached similar values generated for the control station for the remainder of the study.

DISCUSSION

The results of this study appear to indicate that the fishery component of a transplanted eelgrass bed in Mission Bay quickly achieved a high value relative to an adjacent natural bed. Similarly, species which composed the majority of fish caught at the transplant station were almost identical to those species at the control site. While a statistical analysis of the data only indicated significance between the transplant and control stations for the January 5, 1988, sample, the other parameters analyzed clearly showed a trend which suggested that habitat equivalency between the two stations did not occur for at least one year.

The extremely high percentage of the fish community composed of A. affinis at the transplant station during the first three quarterly sampling efforts is particularly interesting, since similar high values have also been seen at non-vegetated stations in San Diego Bay (Hoffman, 1986). This would suggest that during the early growth phase of an eelgrass transplant, the fishery component closely mimics a non-vegetated site.

Some qualification of these conclusions is warranted given the differences between the two stations. First, the difference in slopes between the two stations confounds any analysis. Although there is no evidence which would suggest that any of the bay associated fish species specifically select for one slope over another, assuming all other environmental factors are the same, this variable cannot be definitively ruled out as factor contributing to

NUMBER OF SPECIES CAPTURED

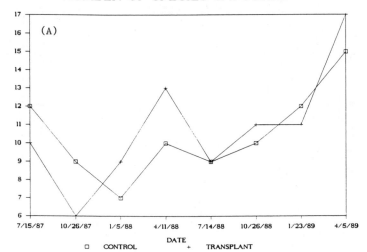

SPECIES DIVERSITY

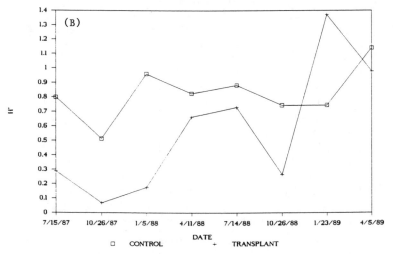

Figure 4. Time series analysis of number of species captured (A) and species diversity (B).

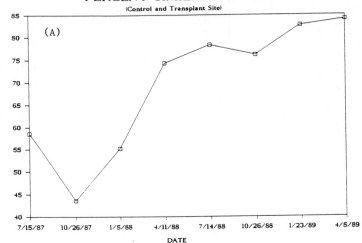

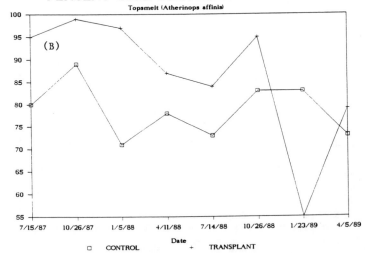

Figure 5. Time series analysis of percent similarity between control and transplant site (A) and percent composition of sample by transplant (B).

results found in this study. Second, beach seine sampling of dense eelgrass beds unquestionably has certain inefficiencies associated with the use of that sampling technique. Observations made of the seine as it moved to shore clearly indicated that the net did not remain on the bottom but tended to ride over the eelgrass. As a consequence, an undetermined number of individuals of the more demersal species were probably not captured. Therefore, the results for the control site are likely to underestimate those species relative to the transplant site. However, this situation only existed during the early stages of eelgrass growth in the transplant site. As the transplant site more closely resembled the control area in terms of density of eelgrass coverage (i.e., approximately one year after transplant), any differences in sampling efficiencies between stations was assumed to be minimal.

The results of this study appear to provide supporting evidence that eelgrass transplanting is an effective mitigation measure to replace beds lost from construction activities. Not only does the vegetative cover become reestablished, but the fishery values associated with natural beds also recover rapidly in transplanted beds.

ACKNOWLEDGMENTS

This study could not have been completed without the field assistance of the following individuals: Richard Nitsos from the California Department of Fish and Game, Keith Merkel from Pacific Southwest Biological Services, and Martin Kenney from the U.S. Fish and Wildlife Service.

Literature Cited

Hoffman, R.S. 1986. Fishery utilization of eelgrass (Zostera marina) beds and non-vegetated shallow water areas in San Diego Bay. NMFS, SWR Admin. Rept. SWR-86-4, 29 p.

Homziak, J., M.S. Fonseca, and W.J Kenworthy. 1982. Macrobenthic community structure in a transplanted eelgrass (Zostera marina) meadow. Mar. Ecol. Prog. Ser. 9: 211-211.

Horn, M.H. 1980. Diel and seasonal variation in abundance and diversity of shallow-water fish populations in Morro Bay, California. Fish. Bull. 78(3): 759-770.

Kenworthy, W.J., G.W. Thayer, and M.S. Fonseca. 1988. The utilization seagrass meadows by fishery organisms. In D.D. Hook et al. (editors), The ecology and

30   COASTAL WETLANDS

Roy, C.P. and C. McMillian. 1977. Structure, function
and classification seagrass communities. In C.P. McRoy
and C. Hefferich (editors), Seagrass ecosystems, a
scientific perspective, p. 52-58. Marvel Dekker, N.Y.

Phillips, R.C. 1984. The ecology of eelgrass meadows in
the Pacific Northwest: A community profile. U.S.
Fish Wildl. Serv. FWS/OBS-84/23. 85pp.

Smith, I., M.S. Fonseca, J.A. Rivera, and K.A. Rittmaster.
1989. Habitat value of Natural versus recently
transplanted eelgrass, Zostera marina, for the bay
scallop, Argopecten irradians. Fishery Bulletin
(1):189-196.

Thayer, G.W., S.M. Adams, and M.W. LaCroix. 1975.
Structural and functional aspects of a recently
established Zostera marina community. In L.E. Cronin
(editor), Estuarine research, vol. 1, p. 517-540.
Acad. Press, N.Y.

Whittaker, R. H., and C.W. Fairbanks. 1958. A study of
plankton copepod communities in the Columbia Basin,
southeastern Washington. Ecology 39:46-65.

Zieman, J.C. 1982. The ecology of the seagrasses of south
Florida: A community profile. U.S. Fish Wildl. Serv.
Biol. Serv. Program. FWS/OBS-82/85. 158pp.

WET-its efficacy in wetland functional
assessment

M. Frances Eargle [1]

Abstract

    The acquisition of knowledge related to wetland
functional attributes provides an integral database for
the management of coastal ecosystems. Inasmuch as the
knowledge of functional attributes is a prerequisite
for wise management of these ecosystems, the
substantive ability to gain knowledge of wetland
functions remains complex. Predictive models can be
used to explore relationships between variables and to
derive databases for management and policy initiatives.
This study evaluates the Wetland Evaluation Technique
(WET) as an appropriate methodology for deriving
wetland functional attributes among various coastal
wetland ecotypes. The predictive validity of the
methodology was analyzed by comparing long-term
monitoring data and field observations with the
predictions derived from the WET analysis.
Sensitivities and specific tasks which affect the
performance of the analysis in these wetland ecosystems
is discussed.

Introduction

    The concept of 'sustainable' development has been
a prevalent axiom and driving force in resource
management and environmental protection. Regardless of
how one defines 'sustainable' development, an
underlying implication remains that an ecological
carrying capacity should be factored within
economic-environmental trade-offs. The speed, overall
acceptance, and likelihood of development interests in
the coastal zone escalates the need to gain substantive
information on these wetland ecosystems. Without
knowledge of ecological relationships, 'wise'
decisions cannot be weighed and trade-offs assessed

---

[1]Sea Grant Fellow, U.S. House of Representatives, Merchant Marine & Fisheries Committee, Washington, D.C.

against the growing consequences of economic activity.
Along with these catapulting economic and social forces
determining the course of management, there will be
equal demand to increase ecological information on
these fragile ecosystems.

The mosaic, an ancient artform, provides an
important analogy for understanding the issues involved
in coastal wetlands management. The consummate image of
the mosaic is formed not only by the juxtaposition of
the individual pieces, known as tessara, but also the
grout between them. The complexity of the mosaic can
be examined with closer detail whereby discreet
patterns emerge within the consummate image.
Similarly, within a wetland ecosystem, microtopography,
hydrology, and biota are tightly coupled to form a
consummate "image", we call wetland function.

Effective management of these fragile ecosystems
currently requires that the physical and biological
integrity of these systems remain intact, while
maintaining a "sustainable" level of development.
Drawing again on the mosaic analogy, coastal resource
agencies are forced with the dilemma and ultimately
faced with decisions, such as- which "pieces" can be
sacrificed without jeopardizing the integrity of the
system?.

Gathering data on the 'pieces' (wetland function)
and its role in maintenance of these systems has been
costly and time-consuming. Predictive models, however,
can be used to approximate and extrapolate wetland
functional attributes and its role in maintenance of
coastal ecosystems.

This study analyzed one such of these models the
Wetland Evaluation Technique (WET). WET ranks the
probability of a particular function using HIGH,
MODERATE, or LOW categories of measurement within
various categories of wetland function. For a model to
be an effective predictor and therefore an integral
component of wetlands management, the model must
possess both scientific accuracy and reproducibility
among investigators.

The objective this analysis was two-fold: the
model was tested for its scientific efficacy to predict
wetland function among a sub-set of complex wetland
systems, estuarine and bottomland hardwoods (blh), as
representative ecosystems found in coastal
environments. Secondly, the model was tested for its
reproducibility among various investigators for
consistency among individual results.

The scientific validity of the WET methodology was tested by comparing the predictions derived from the model with long-term monitoring databases and field observations. The two bottomland hardwood sites were compared/contrasted by using the regionalized version of WET, known as the WET-BLH.

Individual data sets were compared and contrasted among a graduate wetlands ecology class at the University of South Carolina to test for reproducibility of the model. The results of the individual analyses were run through an overlay analysis within the model to compare and contrast results.

Description of the Model

The WET method developed by Paul Adamus (1987) evaluates groundwater recharge/discharge, flood storage, shoreline anchoring and dissipation of erosive forces, sediment trapping, nutrient retention and removal, food chain support (production export), and habitat for fish and wildlife.

The technique uses predictors or integrated variables which are assumed to be correlated with the biological, physical, and chemical characteristics of the wetland. The technique poses a series of questions related to each predictor, whereby the investigator responds with an answer of YES, NO, OR UNCERTAIN. Responses to the questions are analyzed through a series of interpretive keys which relate the predictors to particular wetland functions. The interpretative analysis, (similar to a taxonomic key) assigns a rating of HIGH, MODERATE, and LOW for each category of function. Functional attributes are assessed for both the EFFECTIVENESS and OPPORTUNITY. EFFECTIVENESS is defined as the capability of the wetland to perform a function due to its physical, chemical, or biological characteristics. OPPORTUNITY assesses the opportunity of a wetland to perform a function to its level of capability. For example, OPPORTUNITY may be viewed in this example. Given the location of a wetland in a particular watershed, does the wetland possess the ability to provide a floodflow function? An example of EFFECTIVENESS, could be viewed- given the morphology and hydrologic regimes of a particular wetland, how well does it perform that function in intercepting peak high flooding events downstream of the assessment area? The effectiveness rating usually would be viewed as the most important rating within the evaluation and represents the "bottomline" of the assessment.

Study Sites

The model was evaluated among three wetland types-
two bottomland hardwoods and one marsh/estuarine
ecosystem.   Two bottomland hardwood ecosystems were
selected   for   the   analysis   which   represent   two
hydrologically contrasting areas, both which originate
in the Atlantic Coastal Plain physiographic province.
The marsh/ estuarine site was selected as a control and
to determine the effectiveness in evaluating function
estuarine functional values.

All study sites were selected because they provide
pristine, idealized wetlands of known high value. The
assumption being that if WET could predict a high
rating of known high quality wetlands, it could also
predict levels of function with disturbed conditions.

North Inlet, located along the central Coastal
Plain of the Atlantic Seaboard is the selected example
of pristine marsh-estuarine ecosystem.   North Inlet
consists of approximately 5,200 acres of pristine salt
marsh (See Figure 2). This site has been established
as a research facility operated by the University of
South Carolina and Clemson University. This area has
also been designated by the National Science Foundation
as a long-term monitoring site (LTR).

The North Inlet Estuary is a vertically mixed
marsh-estuary system with a single outlet to the
Atlantic Ocean.   The estuary has three major tidal
creeks, (Debidue, Town, and Jones) with numerous small
tributaries. The salinity in these tidal creeks seldom
fall below 32 parts per thousand.   High salinity
supports dominant communities of smooth cordgrass
(Spartina   alterniflora)   with   S.   patens,   S.
cyansuroides, and Juncus roemerianus. Also occurring
within various higher salinity zones are dominant
halophytes such as Salicornia, which occur in high
abundance.

The two bottomland hardwood sites are formed by a
complex mosaic of microtopography, hydrology, and
vegetation. The Congaree Swamp represents one of the
last examples of virgin bottomland hardwoods along the
major floodplains of the southeastern United States.
Congaree Monument has recently been designated by
UNESCO as part of a network of biosphere reserves.
Congaree Swamp is noted for its record large trees--5
national record trees and 17 state records.

The Congaree Swamp is a floodplain forest formed
just below the fall line originating in the upper
portion of the coastal plain.   The swamp is formed

along the Congaree River, part of the Santee River Basin, whose watershed originates along the eastern Blue Ridge Mountains in western North Carolina and extends for approximately 300 miles southeasterly to the coast of South Carolina between Charleston and Winyah Bay. The Congaree National Monument consists of approximately 15,000 acres of riverine bottomland hardwoods. This site provides an example of virgin bottomland hardwoods, which originally existed along major floodplains in the southeastern United States. (See Figure 2)

Flooded periodically, the hydrology become the driving force in the maintenance of floodplain dynamics. The close coupling of the river and floodplain, allows an open exchange of water between the river and floodplain. Other sources of water contributing flow into the swamp originate from: overland flow, tributaries, groundwater, and direct precipitation.

Vegetative communities at Congaree respond to the diverse landforms and hydrologic regimes formed within braided sloughs, guts, ridges, and levees. Gaddy et al. (1973) identified 11 plant communities within the AA, which are determined by subtle changes in landform, elevation, and hydroperiod variation.

Francis Beidler Forest which consists of a 5,000 acre tract, located at Four Holes Swamp, preserves the largest stand of virgin cypress-tupelo swamp in the world. The site is jointly managed in perpetuity by the National Audubon Society and the Nature Conservancy.

Four Holes Swamp is a low-flow blackwater river swamp that originates in the Lower Coastal Plain and runs approximately 60 miles. The swamp has confluence with the Edisto River and eventually runs into the Atlantic Ocean in the southern portion of the state.

The Four Holes Swamp differs hydrologically from the Congress swamp, in being primarily a stream-fed system formed by a network of springs, overland run-off and precipitation. Water flows through various braided meanders with no unbroken channel flow, and exits the assessment area through a constricted outlet.

Beidler plant communities are significantly influenced, by the length of hydroperiod. The plant communities identified by Porcher (1981) include: mixed mesophytic forest, seepage bogs, swamp forest, hardwood bottoms, and ridge bottoms. Because there is a higher dominance of cypress and tupelo, Beidler more closely resembles a southern deepwater swamp.

ASSESSMENT PROCEDURES

Background

The WET analysis consists of a series of questions related to predictors or integrated variables which relate to wetland function. The responses to these questions are submitted through a matrix of threshold criteria or interpretative keys which yield a HIGH, MODERATE, LOW, or UNCERTAIN rating for each function. WET utilizes a series of steps to conduct the analysis which organizes the available databases and specifies the parameters to study.

A critical step in the analysis is "bounding" the assessment area (AA). Drawing the "red line" is probably the most critical portion of the procedures, and yet the most problematic (see discussion and conclusion). Hydrologic connectivity is the critical element in "bounding" the assessment area. However, if the hydrology criteria is met, geopolitical boundaries can be used to bound the assessment area.

Long-term monitoring data, on-site field observations, GIS data, cartographic data, and expert opinion was utilized to conduct the WET analysis. A computerized analysis was utilized to run the derive the functional ratings, provided in the model. The predictions derived from the WET analysis were then compared and contrasted with actual data collections.

The bottomland hardwood sites were also compared using a regionalized version (WET-BLH) (Adamus 1987), specifically designed for assessment in blh tracts. The results of these evaluations were also compared and contrasted.

Wetland functions were summarized in the following output of the WET analysis in Figures 3,4, and 5.

Figure 3. Summary of evaluation results for Congaree

Figure 4. Summary of evaluation results for Beidler

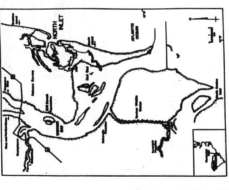

Figure 2. Location of North Inlet Estuary in Georgetown, SC

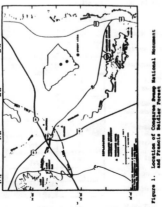

Figure 1. Location of Congaree Swamp National Monument and Francis Beidler Forest

```
•••••••••••••••••••••••••••••••••••••••••••••••••••
        Summary of Evaluation Results for "biNinlet"
•••••••••••••••••••••••••••••••••••••••••••••••••••

                                    Social
                                 Significance  Effectiveness  Opportunity

Ground Water Recharge                  L             •             •
Ground Water Discharge                 L             L             •
Floodflow Alteration                   L             L             •
Sediment Stabilization                 M             •             •
Sediment/Toxicant Retention            M             M             M
Nutrient Removal/Transformation        L             M             M
Production Export                       •             M             •
Wildlife Diversity/Abundance           M             •             •
Wildlife B/A Breeding                   •             L             •
Wildlife B/A Migration                  •             M             •
Wildlife B/A Wintering                  •             M             •
Aquatic Diversity/Abundance            M             M             •
Uniqueness/Heritage                    M             •             •
Recreation                             M             •             •

Note:  "H" = High, "M" = Moderate, "L" = Low, "U" = Uncertain, and
"•"'s identify conditions where functions and values are not evaluated.
```

Figure 5. Summary of evaluation results for North Inlet

FUNCTION	North Inlet NI	Beidler BE	Congaree CO
Groundwater Recharge			
WET	L	L	L
Actual Data	L	L	L
Groundater Discharge			
WET	L	M	M
Actual Data	L	M	M
Floodflow Alteration			
WET	L	M	M
Actual Data	M	M	M
Sediment Stabilization			
WET	M	M	M
Actual Data	M	M	M
Sediment/Toxicant Retent.			
WET	M	M	M
Actual Data	M	M	M
Nutrient Retention/Trans.			
WET	M	M	M
Actual Data	M	M	M
Production Export			
WET	M	M	M
Actual Data	M	M	M
Wildlife Diversity			
Breeding			
WET	L	M	M
Actual Data	L	M	M
Migration			
WET	M	M	M
Actual Data	M	M	M
Wintering			
WET	M	M	M
Actual Data	M	M	M
Aquatic Diversity			
WET	M	M	M
Actual Data	M	M	M

H=HIGH   M=MODERATE   L=LOW

Figure 6. Comparison of WET ratings with actual data

Figure 6. WET-BLH summary evaluation for Francis Beidler

Enter ratings (H, M, or L) in all boxes below.

	Significance	Effectiveness	Opportunity
Ground Water Discharge	M	H	
Flood Flow Alteration	L	H	L
Sediment Stabilization	L		L
Sediment/Toxicant Retention	L	H	H
Nutrient Retention/Transformation	L	H	H
Production Export		H	
Shellfish Habitat	M	H	
Finfish Habitat	M	M	
Wildlife Habitat	H	H	
Uniqueness/Heritage	M		

Figure 7. WET-BLH summary evaluation for Congaree Swamp

Groundwater Recharge

Peizometric data collected by Patterson et al (1985) at the Congaree site indicated that recharge may be occurring along stream levees, where low permeability sediments are breached. Overall, the areal extent of these levees constitute a small portion of the total AA, therefore supporting the LOW rating. Literature for bottomland hardwoods also supports a low recharge potential because of impeding permeability of most blh sediments.

The peizometric data suggest that groundwater recharge in at least one bottomland hardwood site and the estuarine-marsh site, is spatially segregated within particular zones of the assessment area.

Groundwater Discharge

The WET analysis yielded a LOW rating for the North Inlet site which is supported by peizometric data. The data indicates that a shift in pressure head occurs during tidal inundation preventing large contributions of discharge.

The bottomland hardwood results yielded a HIGH rating for Congaree and a MODERATE rating for Beidler. The discharge data for these study sites indicate a similar pattern, whereby this function is distributed disproportionately throughout segments of the assessment areas.

Floodflow Alteration

The North Inlet site received the lowest rating for this function. There is an inherent assumption built into model which rates all tidal marshes as LOW. The underlying assumption is that marshes have the ability to buffer low energy storm surges only during low tide events (Adamus et al 1987). This assumption may need further investigation given the broad diversity of morphological settings for estuaries; i.e. Type I, II, III geomorphological settings described by Odum (1980).

The bottomland hardwood sites rated HIGH for floodflow alteration for all sites. Hydrographic data for the Congaree swamp support WET predictions by indicating peak flow velocities decreasing throughout passage in the floodplain. Flow rates were not available for the Beidler site, but examination of high altitude infrared photographs (NHAP) suggest high storage potential in the braided morphology of the basin.

Sediment Stabilization

Sediment export at North Inlet site indicates that during neap tides, sediment is exported annually in the amount of 40 x 10 kg⁻2/ yr⁻2 (Dame et al. 1986). The WET analysis predicted a MODERATE rating which can be explained by average calculations throughout the year.

Both bottomland hardwood sites were rated MODERATE for this function. Total suspended solids (TSS) decrease throughout passage of the Congaree floodplain (Birch 1983). However, TSS levels fluctuate considerably throughout locations within the floodplain. In fact, TSS levels increase at confluence areas with higher order streams. A dam structure located upstream of the AA produces artificial fluctuations in water levels of the Congaree River which has significant effects on the natural ability of the floodplain to decrease flow velocities and TSS. The method clearly was sensitive to the anthropogenic sources of sediment, rating the Congaree with a MODERATE.

TSS levels were not available for the Beidler site, however it is hypothesized to have protracted ability to decrease TSS to the low flows and constricted nature of the floodplain. However, WET predicted a MODERATE-lower than would be expected. However, the WET-BLH evaluation, reflected a finer level of distinction and reflected distinctions in the two blh tracts, and rated Beidler, HIGH.

Sediment/Toxicant Retention

All three study sites were rated HIGH for toxicant retention. Water quality data from the Congaree Swamp indicates that the floodplain is indeed a sink for chromium, tin, lead, and copper (Birch 1983). However, analysis of data indicate that the sediment/toxicant retention function tends to be concentrated with particular zones of the floodplain. When in fact, some areas of the floodplain were actually showed elevated levels of these contaminants, suggesting a possible source from overland run-off into the wetland.

North Inlet data indicates that the ability of the marsh to sequester heavy metals and provide a toxicant retention function is dependent on the sediment type and ultimately the geologic age of the marsh sediments (Gardner 1973; Morris pers. comm). The Congaree data, as well, indicates a spatial segregation within the floodplain of this particular function. So that, some areas of the AA may be contributing a higher degree of efficiency in removing toxicants.

Nutrient Removal/Transformation

Nutrient retention was rated HIGH for all three study sites. North Inlet, at least seasonally, is a transformer of nutrients and a source of nutrient source to the receiving waters of the estuary and Atlantic Ocean (Dame et al. 1986).

Water quality data for Congaree indicates that the floodplain is a sink to phosphorus and a transformer of nitrogen. However, the ability of the floodplain to remove or transform certain nutrients is a function of the location within the floodplain. WET predictions clearly reflect the overall value of the swamp to the Congaree river and downstream ecosystem. Likelihood exists that this function is also spatially segregated within the assessment area.

Production export

Productivity studies at North Inlet indicate that total carbon exported from North Inlet was 453 $m^{-2}yr^{-1}$ about 42% of above-ground and 25% of the total primary productive from the system. The actual measured level compared higher than five other marshes, including the Great Sippesiwisset and Flax Pond marshes, which supports the HIGH rating.

The Congaree and Beidler sites were rated HIGH for production export. Actual data and literature sources concurred with this prediction.

Wildlife/Diversity/Abundance

WET evaluates habitat for wetland-dependent bird species for categories of breeding, overwintering, and migration. The two bottomland hardwood sites rated HIGH for breeding and LOAW for North Inlet. All study sites rated HIGH for over-wintering and migration.

Breeding bird surveys at Congaree substantiate clearly the HIGH rating derived from WET for all categories of wildlife diversity. A count of 118 species utilized the swamp and 84 species are known to nest (Dennis 1963). The North Inlet data also supported the LOW rating for breeding and HIGH for over-wintering and migration. The results comparing wildlife diversity for breeding and overwintering habitat at the North Inlet site proved very sensitive to these particular sub-categories of function.

Bildstein et al. (1982) listed 94 species which utilize
the marsh with only five species which are known to
nest.

Aquatic Diversity

The aquatic diversity category combines habitat
for finfish, shellfish, and macroinvertebrates. All
three study sites rated HIGH. Trawl data from North
Inlet substantiates this HIGH rating indicating that 94
species of fish and 12 species of macroinvertebrates
utilize the tidal creeks (Ogburn, Dennis, and Michener
1987).

Habitat for finfish in the bottomland hardwood
sites, however, is limited to the permanent stream
reaches and oxbow lakes. One possible explanation for
the blh sites rating HIGH is that the category combines
the shellfish and finfish categories, whereby
macroinvertebrate densities could have resulted in the
HIGH rating for habitat. The WET-BLH version breaks
down this category into two separate evaluations, which
rated Beidler slightly lower. Since Congaree provides
a deepwater, riverine habitat, as opposed to Beidler
with no channel present, the WET-BLH predicted with
finer distinctions these differences.

Summary and Conclusion

The WET method of wetland functional assessment
provides an adequate predictive tool to gain baseline
information of wetland function in coastal wetland
ecosystems. This study evaluated the predictions
derived from the WET analysis and compared these
results with a synthesis of collected field data and
from long-term monitoring of the sites, existing
literature, and expert opinion. From these
evaluations, the study was able to conclude that WET
provides relative degrees of scientific efficacy in
predicting wetland function. The WET analysis compared
favorably with summations of wetland function derived
from actual data collections, existing literature, and
expert opinion. However, the two bottomland hardwood
sites yielded similar ratings among most categories of
function. The similar ratings were expected, given the
similar settings. However, the WET-BLH version
predicted finer distinctions in wetland function, which
allowed comparison between the two blh tracts. Several
observations were made regarding the sensitivities and
efficacy of using this type of predictive model.

One such of the sensitivities arises from
evaluating wetland function within artificial
components of study. When conducting an analysis or

when forced to "draw the red line", an investigator is forced to make broad generalizations across a diverse "mosaic" of topographic, hydrologic, and biotic features. Bounding of the assessment area is likely to create artificial compartments of wetland function, when in fact, wetland function tends to be segregated within particular zones. North Inlet data implies that among some categories of function, substrate type and age of marsh may be determining factors in levels of functioning. Likewise, in bottomland hardwoods actual data implies segregation of wetland function within particular zones of the floodplain. It is not known how important these variables become in the overall analysis until they are tested among a wide variety of study sites.

Lastly, the conclusion regarding scientific efficacy is that the regionalized version designed for evaluation of bottomland hardwoods yielded more precise ratings and allowed for comparison between systems. A similar regionalized version for use in evaluating estuarine and marsh ecosystems could more precise predictions.

Reproducibility

The analysis for reproducibility tested the WET method among various investigators. When results taken from 8 trained graduate student analyses with various ecological backgrounds, indicated that the range in questions answered differently between 94-104 questions. However, those differences in responses to questions did not significantly effect the overall rating of the function. One possible explanation is there are numerous checks and balances within the method to take into account differences in responses.

Applications of the methodology

The methodology presents various options for use within regulatory contexts, in managing individual estuaries and wetland systems. The most obvious need for a wetland assessment technique is within the scale of individual site-specific permit reviews. Within this context, various sensitivities regarding its application have been drawn. Understanding the sensitivities of the model provides some framework for which to evaluate a site. For example, within the AA when obvious diverse substrates and hydrology exist, boundaries could be established using additional criteria. Boundaries in areas where heterogeneous substrates and hydrology exist, the area could be broken into several AA's.

The methodology gives simple "rules of thumb"
approaches to wetland assessment when the direct
evidence or actual data is lacking. The most valuable
asset of the model is in providing a scientific
framework from which to evaluate sites. The method
requires an investigator to organize the data, define
the study area, and provides a scientific framework to
derive predictions. The results are also expressed in
terms of probability which gives greater scientific
validity.

Further Study

WET predicted with fair accuracy wetland function
among the selected study sites. This study identified
several factors which could influence the efficacy fo
the application in predicting wetland function.
"Bounding" of the assessment area offers intrinsic
complex parameters which may have significant effect on
the accuracy of the predictions. Future study is
needed to resolve the effects of "bounding" the AA when
zonations occur in the study site. This may be
particularly problematic when involving estuarine and
blh tracts.

Acknowledgements

The author expresses appreciation to the U.S. Park
Service, Congaree National Monument staff, National
Audubon Society, for permission to use the parks as
study sites.

Special thanks is extended to J. Dean, M. Harvley,
E. Laban, and S. Emanuel.

REFERENCES

Adamus, P.R. 1987. Wetland evaluation technique (WET), Volume II, Methodology (operational draft). Technical Report 1-87. U.S. Army Corps of Engineers Waterways Experiment Station, Vicksburg, Mississippi.

Bildstein, K.L., R.L. Christy, and P. DeCoursey. 1982. Energy flow through a South Carolina saltmarsh shorebird community. Water Study Group Bull. 43:35-37.

Birch, J.B. 1983. Water quality of the Congaree National Monument. University of Georgia, Athens, Georgia.

Dame, R., T. Chrzanowski, K. Bildstein, B. Kjerve, H. McKellar, D. Nelson, J. Spurrier, S. Stanck, H. Stevenson, J. Vernberg, and R. Zingmark. 1986. The outwelling hypothesis and North Inlet, South Carolina. Marine Ecology 33:217-229.

Gaddy, L.L. and G.A. Smathers. 1980. The vegetation of Congaree Swamp National Monument. Veroff. Geobot. Inst. ETH. Stiflung Rubel, Zurich 69, Germany. Heft (1980): 171-182.

Gardner, L.R. 1973. The effect of hydrologic factors on pore water chemistry of intertidal marsh sediments. Southeastern Geology 15: 17-28.

Morris, J. (personal communication). Department of Biology, University of South Carolina, Columbia, South Carolina.

Odum, E.P. 1980. The status of three ecosystems-level hypotheses regarding salt marsh estuaries: tidal subsidy, outwelling, and detritus-based food chain. In: Estuarine perspectus. Academic Press, N.Y.

Patterson, G.G., G.K. Speiran and B.H. Whetstone. 1985. Hydrology and its effects on distribution of vegetation in Congaree Swamp National Monument, South Carolina. U.S. Geological Survey Water Resources Investigation Report 83-4256.

Porcher, R.D. 1981. The vascular flora of the Francis Beidler Forest in Four Holes Swamp, Berkeley and Dorchester Counties, South Carolina. Castanea. 46:248-280.

HABITAT DEGRADATION AND FISHERY DECLINES IN THE U.S.

James R. Chambers[1]

Abstract

Habitat degradation and loss are adversely affecting coastal, estuarine and riverine ecosystems which are essential for spawning, feeding, growth or as migratory routes for a majority of the Nation's living marine resources. Primary habitat threats are due to wetland loss and degradation, toxic chemical releases, alteration of freshwater flows and nutrient over-enrichment. Landings of most species fished both commercially and recreationally are at historic low levels. Projected human growth trends indicate that future development and its associated effects will occur in coastal regions where living marine resources' estuarine-dependency is highest. This adds urgency to the need for added protection of remaining habitats and an increased consideration of the importance of habitat in national policy and land use decision-making.

Estuarine-Dependency of Living Marine Resources

Of U.S. commercial fishery landings in 1985, about 77% by weight and 71% by value were estimated by the National Marine Fisheries Service (NMFS), National Habitat Conservation Program to be composed of estuarine-dependent species, i.e., those dependent for reproduction, as nursery areas, for food or as migratory pathways (Chambers, 1991). The estuarine-dependency of recreational species has not yet been quantified, due to the lack of comparable catch statistics, but is assumed to be equally extensive. By

---

[1]Office of Protected Resources, National Marine Fisheries Service, 1335 East-West Highway, Silver Spring, Maryland 20910.

region, NMFS estimates estuarine-dependency of the U.S.
commercial landings to be as follows:

98%	Gulf of Mexico	94%	Southeast Atlantic
78	Chesapeake	76	Alaska
52	Pacific Northwest	41	Northeast Atlantic
18	California	1	Pacific Islands

Economic Value of Estuarine-Dependent Fisheries

Estuarine ecosystems are critical to the survival of
living marine resources which have a cumulative value to
society approaching $14 billion per year. Marine
recreational species accounted for 30% of the total
domestic harvest of edible marine fish used for food and
an estimated total economic impact of over $8.2 billion in
1985 (Prosser et al., 1988). In 1986, commercial landings
of estuarine-dependent species contributed approximately
$5.5 billion to the U.S. economy (Chambers, 1991).

National Trends in Estuarine-Dependent Resources

Populations of virtually all estuarine-dependent
species off the Atlantic, Pacific and Gulf of Mexico coasts
which have been exploited by both commercial and
recreational fisheries have been reduced to all time low
levels of abundance. These include: Atlantic salmon,
striped bass, weakfish, spotted seatrout, summer flounder,
winter flounder, American shad, hickory shad, croaker, king
mackerel, spanish mackerel, greater amberjack, red drum,
black drum, snook, California halibut, English sole and
many races of chinook salmon, coho salmon, sockeye salmon,
chum salmon, and steelhead. Following is a review of
information on trends of estuarine-dependent species and
their essential habitats. In most of these cases, observed
declines in living marine resources are believed by the
author to be due to the combined effects of fishery
harvests and habitat degradation and loss, in addition to
mortality caused by natural factors.

Southeast Atlantic and Gulf of Mexico Coasts

Since 1982, commercial landings of fish and shellfish
along the Southeast Atlantic and Gulf of Mexico coasts have
decreased by 42%. Virtually all landings are composed of
estuarine-dependent species. The region contains over 300
estuaries composed of an estimated 17.2 million ac (68,800
km[2]) of coastal marsh. Extensive regional losses of coastal
fishery habitats associated with thousands of Federal
projects and permit approvals have been documented by NMFS'
Habitat Program scientists involved in the decision-making
processes (Lindall and Thayer, 1982; Mager and Thayer,
1986). NMFS scientists have also demonstrated, in

preliminary terms for the Mid-Atlantic coastal region, that
estuarine wetland productivity is essential for support of
offshore fishery biomass (Peters and Schaaf, 1983). This
relationship is based in large part on a short, direct food
chain (Lewis and Peters, 1984) involving coastal wetlands,
menhaden - the primary forage fish species which we have
recently discovered can digest plant detritus - and many
of the marine fish species sought by commercial and sport
fishermen.

National Wetland Losses

     Both the area of wetlands currently remaining in the
U.S. and that already lost is very poorly understood. Of
the Nation's estimated orriginal 226 million ac (915,000
km^2) of wetlands in the lower 48 states (including both
freshwater and coastal), about 54% had been lost by the
mid-1970s (Tiner, 1984). Losses have been most pronounced
during the mid-1950s to mid-1970s, averaging an estimated
400-500,000 ac/yr (1,600-2,000 km^2/yr) (Frayer et al.,
1983). Regional losses of both freshwater and marine
wetlands have been extensive (Tiner, 1984) with estimated
declines greater than 70% experienced by 10 states - Ohio,
Indiana, Illinois, Missouri, Iowa, Kentucky, California,
Connecticut, Maryland and Arkansas (Dahl, 1990).

     Coastal wetlands are relatively scarce, making up only
about 5% of the total national wetland acreage (Frayer et
al., 1983). However, these habitats (including upstream
transitional habitats) are among the most productive
ecosystems on earth and have high value as habitat for
living marine resources. Coastal wetland losses were
estimated to have averaged 18,000 ac/yr (73 km^2/yr) between
the mid-1950s and mid-1970s (Frayer, et al., 1983), and are
thought to have been greatest in Florida, Louisiana, Texas
and California. There is evidence that more recent annual
losses may actually be much higher.

Louisiana Wetland Losses

     Approximately 80% of the lower 48 states' coastal
wetland loss occurs in Louisiana. The State is thought to
be losing coastal wetlands at a rate of about 35,200 ac/yr
(142 km^2/yr) or 55 mi^2/yr (from Gagliano et al., 1981 and
Gosselink et al., 1979). These losses are due primarily
to (1) canal dredging through marsh for oil and gas
development and for navigation channels, (2) flood control
levee construction resulting in direct loss of wetlands and
loss of sediment re-nourishment over the deltaic plain to
replace its natural subsidence and (3) levees and water
control structures for marsh management, intended to
preserve wetlands. The latter practice, which blocks
access by living marine resources to their nursery areas

and can actually accelerate wetland loss, has been proposed recently for more than 600 mi^2 (1,500 km^2) of coastal Louisiana. The U.S. Army Corps of Engineers estimates that by the year 2040, nearly 1 million ac (4,000 km^2) of additional Louisiana wetlands will be lost for a total, then, of 2.4 million ac (9,700 km^2). If this rate is not reduced, the Gulf shoreline may advance inland as much as 33 mi (53 km) in some areas (U.S. Army Corps of Engineers and State of Louisiana). Any relative sea level rise attributable to global climate change or other factors will magnify this shoreline retreat and habitat loss.

Louisiana produces about 30% of the lower 48 states' commercial fishery landings (by weight); essentially all are estuarine-dependent. In fact, the productivity of the valuable Gulf of Mexico shrimp fishery has been shown to be directly proportional to its total area of intertidal marsh habitat (Turner, 1979). NMFS scientists hypothesize that extensive marsh deterioration in parts of Louisiana and Texas may be resulting in a combination of increased "edge," more extensive flooding of tidal marsh, and consequently greater access to rich feeding areas (Zimmerman and Minello, manuscript in preparation). These factors may have increased the production and survival of juvenile fish and shellfish observed since 1960. However, with continued regional wetland loss and degradation, a precipitous decline in the Gulf of Mexico's living marine resources may be forthcoming. We think we may be at or near the "peak of the curve" in those areas, such as Barataria Bay, experiencing the greatest losses.

California Wetland Losses

By 1977, California had lost over 90% of its original wetlands, which are thought to have totaled over 5 million ac (20,000 km^2) (U.S. Fish and Wildlife Service, 1977), and today it is believed to have lost 91% of its original 3.5 million ac (14,000 km^2) of coastal wetlands (J. Zedler, personal communication). San Francisco Bay wetlands, which once covered an estimated 303 mi^2 (784 km^2), have declined approximately 85% (Dedrick, 1989).

Chesapeake Bay Nutrient Over-enrichment

During the past three decades, landings of migratory species from Chesapeake Bay and its tributaries declined as follows: hickory shad (96%), alewife and blueback herring (92%); striped bass (70%), yellow perch (72%), American shad (66%) and white perch (51%) (Chesapeake Bay Program, 1988). Maryland's oyster harvest has declined 90% from levels a century ago, due primarily to disease, over-exploitation and habitat degradation. Seed oyster beds have experienced equivalent losses throughout the Bay.

Submerged aquatic vegetation (a primary nursery habitat)
declined 65% between 1971-1979 (Stevenson et al., 1979),
but may have increased, recently. Losses are thought to
be due primarily to an increase in shading by microscopic
plant growth (on the leaves and in the water column above)
resulting from increased nutrient and sediment discharges
from municipalities and agricultural areas throughout the
drainage basin. The Bay has also been affected by dams on
most major tributaries and particularly by upstream water
diversions, an under-appreciated problem discussed below
for other estuarine systems.

California Water Diversions

In California, natural reproduction of salmon has been
reduced by 65% over the past 20 years (California
Department of Fish and Game, 1983). Approximately 90% of
the State's historic chinook salmon spawning habitat has
been lost due to extensive water project developments
(dams, pumping plants and aqueducts) throughout the Central
Valley which effectively have eliminated spawning in whole
river basins (e.g., San Joaquin). In 1982, the chinook
salmon population was reported to have declined to 30% of
its levels in the 1960s (Kjelson et al., 1982). Sacramento
River winter run chinook salmon have declined from an
average of 86,500 fish in 1967-69 to only 500 in 1990, a
99% reduction due entirely to habitat loss and degradation
from Federal and State water projects (specifically, Shasta
Dam and the Red Bluff Diversion Dam). In 1990, NMFS listed
this run as a threatened species under the Endangered
Species Act (ESA). The other anadromous races, including
steelhead and both late fall and spring run chinook salmon,
are also at very low population levels.

Survival of young striped bass has declined to levels
approaching zero in the two primary spawning areas of San
Francisco Bay. At the same time, adult populations in the
Bay have declined 60-80% (Stevens et al., 1985). In
addition, the high pollutant body burdens of the Bay's
striped bass population has been shown by NMFS' scientists
to be correlated with generally poor condition, growth rate
and a compromised reproductive capacity, as compared to
populations elsewhere in the U.S. (Whipple et al., 1987).

The survival of juvenile striped bass and salmon has
been shown to be correlated directly with freshwater
inflows, and both sport and commercial catches are related
to volumes of spring flows experienced by the young
(Rozengurt et al., 1987). By 1980, more than 62% of the
annual historic freshwater inflows to San Francisco Bay had
been diverted for agriculture in the Central Valley and for
municipal uses in southern California. Planned diversions
will increase the annual loss of freshwater to 71% by the

year 2000 (Nichols et al., 1986), cutting off more of its "lifeblood" and threatening the system's survival.

The effects of freshwater diversions on estuarine ecosystems was examined in 1980 at the National Symposium on Freshwater Inflow to Estuaries. Its summary and recommendations stated:

"Published results regarding water development in rivers entering the Azov, Caspian, Black and Mediterranean Seas in Europe and Asia all point to the conclusion that no more than 25-30 percent of the historical river flow can be diverted without disastrous ecological consequences to the receiving estuary. Comparable studies on six estuaries by the Texas Water Resources Department showed that a 32 percent depletion of natural freshwater inflow to estuaries was the average maximum percentage that could be permitted if subsistence levels of nutrient transport, habitat maintenance, and salinity control were to be maintained" (Clark and Benson, 1981).

Dams in the Pacific Northwest

Columbia River Basin salmon and steelhead runs have declined 75-84% from estimated historic levels of 10-16 million fish to 2.5 million fish (from Northwest Power Planning Council, 1986). Moreover, approximately 66% of the remaining population of 2.5 million fish are produced in hatcheries, as partial mitigation for continuing losses resulting from an extensive system of hydroelectric power-generating dams constructed throughout the Columbia River Basin.

More than 55% of the Columbia River Basin formerly accessible to salmon and steelhead has been blocked by dams (Thompson, 1976). Access to over 1,000 mi (1,609 km) of the upper Columbia River and its tributaries was blocked by completion of the Grand Coulee Dam in 1941. Chinook salmon once migrated 600 miles up the Snake River to Shoshone Falls, Idaho; Hells Canyon Dam now limits fish to the lower 250 mi of this river. An estimated 3,600 mi (5,700 km) of habitat has been lost in the Snake River Basin.

The Snake River run of coho salmon is thought to be extinct. Populations of Snake River sockeye salmon; Snake River races of spring, summer and fall chinook salmon; and lower Columbia River coho salmon are so low now that petitions have been received by NMFS to list them also as threatened or endangered species under the ESA.

Water management schemes for dams throughout the
Columbia River Basin have often resulted in inadequate
springtime flows, which combined with turbine-related
mortality, predation and other factors associated with the
dams, have been responsible for mortality rates of 70-90%
for juveniles migrating downstream to the ocean. NMFS
estimates that ten times the number of juveniles are now
being contributed to the system from hatcheries as were
ever produced naturally (J. Williams, personal
communication). However, the hatchery fishes' survival
must be much lower than that of wild fish; only 10-20% of
the juveniles produced in the Columbia River system are
wild, whereas 33% of the adult returns are wild. Dwindling
stocks of wild fish intermingle as returning adults with
larger numbers of hatchery-produced fish but are
indistinguishable to fishermen who unknowingly contribute
to additional weakening of the wild races prior to
spawning. Thus, we are also losing the genetic diversity,
characteristic of the many wild races, which may be
essential to these species' long-term survival.

Coastal Pollution

Contaminants (e.g., organic chemicals and trace metals)
have been found by NMFS' National Benthic Surveillance
Project in high concentrations in fish and sediment from
urbanized and industrialized areas where disease also
occurs (Varanasi et al., 1989; Zdanowicz et al., 1986; and
Hanson et al., 1986). Cancerous liver tumors have been
observed in as many as 24% of the bottom-dwelling English
sole (Myers et al., 1987) and winter flounder (Murchelano
and Wolke, 1985) sampled from Puget Sound and Boston
Harbor, respectively. These tumors and other abnormalities
(called lesions) are believed by our scientists to be
biological responses to persistent exposure to contaminants
in sediments and their food.

English sole landings from Puget Sound have declined
in recent years from a high of 2.4 million lbs (1 million
kg) to a low in 1987 of 0.7 million lbs (0.3 million kg)
(Bargmann, 1988). Certainly, harvesting practices have had
a pronounced effect on the population. However, almost 40%
of the female English sole NMFS has tested from Eagle
Harbor and the Duwamish Waterway of Puget Sound failed to
mature sexually (Johnson et al., 1988). Sediments in both
areas have high levels of aromatic hydrocarbons, which we
have shown can seriously compromise fishes' health (Malins
et al., 1984; 1985) and can induce a variety of lesions
leading eventually to the development of cancerous tumors
(Varanasi et al., 1987). Fish with such contaminant-
induced tumors also show evidence of organ malfunction
(Casillas et al., 1983) indicating the fishes' health is
also being affected.

NMFS scientists have demonstrated that contaminants can constitute a kind of "triple threat" to successful reproduction by English sole. First, normal egg maturation is inhibited in fish from contaminated areas (Johnson et al., 1988). Secondly, sexually mature fish are less likely to spawn when induced in the laboratory, if they were collected from contaminated sites. Thirdly, adults from contaminated sites which do successfully spawn, produce a high percentage of deformed larvae in comparison to those from cleaner areas (Johnson et al., 1989). Thus, prior exposure of adults to contaminants (1) can cause failure of egg development in some of the fish, (2) of those which do produce eggs, it can interfere with the timing of their spawning, and (3) of those which do spawn on time, it can result in producing deformed young.

The distribution of serious contaminant-related lesions in fish (Harshbarger and Clark, 1990) overlaps coastal urbanization and the pollution from about 1,900 major industrial and municipal dischargers (Office of Technology Assessment, 1987) and countless smaller sources. From NMFS' annual sampling of coastal waters, the areas having high concentrations of contaminants are: Puget Sound (WA), San Francisco Bay (CA), Southern California (Los Angeles and San Diego Bay), Galveston Bay (TX), Mobile Bay (AL), Tampa Bay (FL), Biscayne Bay (FL), the St. Johns River (FL), urbanized areas of Chesapeake Bay (VA, MD), Delaware Bay (DE, NJ), the Hudson-Raritan estuary (NJ, NY), Western Long Island Sound (NY, CT), Narragansett Bay (RI), Boston Harbor (MA), Salem Harbor (MA), Casco Bay (ME), and Penobscot Bay (ME). However, agricultural areas can also contribute high levels of toxic pesticides to adjacent coastal ecosystems. In priority order, those coastal areas estimated (Pait, et al. 1989) to be at highest risk from agricultural pesticide use are: Albemarle Sound (NC), Chesapeake Bay, Laguna Madre (TX), Pamlico Sound (NC), Winyah Bay (SC), Delaware Bay, Cape Fear River (NC), Hudson-Raritan estuary, St. Johns River and Puget Sound.

Shellfish bed closures - an indicator of the loss of good water quality - occur coast-wide. On any given day one-third of the nation's remaining 16 million ac (65,000 km^2) of shellfish waters are closed. In the Gulf of Mexico, 74% of our shellfish waters are restricted (U.S. Department of Commerce, 1985) due to inadequate septic systems, sewage discharges and urban runoff.

NMFS' National Habitat Conservation Program

NMFS is the Federal steward of the Nation's living marine resources throughout their range. This stewardship extends from 200 mi at sea, through our estuaries, and inland as far as salmon and other anadromous species swim

for their spawning migrations. Therefore, in addition to
any activities affecting marine waters, NMFS' National
Habitat Conservation Program is equally concerned with
activities in drainage basins far inland, which may damage
habitats important to living marine resources. NMFS'
Regional Habitat Conservation Divisions assess the effects
and make recommendations to other agencies on programs,
policies and approximately 10,000 proposed projects which
would adversely affect over 300,000 ac (1,200 km^2) of
habitat per year. Program scientists in NMFS' Fishery
Centers conduct supporting research on the importance of
habitats to living marine resources, and the effects of
habitat degradation and loss. The Habitat Program's field
organization consists of approximately 50 Regional
ecologists, 265 Center scientists and support staff located
in each of NMFS' 5 Regions, 5 Centers and most of its 22
laboratories, with personnel in nearly every coastal state.

Authorities for this Program include the Fish and
Wildlife Coordination Act, which requires any Federal
agency seeking to affect "waters of the U.S." to consult
first with NMFS and the other resource agencies (U.S. Fish
and Wildlife Service and state fish and game agency). The
Magnuson Fishery Conservation and Management Act provides
broad authority to conserve and manage fisheries within the
U.S. Exclusive Economic Zone and to conduct research on the
effects of man's activities on these resources. Other
authorities include the Clean Water Act (FWPCA); National
Environmental Policy Act; Superfund (CERCLA); Federal Power
Act; Pacific Northwest Electric Power Planning and
Conservation Act; Mitchell Act; ESA; Marine Mammal
Protection Act; and Marine Protection, Research and
Sanctuaries Act. NMFS' Habitat Policy was published in the
Federal Register (Vol. 48, No. 228) on November 25, 1983.

Demographic Trends

The coastal areas which are experiencing stress are
also those most densely populated. People are moving to
the coasts. Most of this anticipated movement is toward
the "Sun Belt" with growth in coastal counties averaging
four times that of the national average (Culliton, et al.,
1990). For example, at a current rate of 900 new
Floridians a day, the State is expected to increase from
a current population of 10 million to approximately 15
million by the year 2000. NMFS' estimates of regional U.S.
population increases from 1980 to 2010 for counties within
50 mi (80 km) of the coast, are (from Edwards, 1989a):

73%	South Atlantic	53%	California
46	Gulf of Mexico	28	Pacific Northwest
16	Northeast	-3	Great Lakes

Several states, which still have extensive estuarine and coastal systems, are projected to experience substantial additional growth in coastal counties, as follows: Virginia (50%), South Carolina (55%), Georgia (64%), Florida (81%), Mississippi (46%), Texas (68%) and Alaska (81%). Overall, the coastal population of the U.S. is projected to increase by 27% between 1980 and 2010, from 119 to 151 million, and as a proportion of the national population, from 52.6 to 53.6% (Edwards, 1989b).

Conclusions

Demographic trends may have serious implications for the Nation's living marine resources. This is particularly true for the Southeast Atlantic and Gulf of Mexico coasts where recreational and commercial species are almost completely dependent upon estuarine and coastal habitats for their survival. Habitat degradation and loss can have long-term adverse effects on living marine resources, in addition to those losses attributable to commercial and recreational harvesting practices. The importance of living marine resources' habitat must be elevated in priority both in decisions on projects by responsible Federal and state agencies and in the level of support given to agencies responsible for stewardship of such resources. Only then will we be able to significantly reduce those losses which we are now beginning to see as a result of past development practices and policies. Providing a scientific understanding of the effects of such development in Federal and state decision-making processes is the responsibility of NMFS' National Habitat Conservation Program - the Federal steward of the nation's living marine resources and their habitats.

Acknowledgements

The author gratefully acknowledges the substantial contributions of the NMFS National Habitat Conservation Program's field staff for its involvement in determining estuarine-dependency of U.S. commercial fishery landings data and for providing specific information on both habitat issues and resource trends. In this regard, particular recognition is extended to Bruce McCain (contaminants), Roger Wolcott (California water issues), and Merritt Tuttle, Monique Rutledge and Nicholas Iadanza (Columbia River salmon issues).

References

Bargmann, G.G. 1988. Trends in abundance of economically important marine fish in Puget Sound. Proceedings of the 1st Ann. Mtg. on Puget Sound Research, Puget Sound Water Qual. Auth. Seattle, WA. pp. 72-76.

California Department of Fish and Game (Anadromous
Fisheries Branch). 1983. Salmon Management in California.
Prepared for the Pacific Fisheries Management Council.

Casillas, E., M.S. Myers and W.E. Ames. 1983.
Relationship of serum chemistry values to liver and kidney
histopathology in English sole (Parophrys vetulus) after
exposure to carbon tetrachloride. Aquat. Toxicol. 3:61-
78.

Chambers, J.R. 1991. Coastal habitat degradation and
fisheries declines. In: R.H. Stroud (editor). Proceedings
of the Symposium on Coastal Fish Habitat Conservation.
Baltimore, MD (in press).

Chesapeake Bay Program. 1988. Draft strategy for removing
impediments to migratory fishes in the Chesapeake Bay
watershed. Agreement Commitment Report.

Clark, J. and N.G. Benson. 1981. Symposium Summary and
Recommendations. In: R.D. Cross and D.L. Williams
(editors). Proceedings of the National Symposium on
Freshwater Inflow to Estuaries, Vol. II. Washington, D.C.,
U.S. Department of the Interior. pp. 523-528.

Culliton, T.J., M.A. Warren, T.R. Goodspeed, D.G. Remer,
C.M. Blackwell and J.J. McDonough. 1990. 50 years of
population change along the Nation's coasts. NOAA,
Rockville, MD. 41 p.

Dahl, T.E. 1990. Wetland Losses in the United States,
1780s to 1980s. U.S. Department of the Interior, Fish and
Wildlife Service, Washington, D.C.

Dedrick, K.G. 1989. San Francisco Bay tidal marshland
acreages: recent and historic values. In: Coastal Zone
'89. Proceedings of the Sixth Symposium on Coastal and
Ocean Management. Amer. Soc. Civ. Eng. 1:383-398.

Edwards, S.F. 1989a. Estimates of future demographic
changes in the Coastal Zone. Coastal Management. 17:229-
240.

Edwards, S.F. 1989b. What percent of the population will
live in coastal zones? In: Coastal Zone '89: Proceedings
of the Sixth Symposium on Coastal and Ocean Management.
Amer. Soc. Civ. Eng. 1:114-126.

Frayer, W.E., T.J. Monahan, D.C. Bowden and F.A. Graybill.
1983. Status and trends of wetlands and deepwater habitats
in the conterminous United States, 1950s to 1970s. Dept.
Forest and Wood Sci., Colo. St. Univ., Ft. Collins. 32 p.

Gagliano, S.M., K.J. Meyer-Arendt and K.M. Wicker. 1981.
Land loss in the Mississippi River Deltaic Plain. Trans.
Gulf Coast Assn. Geol. Soc. 31:295-300.

Gosselink, J.G., C.C. Cordes and J.W. Parsons. 1979. An
ecological characterization study of the Chenier Plain
Coastal Ecosystem of Louisiana and Texas. FWS/OBS-78/9-
78/11 (3 vols.), Office of Biological Services, U.S. Fish
and Wildlife Service, Slidel, LA.

Hanson, P.J., J.A. Wells and M.W. Newman. 1986.
Preliminary results of the 1984-85 National Benthic
Surveillance Project: Southeast Atlantic and Gulf of Mexico
Coasts. In: Oceans '86 Conference Record, Vol. 2,
(Washington, D.C.: Marine Technology Society and IEEE Ocean
Engineering Society, 1986), pp. 572-577.

Harshbarger, J.C. and J.B. Clark. 1990. Epizootiology of
neoplasms in bony fish of North America. Sci. Total Env.
94:1-32.

Johnson, L.J., E. Casillas, T.K. Collier, B.B. McCain and
U. Varanasi. 1988. Contaminant effects on ovarian
development in English sole (Parophrys vetulus) from Puget
Sound, Washington. Can. J. Fish. Aquat. Sci. 45:2,133-
2,146.

Johnson, L.L., E. Casillas, D. Misitano, T.K. Collier, J.E.
Stein, B.B. McCain and U. Varanasi. 1989. Bioindicators
of reproductive impairment in female English sole
(Parophrys vetulus) exposed to environmental contaminants.
In: Oceans '89 Conference Record. (Seattle, WA: Marine
Technology Society and IEEE Ocean Engineering Society,
1989). pp. 391-396.

Kjelson, M.A., P.F. Raquel and F.W. Fisher. 1982. Life
history of fall-run juvenile chinook salmon, Oncorhynchus
tshawytscha, in the Sacramento-San Joaquin estuary,
California. In: V.S. Kennedy (editor). Estuarine
Comparisons. New York. Academic Press. pp. 393-411.

Lindall, W.N. and G.W. Thayer. 1982. Quantification of
National Marine Fisheries Service habitat conservation
efforts in the southeast region of the United States. Mar.
Fish. Rev. 44(12):18-22.

Lewis, V.P. and D.S. Peters. 1984. Menhaden - a single
step from vascular plant to fishery harvest. J. Exp. Mar.
Biol. Ecol. 84:95-100.

Mager, A. and G.W. Thayer. 1986. National Marine Fishery
Service habitat conservation efforts in the southeast

58   COASTAL WETLANDS

region of the United States from 1981 through 1985. Mar.
Fish. Rev. 48(3):1-7.

Malins, D.C., B.B. McCain, D.W. Brown, S-L. Chan, M.S.
Myers, J.T. Landahl, P.G. Prohaska, A.J. Friedman, L.D.
Rhodes, D.G. Burrows, W.D. Gronlund and H.O. Hodgins.
1984. Chemical pollutants in sediments and diseases in
bottom-dwelling fish in Puget Sound, Washington. Environ.
Sci. Technol. 18:705-713.

Malins, D.C., M.M. Krahn, D.W. Brown, L.D. Rhodes, M.S.
Myers, B.B. McCain and S-L. Chan. 1985. Toxic chemicals
in marine sediment and biota from Mukilteo, Washington:
relationships with hepatic neoplasms and other hepatic
lesions in English sole (Parophrys vetulus). J. Natl.
Cancer Inst. 74:487-494.

Murchelano, R.A. and R.E. Wolke. 1985. Epizootic
carcinoma in winter flounder (Pseudopleuronectes
americanus). Science. 228:587-589.

Myers, M.S., L.D. Rhodes and B.B. McCain. 1987.
Pathologic anatomy and patterns of occurrence of hepatic
neoplasms, putative preneoplastic lesions and other
idiopathic hepatic conditions in English sole (Parophrys
vetulus) from Puget Sound, Washington, U.S.A. J. Natl.
Cancer Inst. 78(2):333-363.

Nichols, F.H., J.E. Cloern, S.N. Luoma and D.H. Peterson.
1986. The modification of an estuary. Science. 231:567-
573.

Northwest Power Planning Council. 1986. Council staff
compilation of information on salmon and steelhead losses
in the Columbia River Basin. Columbia River Basin Fish and
Wildlife Program.

Office of Technology Assessment. 1987. Wastes in Marine
Environments, OTA-O-334. U.S. Congress, U.S. Government
Printing Office, Washington, D.C. 313 p.

Pait, A.S., D. Farrow, J.A. Lowe and P.A. Pacheco. 1989.
Agricultural pesticide use in estuarine drainage areas.
In: Coastal Zone '89. Proceedings of the Sixth Symposium
on Coastal and Ocean Management. Amer. Soc. Civ. Eng.
1:452-466.

Peters, D.S. and W.E. Schaaf. 1983. Atlantic marsh-
estuarine nearshore detrital system (AMENDS) model, p. 173-
177. (Abstract). In: K.W. Turgeon (editor), Marine
ecosystem modeling, proceedings from a workshop held April
6-8, 1982, Frederick, MD. NOAA, NESDIS, Washington, D.C.

Prosser, N.S., G.C. Radonski and D.B. Rockland. 1988. A review of state agencies' capabilities to manage marine recreational fisheries. 118th Ann. Mtg., Amer. Fish. Soc., Toronto, Canada. 11 p.

Rozengurt, M.A., M.J. Herz and M. Josselyn. 1987. The impact of water diversions on the river-delta-estuary-sea ecosystems of San Francisco Bay and the Sea of Azov. In: D.L Goodrich (editor). San Francisco Bay: issues, resources, status, and management. NOAA Estuary-of-the-Month Seminar Series No. 6. U.S. Department of Commerce, Washington, D.C. pp. 35-62.

Stevens, D.E., D.W. Kohlhorst, L.W. Miller and D.W. Kelley. 1985. The decline of striped bass in the Sacramento-San Joaquin Estuary, California. Trans. Amer. Fish. Soc. 114:12-30.

Stevenson, J.C., N. Confer and C.B. Pieper. 1979. The decline of submerged aquatic plants in Chesapeake Bay. U.S. Fish and Wildlife Service. FWS/OBS- 79/24. 12 p.

Thompson, K. 1976. Columbia Basin fisheries, past, present, and future. Columbia River Fisheries Project Report. Pacific Northwest Regional Commission. 41 p.

Tiner, R.W. 1984. Wetlands of the United States: current status and recent trends. U.S. Fish and Wild. Ser. 59 p.

Turner, R.E. 1979. Louisiana's coastal fisheries and changing environmental conditions. In: J.W. Day, Jr., D.D. Culley, Jr., R.E. Turner and A.J. Mumphrey, Jr. (editors). Proceedings of the Third Coastal Marsh and Estuary Management Symposium. March 6-7, 1978. Louisiana State University. pp. 363-370.

U.S. Army Corps of Engineers and State of Louisiana. Crisis on Louisiana's coast...America's loss. Louisiana Comprehensive Coastal Plan.

U.S. Department of Commerce, National Oceanic and Atmospheric Administration; and Department of Health and Human Services, Food and Drug Administration. 1985. 1985 National Shellfish Register of Classified Estuarine Waters. Washington, D.C. 19 p.

U.S. Fish and Wildlife Service. 1977. Concept plan for waterfowl wintering habitat preservation, Central Valley, California. Region 1, Portland, Oregon. 116 p.

Varanasi, U., D.W. Brown, S-L. Chan, J.T. Landahl, B.B. McCain, M.S. Myers, M.H Schiewe, J.E. Stein and D.D. Weber.

1987. Etiology of tumors in bottom-dwelling marine fish.
Final Rept. Natl. Cancer Inst.   250 p.

Varanasi, U., S-L. Chan, B.B. McCain, J.T. Landahl, M..
Schiewe R.C. Clark, .W. Brown, M.S. Myers, M.K. Krahn, W.D.
Gronlund and W.D. McLeod.   1989.   National Benthic
Surveillance Project: Pacific Coast. Part II, Technical
Presentation of the Results for Cycles III (1984-86). NOAA
Tech. Memo. NMFS NWC-170.   204 p.

Whipple, J.A., R.B. MacFarlane, M.B. Eldridge and P.
Senville, Jr. 1987. The impacts of estuarine degradation
and chronic pollution on populations of anadromous striped
bass in the San Francisco Bay-Delta: a summary.   In: D.L.
Goodrich (editor). San Francisco Bay: Issues, Resources,
Status and Management. NOAA Estuary-of-the-Month Seminar
Series No. 6.   Washington, D.C., U.S. Department of
Commerce, NOAA Estuarine Programs Office.   pp. 77-106.

Zdanowicz, V.S., D.F. Gadbois and M.W. Newman.   1986.
Levels of organic contaminants in sediments and fish
tissues and prevalences of pathological disorders in winter
flounder from estuaries of the Northeast United States,
1984.   In: Oceans '86 Conference Record, Vol. 2.
(Washington, D.C.: Marine Technology Society and IEEE Ocean
Engineering Society, 1986).   pp. 578-585.

Zimmerman, R.J., E.F. Klima, T.J. Minello and J. Nance.
Sea level rise, marsh deterioration and stimulation of
coastal fisheries. NOAA, NMFS, Galveston, TX (manuscript
in preparation).

Eutrophication and its Effects on Coastal Habitats

R. Eugene Turner[1] and Nancy N. Rabalais[2]

Abstract

Coastal eutrophication occurs worldwide. The causes are varied, but increased fertilizer use and consumption, as well as inadequate disposal/treatment, are driving factors leading to declines in estuarine water quality. The potential and observed consequences of eutrophication for coastal habitats include altered food chains, noxious blooms, hypoxia, increased stress on coral reefs and seagrasses, and, altered habitat biodiversity, if not loss. The few documented examples of reduced nutrient loading show partial recovery within a few years. The present national policy of water quality treatment in freshwaters (mostly phosphorus control) is not sufficiently responsive to the realities of coastal water quality problems. Limitation of use, as opposed to treatment after use, is a more fundamental solution to resolving water quality problems.

Introduction

Eutrophication is a worldwide phenomena of aesthetic, social, health, environmental, and economic concern. It has been recognized as such for many years, and is the subject of numerous scientific meetings like this one. In this paper we briefly review the worldwide occurrence of coastal eutrophication and its causes, and focus on how eutrophication affects coastal habitats.

Occurrence of Coastal Eutrophication

The rise in nutrient loading into coastal habitats is widespread (Nehring 1984, Fransz and Verhagen 1985, Rosenberg 1985, Lancelot et al. 1987, Andersson and Ryberg 1988, Wulff and Rahm 1988, Turner and Rabalais 1991). An example of water quality changes in the U.S. is represented by the nitrate and silicate concentrations in the Mississippi River, which drains 40% of the U.S. and is the dominant freshwater inflow source into the Gulf of Mexico. The average annual nitrate-nitrogen concentration doubled after 1950, and the silicate concentration was reduced by half (Figure 1). The average annual nitrate concentration is positively related to nitrogen fertilizer use; the average

---

[1]Coastal Ecology Institute, Center for Wetland Resources, Louisiana State University, Baton Rouge, Louisiana 70803.
[2]Louisiana Universities Marine Consortium, Cocodrie, Louisiana 70344.

annual silicate concentration in the river is inversely related to phosphorus fertilizer use, presumably because phosphorus stimulates diatom growth and diatom tests sink to lake bottoms, thus storing silica that would otherwise go into the water column and downstream.

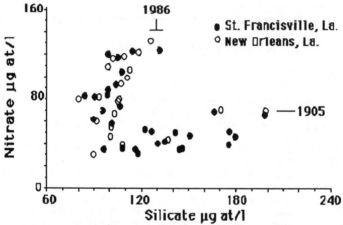

Figure 1. Average annual nitrate and silicate concentration in the lower Mississippi River. The data are from U.S. Geological Survey and the New Orleans Water Board. Further details are in Turner and Rabalais (1991).

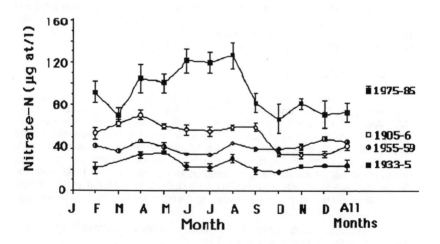

Figure 2. Seasonal nitrate concentrations in the lower Mississippi River. A +/- 1 standard error of the mean is shown.

The widespread occurrence of changes in nutrient loading to coastal zones has also changed the annual and seasonal variability of nutrient concentrations. Fertilizers, a major contributor to the nitrogen and phosphorus budget, are usually applied in the spring, just before or during the planting period. An example is shown for the lower Mississippi River at St. Francisville, and New Orleans, Louisiana (Figure 2). A spring rise in nitrate concentration is now evident.

These changes are important to understand, if only because nitrogen is commonly thought to be limiting phytoplankton growth in coastal and oceanic waters (e.g. Harris 1986, Valiela 1984). However, not all coastal systems are nitrogen limited (e.g., the Huanghe in China is phosphorus limited; Turner et al. 1990), nor is changing nutrient loading the only factor influencing phytoplankton growth (Skreslet 1986). Marine phytoplankton may also respond differently to nutrient additions if introduced gradually or suddenly, with changing flushing rates or salinity, and with cell density (Sakshaug et al. 1983, Sommer 1985, Suttle and Harrison 1986, Turpin and Harrison 1980).

Causes of Eutrophication

A major global source of the new nitrogen and phosphorus entering estuaries is originally from fertilizer applications, and, to a lesser extent, the combustion of fossil fuels. Sewage outfalls are a dramatic, but only symptomatic fruition, of the more basic rise in consumption of basic commodities, and of the need to dispose of the undesirable by-products. There would be no need for disposal if the products were not grown, produced and consumed.

World fertilizer consumption has been increasing since World War II, when industrial processes evolved for fixing atmospheric nitrogen into fertilizers, and mining of P-rich mineral deposits expanded rapidly (Figure 3). Additional nutrients are released through de-vegetation, farming, soil erosion, weathering, etc. A linear relationship between fertilizer application and water quality is not always expected, because of the interaction of various ecosystem components, and subsequent adjustments by microorganisms, in particular (see for example, Aber et al. 1989).

This new industrially-produced nitrogen is recycled through the atmosphere, and into water supplies. The worldwide rise in the concentration of nitrogen in rainfall and snowfall is symptomatic of the extent of these human activities on the environment. The importance of atmospheric inputs of essential nutrients has only recently been understood, but is clearly important and changing. For example, the inputs from precipitation and river runoff are equal in Chesapeake Bay (Correll and Ford 1982), an estuarine system often described as eutrophic. The nitrogen concentration in precipitation has recently increased in the eastern U.S. and in Europe (e.g. Likens and Borman 1979; Brimblecombe and Pitman 1980). These changes are so pervasive as to appear in snowfall in the Antarctic (Figure 4.). It is therefore likely that most landscapes are subject to significantly increased nitrogen loading this century. In addition, we have observed a change this century from viewing eutrophication as a local and perhaps point-source problem manageable on a regional scale, to a phenomena that is the cumulative result of many small actions throughout the world, and whose scale of management is vastly expanded (and more expensive).

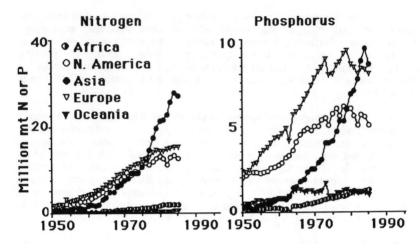

Figure 3. World and nitrogen and phosphorus fertilizer consumption since 1950.

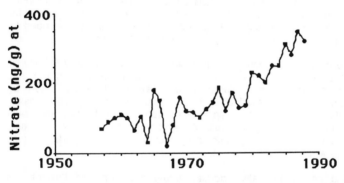

Figure 4. Recent changes in nitrate in snowfall in Antarctic (adapted from a figure in Turner 1991, whose data source is Mayewski and Legrand 1990).

Some Consequences of Eutrophication in Coastal Habitats

There are several major categories of ways that eutrophication has affected coastal habitats: through changes in food source quantity, food source quality, habitat suitability, alterations of predator-prey relationships, catastrophic events, ecosystem predictability, and intraspecific competition. We know more about what is on this list, which is seemingly all-encompassing, than of the scientific understanding of the subtle interactions. We also have limited experience treating the symptoms. However, we do have the experience to recognize the validity of the list, although it has not really been compiled for coastal systems in a way comparable to what has been done for freshwater systems. Some brief examples are provided below to illustrate the major points.

The abundance of coastal diatoms is influenced by the silicon supplies, whose Si:N atomic ratio is about 1:1 (the Redfield ratio). Diatoms out-compete other algae in a stable and illuminated water column of favorable silicate concentration. When nitrogen increases and silicate decreases, flagellates may increase in abundance (Officer and Ryther 1980) and form blooms. In particular, noxious blooms of flagellates are becoming increasingly common in coastal systems (Table 1; many other examples are in Shumway 1990). Zooplankton, the main consumers of whole diatoms and a staple of juvenile fish, are thus affected by these nutrient changes in a cascading series of interactions. Furthermore, where eutrophication occurs, hypoxia often follows (Table 2), presumably as a consequence of this increase in organic loading. Supportive evidence of this benthic-pelagic coupling is the observations of Cederwall and Elmgren (1980) who demonstrated a rise in macrobenthos around the Baltic islands of Gotland and Oland, which they attributed to eutrophication, a known event (Nehring 1984).

Table 1. Examples of the appearance of noxious blooms in coastal waters.

Area	Species	Year(s)	Reference
East coast of Florida	*Synechoccus* sp.	1972, 73	Mitsui et al. 1989
Laguna Madre, TX	"brown" tide	1990	Stockwell and Busley, pers. comm.
Long Island bays	brown tide	1985, 86	Cosper et al. 1987
Narraganset Bay, RI	*Aureococcus*	1985	Tracey 1988
North Carolina	*Ptychodiscus brevis*	1987	Summerson and Peterson 1990
Swedish west coast	*Prorocentrum minimum* and *Gyrodinium aureolum*	since 1980	Graneli, et al. 1986
Texas coast	*Gymnodinium splendens, Pthychodiscus brevis*	1955, 1984, 86	Riley et al. 1989; Harper and Guillen 1989
Norway	*Chrysochromulina polyepis*	1988	Cherfas 1990
West coast of Florida	*Gymnodinium breve*	common	Eng-Wilmont and Martin 1977

Table 2. Examples of coastal eutrophication and its effects.

Area	Probable or Observed Effect[1]	Reference
Adriatic Sea	ox.; turb.; food chain	Krstulovic´ and Solic´ Justic´ 1991; Faganeli, 1985; Stachowitsch 1986
Baltic Sea, incl. Kattegat and Skagerrak	ex. al.; food chain; ox., macrophyt. gain; incr. sec. prod.	Rosenberg 1985, 1986; Rosenberg and Loo 1988; Cederwall and Elmgren, 1980; Andersson and Rydberg 1988; Ankar 1980
Bayou Texar, FL	ox., food chain, health, nox.	Moshiri et al. 1981
Chesapeake Bay	food chain, incr. prod., ox.	Seliger et al. 1985
Great South Bay, Long Island, NY	ex. al.	Ryther 1954
Lac de Tunis, Tunis, Tunisia	ex. al.; ox.; macrophyt. loss; macrophyte gain; food chain	Kelly and Naguib 1984
Mississippi R. Delta Bight, U.S.	changes in area and extent of hypoxia	Rabalais et al. 1991
Southern Bight of the North Sea	ex. al.; incr. prod., food chain; nox.; incr. sec. prod.	van Bennekom et al. 1975; Beukema and Cadée 1986; Westernhagen and Dethlefson 1983; Lancelot et al. 1987
Tampa Bay, FL	macrophyt. loss; turb; healt.; ox.; nox.; food chain	Johansson and Lewis 1991 Santos and Simon 1980

[1]Key to abbreviations:

ex. al.	= excessive algal growth (including filamentous and attached)
food chain	= food chain alterations affecting important fisheries species, including fish kills, loss of benthic organisms
incr. al. prod.	= increased primary productivity
incr. sec. prod.	= increased secondary productivity, including benthos
nox.	= noxious algal blooms
healt.	= health problems with seafood consumption
macrophyt. loss	= loss of important macrophytes
macrophyt. gain	= gain of macrophytes
macrophyt. inv.	= invasion of undesirable macrophytes
ox.	= low oxygen levels
turb.	= increased turbidity from phytoplankton growth

Habitat suitability is the result of several interacting conditions, including oxygen. Low oxygen levels in the water column, typically below 2 mg/l, are generally not satisfactory fish habitat. Because eutrophication frequently leads to low oxygen levels, an indirect effect of eutrophication may be the contraction of suitable habitat necessary for reproduction and growth. A recent example of this interaction is provided by Coutant and Benson (1991). They examined the summer habitat conditions of the striped bass, *Morone saxatilis*, an important commercial and recreational fisheries species in Chesapeake Bay, where the frequency and duration of low oxygen events appears to be increasing (Officer et al. 1984; Seliger et al. 1985). They estimated habitat suitability as the thickness of the July water column that had temperature and oxygen conditions suitable for striped bass (for the Central Basin); they then compared the changes in the habitat to an empirically-derived estimate of juveniles. Among the annual variations usually attributable to climatic influences is a coincidental decline in both parameters (Figure 5). In effect, the decline in water quality appears to have compromised the habitat suitable for juvenile reproduction. Without the long-term surveys of juvenile abundance, the relationships might have gone unnoticed, because the adults are harvested over a much larger area.

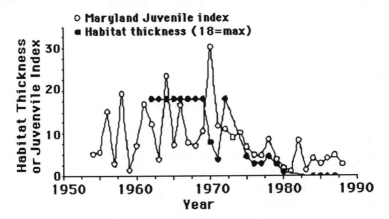

Figure 5. Decline in habitat suitability and the abundance of juvenile striped bass in Chesapeake Bay from 1954 to 1988 (from data in Table 1, in Coutant and Benson 1991). The maximum value for the area of suitable habitat is 18 m, and is based on the water column with a temperature below 25°C and dissolved oxygen above 2 mg/l during July. The juvenile index is based on summer-time beach seining surveys of 100 day-old striped bass at standard locations.

Another aspect of eutrophication has to do with the loss of emergent and submerged macrophytes that limit fisheries species during critical recruitment periods. It is well-established that certain coastal fisheries species seem to require a physical structure to escape from predators while young. Where the area of estuarine macrophytes declines or improves, fisheries harvest is observed to respond proportionally (e.g. Turner and Boesch 1987). The loss of seagrass beds following decreased water clarity is often observed (Cambridge and McComb 1984, Cambridge et al. 1986), leading one to conclude that the potential harvest of dependent fisheries will probably decline too. Such subtle changes are difficult to detect without substantial amounts of long-term data (two examples are in Turner and Boesch 1987).

We could not find examples of investigations of whether coastal emergent marsh communities changed following eutrophication. The most likely communities to change may be the more diverse ones, where a small alteration in a limiting nutrient could determine the plant community composition.

A change in the timing of the spring bloom that fish entering estuaries to feed during critical recruitment periods may also be an important consequence of eutrophication. Townsend and Cammen (1988) point out that the timing of the spring plankton bloom in high latitudes is important to the benthic-pelagic couplings upon which fisheries recruitment is dependent. The timing of a bloom is often very sensitive to light conditions, but declines from nutrient depletion. Changes in the nutrient loading, and timing, could affect recruitment success through a mis-match of larval recruitment and food supply, as well as an altered food chain. Early blooms with a greater sedimentation to the benthos could positively affect demersal fishes, but late blooms positively affect pelagic fisheries through a zooplankton food chain. Zooplankton graze on diatoms, extensively, so that changes in nutrient loading and the ratio of nutrients could affect the balance and timing of the demersal and pelagic food webs.

Reversal of Coastal Eutrophication

There are a few examples of what happens when eutrophication is reversed. Four are discussed here: (1) Bayou Texar, Florida, (2) Tampa Bay, Florida, (3) Seto inland Sea of Japan, and, (4) Kaneohe Bay, Hawaii.

The eutrophication of Bayou Texar, near Pensacola, Florida, was studied by Moshiri et al. (1981) to determine the causes and remedies for extensive fish kills (up to five weeks), closure of recreational use, dinoflagellate blooms, (primarily *Ceratium* sp. and *Gymnodinium* sp.), chrysophytes (primarily from the genera *Chrysochromulina, Chromulina, Navicula* and *Cyclotella*), and high algal biomass which contributed to low dissolved oxygen levels. In 1974 a retention reservoir and weirs in the upstream channels were built, and sewage plants repaired. The authors reported an almost total reduction in fish kills, a 90% reduction in phytoplankton primary production, and a virtual elimination of algal blooms. Public use began again.

Johansson and Lewis (in press) documented a decline in water quality and then restoration in Hillsborough Bay, near Tampa Bay, Florida. Tampa Bay was "grossly polluted" from cannery wastes and poorly treated municipal wastewater discharges in the 1960s. A principal concern was the loss of submerged macrophytes (*Halodule wrightii*), presumably because of increased turbidity following eutrophication. Anoxia and high coliform counts were additional concerns. Following improved sewage treatment, the nitrogen loading to Hillsborough Bay was reduced 30%, primarily between 1979 and 1980. It was not until four years later that the ambient chlorophyll *a* concentration decreased substantially, which was coincidental with the decline in a nuisance planktonic blue-green algae, *Schizothrix calcicola sensu* Drouet. Turbidity decreased.

A major change in seagrasses, reduced to 20% of the areal coverage of 100 years ago, occurred following improvements in water quality around 1984. The areal coverage of submerged macrophytes (seagrasses) in Hillsborough Bay and Middle Tampa Bay doubled from 1986 to 1989. However, seagrasses in most shallow areas have not yet recovered and high concentrations of chlorophyll *a* persist. Major sources of nitrogen and phosphorus remain in the fertilizer plants and storage facilities, as well as leakage during loading terminals. This partial recovery has followed partial reduction of nutrient loads, but the system was still responding after five years of the initial nutrient load reduction.

The third example is a report by Cherfas (1990) of a summary by Smetacek. The annual incidence of red tides in the Seto inland sea of Japan increased from 40 to more than 300 annually, from 1965 to 1972 as nutrient loading increased. In 1972 nutrient loading was reduced by half, and the frequency of red tides peaked in 1975 and has been declining ever since.

Smith (1981), described the impacts of nutrient loading and restoration in Kaneohe Bay, Hawaii, a subtropical coral reef/estuary complex. Effluent from a large sewage treatment plant provided 80-90% of the inorganic nitrogen and phosphorus loading to the bay before being diverted to an ocean outfall in 1977-78. The initial sewage addition resulted in enhanced plankton biomass near the outfall (frequently a bloom developed), with only small increases in nutrient concentration, but a significant increase in benthic biomass. Following the diversion to the outfall, the plankton biomass dropped quickly, but the benthic biomass dropped more slowly. The authors explain the difference response times as being a function of the washout characteristics of the two compartments: the phytoplankton have low storage of nutrients, high nutrient turnover, and a fast washout time. In contrast, the benthic community has a slower nutrient recycling rate of the larger nutrient storage pool, hence the biomass remained nearly the same as before the diversion after three years.

These four studies, particularly the discussion by Smith (1981), illuminates the obvious possibilities for recovery following a reduction in nutrient loading. When the causes of eutrophication are reversed, the symptoms may be reversed. But recovery varies with the cycling rate of various stored nutrients, and, the flushing rate of the estuary. Seagrasses, at least those that are inhibited by the reduced light penetration that often accompanies eutrophication, may recover, but only over a decade or more.

Policy Implications

Most aspects of the policy implications of eutrophication are well stated in other places and the basis for numerous legislation. However, some less obvious issues are related to the interrelationship of the national policy of nutrient control in freshwaters and the impact, or lack of impact, on coastal systems. Our point here is that the management of eutrophication on a national scale has not sufficiently integrated freshwater and estuarine systems. The national freshwater policy is to control phosphorus, and is based on the numerous excellent laboratory and field studies of the stimulatory effect of phosphorus on freshwater ecosystems. However, coastal systems are usually thought to be nitrogen limited.

A national policy in common to both freshwater and coastal systems is sewerage treatment, in general. But, as is shown for the Mississippi River (Turner and Rabalais 1991), the terrestrial system is very leaky, and treatment does not mean a reduction of loading to the estuary via water and precipitation. So a second understated issue, therefore, is that sewerage treatment upstream does necessarily equate to controlling nutrient loading to downstream estuaries.

Third, the minimization and mitigation of uses seems a less prudent management policy, than an outright reduction in use. The ecosystem is simply too leaky to control all flows of important nutrients from use to arrival in the estuary. Of course, getting people to accept that reduction in use is another issue.

References

Aber, J.D., K.J. Nadelhoffer, P. Streudler and J.M. Melillo. 1989. Nitrogen saturation in northern forest ecosystems. Bioscience 39:378-386.

Andersson, L. and L. Rydberg. 1988. Trends in nutrient and oxygen conditions within the Kattegat: Effects on local nutrient supply. Estuarine, Coastal Shelf Sci. 26:559-579.

Ankar, S. 1980. Growth and production of Macoma balthica (L.) in a northern Baltic soft bottom. Ophelia, Suppl. 1:31-48.

van Bennekom, J.A., W.W.C. Gieskes and S.B. Tijssen. 1975. Eutrophication in Dutch coastal waters. Proc. R. Soc. London B. 189:359-374.

Beukema, J.J., and G.C. Cadée. 1986. Zoobenthos response to eutrophication of the Dutch Wadden Sea. Ophelia 26:55-64.

Borman, F.H. 1982. The effects of air pollution on the New England landscape. Ambio 11:338-346.

Borman, F.H., and G.E. Likens. 1979. Pattern and Process in a Forested Ecosystem. Springer-Verlag, New York.

Brimblecombe, P., and J. Pitman. 1980. Long-term deposit at Rothamsted, southern England. Tellus 32:261-267.

Cambridge, M.L., and A.J. McComb. 1984. Loss of seagrass in Cockburn Sound, Western Australia. II. The time course and magnitude of seagrass declines in relation to industrial development. Aquatic Botany 20:229-243.

Cambridge, M.L., A.W. Chiffings, C. Britton, L. Moore and A.J. McComb. 1986. Loss of seagrass in Cockburn Sound, Western Australia. II. Possible cause of seagrass decline. Aquatic Botany 24:269-285.

Cederwall, H. and R. Elmgren. 1980. Biomass increases of benthic macrofauna demonstrates eutrophication of the Baltic Sea. Ophelia, Suppl.1:287-304.

Cherfas, J. 1990. The fringe of the ocean - under siege from land. Nature 248:163-165.

Correll, D.L., and D. Ford. 1982. Comparison of precipitation and land runoff as sources of estuarine nitrogen. Estuarine, Coastal and Shelf Sci. 15:45-56.

Cosper, E.M., W.C. Dennison, E.J. Carpenter, V.M. Bricelj, J.G. Mitchell, S.H. Kuenstner, D. Colflesh and M. Dewey. 1987. Recurrent and persistent brown tide blooms perturb coastal marine ecosystem. Estuaries 10:284-290.

Coutant, C.C., and D. L. Benson. 1991. Summer habitat suitability for striped bass in Chesapeake Bay: reflections on a population decline. Trans. Am. Fish. Soc. 119:757-778.

Eng-Wilmont, D.L. and D.F. Martin. 1977. Growth response of the marine blue-green alga, Gomphosphaeria aponina, to inorganic nutrients and significance to management of Florida red tide. Microbios. 19:167-179.

Faganeli, J., A. Avcin, N. Fanuko, A. Malej, V. Turk, P. Tusnik, B. Vriser and A. Vukovic´. 1985. Bottom layer anoxia in the central part of the Gulf of Trieste in the late summer of 1983. Marine. Poll. Bull. 16:75-78.

Fransz, H.G., and J.H.G. Verhagen. 1985. Modelling research on the production cycle of phytoplankton in the southern bight of the North Sea in relation to riverborne nutrient loads. Netherlands J. Sea Res. 19:241-250.

Graneli, E., H. Persson, and L. Edler. 1988. Connection between trace metals, chelators and red tide blooms in the Laholm Bay south-east Kattegat - an experimental approach. Mar. Envir. Res. 18:61-78.

Harper, D.E., Jr., and G. Guillen. 1989. Occurrence of a dinoflagellate bloom associated with an influx of low salinity water at Galveston, Texas, and coincident mortalities of demersal fish and benthic invertebrates. Contr. Mar. Sci. 31:147-161.

Harris, G.P. 1986. Phytoplankton Ecology: Structure, Function and Fluctuations. Chapman and Hall, New York.

Johansson, J.O.R., and R.R. Lewis, III. 1991. Recent improvements of water quality and biological indicators in Hillsborough Bay, a highly impacted subdivision of Tampa Bay, Florida, U.S.A. In. Prod. Intl. Conf. Marine Coastal Eutrophication, Bologna, 21-24 March, 1990. In press.

Justic´, D. 1991. Hypoxic conditions in the northern Adriatic Sea: Historical development and ecological significance. J. Geol. Soc. London (in press).

Kelly, M., and M. Naguib. 1984. Eutrophication in coastal marine areas and lagoons: a case study of 'Lac de Tunis'. Unesco Reports in Marine Science No. 29., Paris. 54 p.

Krstulovic´, N., and M. Solic´. 1990. Long-term study of heterotrophic bacteria as indicators of eutrophication of the open middle Adriatic Sea. Estuar. Coastal Shelf Sci. 30:611-617.

Lancelot, C., G. Billen, A. Sournia, T. Weisse, F. Colijn, M.J.W. Veldhuis, A. Davies and P. Wassman. 1987. *Phaeocystis* blooms and nutrient enrichment in the continental zones of the North Sea. Ambio 16:38-46.

Likens, G.E., and H. Bormann. 1979. The role of watershed and airshed in lake metabolism. Arch. Hydrobiol. 13:195-211.

Mayewski, P.A., and M.R. Legrand. 1990. Recent increase in nitrate concentration of Antarctic snow. Nature 346:258-260.

Mitsui, A., D. Rosner, A. Goodman, G. Reyes-Vasquez, T. Kusumi, T. Kodama, and K. Nomoto. 1989. Hemolytic toxins in marine cyanobacterium *Synechococcus* sp. In: T. Okaichi, D.M. Anderson, and T. Nemoto (eds.), Red Tides. Biology, Environmental Science, and Toxicology. New York, Elsevier. pp. 367-370.

Moshiri, G.A., N.G. Aumen and W.B. Crumpton. 1981. Reversal of the eutrophication process: A case study. pp. 373-390, In: B.J. Neilson and L.E. Cronin (editors) Estuaries and Nutrients, Humana Press, Inc. Clifton, New Jersey. pp. 373-390.

Nehring, D. 1984. The further development of the nutrient situation in the Baltic proper. Ophelia, Suppl.3:167-179.

Neilson, B.J., and L.E. Cronin (editors). 1981. Estuaries and Nutrients, Humana Press, Inc. Clifton, New Jersey. 643 p.

Officer, C.B. and J.H. Ryther. 1980. The possible importance of silicon in marine eutrophication. Mar. Ecol. Progr. Ser. 3:83-91.

Officer, C.B., R.B. Biggs, J.L. Taft, L.E. Cronin, M.A. Tyler, and W.R. Boynton. 1984. Chesapeake Bay anoxia: origin, development, and significance. Science 223:22-27.

Rabalais, N.N., R.E. Turner, W.J. Wiseman, Jr., and D.F. Boesch. 1991. A brief summary of hypoxia on the northern Gulf of Mexico continental shelf: 1985-1988. J. Geol. Soc. London (in press).

Redfield, A.C. 1958. The biological control of chemical factors in the environment. Am. Sci. 46:205-222.

Riley, C.M., S.A. Holt, G.J. Holt, E.J. Buskey and C.R. Arnold. 1989. Mortality of larval red drum (*Sciaenops ocellatus*) associated with a *Ptychodiscus brevis* red tide. Contr. Mar. Sci. 31:137-146.

Rosenberg, R. 1985. Eutrophication - the future marine coastal nuisance? Marine Pollution Bull. 16:227-231.

Rosenberg, R. (ed.). 1986. Eutrophication in marine waters surrounding Sweden. National Swedish Environmental Protection Board report 3054.

Rosenberg, R. and L.-O. Loo. 1988. Marine eutrophication induced oxygen deficiency: Effects on soft bottom fauna, western Sweden. Ophelia 29:213-225.

Ryther, J.H. 1954. The ecology of phytoplankton blooms in Moriches Bay and Great south Bay long Island, N.Y. Biol. Bull. 106:198-209.

Sakshaug, E., K. Andresen, S. Mkklestad and Y. Olsen. 1983. Nutrient status of phytoplankton communities in Norwegian waters (marine, brackish, and fresh) as revealed by their chemical composition. J. Plankton Res. 5:175-196.

Santos, S.L., and J. L. Simon. 1980. Response of a soft-bottom benthos to annual catastrophic disturbance in a South Florida estuary. Mar. Ecol. Progr. Ser. 3:347-355.

Seliger, H.H., J.A. Boggs, and W.H. Biggley. 1985. Catastrophic anoxia in the Chesapeake Bay in 1984. Science 228:70-73.

Shumway, S.E. 1990. A review of algal blooms on shellfish and aquaculture. J. World Mariculture Society 21:65-104.

Skreslet, S. (editor). 1986. The Role of Freshwater Outflow in Coastal Marine Ecosystems. Springer-Verlag, New York. 453 p.

Smith, S.V. 1981. Responses of Kaneohe Bay, Hawaii, to relaxation of sewage stress. In: B.J. Neilson and L.E. Cronin (editors) Estuaries and Nutrients, Humana Press, Inc. Clifton, New Jersey. pp. 391-410.

Smith, R.A., R.B. Alexander, and M.G. Wolman. 1987a. Analysis and interpretation of water-quality trends in major U.S. rivers, 1974-1981. U.S. Geological Survey Water-Supply Paper 2307. 25 p.

Smith, R.A., R.B. Alexander, and M.G. Wolman. 1987b. Water-quality trends in the nation's rivers. Science 235:1605-1615.

Sommer, U. 1985. Comparison between steady state and non-steady state competition: Experiments with natural phytoplankton. Limnol. Oceanogr. 30:335-346.

Stachowitsch, M. 1986. The Gulf of Trieste: a sensitive ecosystem. Nova Thalassia 8, Suppl. 3:221-235.

Stachowitsch, M. 1984. Mass mortality in the Gulf of Trieste: The course of community destruction. Mar. Ecol. 5:243-264.

Summerson, H.C., and C.H. Peterson. 1990. Recruitment failure of the bay scallop, *Argopecten irradians concentricus*, during the first red tide, *Ptychodiscus brevis*, outbreak recorded in North Carolina. Estuaries 13:322-331.

Suttle, C.A., and P.J. Harrison. 1986. Phosphate uptake rates of phytoplankton assemblages grown at different dilution rates in semicontinuous culture. Can. J. Fish. Aquat. Sci. 43:1474-1481.

Townsend, D.W., and L.M. Cammen. 1988. Potential importance of the timing of spring plankton blooms to benthic-pelagic coupling and recruitment of juvenile demersal fishes. Biol. Oceanogr. 5:215-229.

Tracey, G.A. 1988. Feeding reduction, reproductive failure, and mortality in *Mytilus edulis* during the 1985 "brown tide" in Narraganset Bay, Rhode Island. Mar. Ecol. Prog. Ser. 50:73-81.

Turner, R.E. 1991. Fertilizer nitrogen in Antarctic snowfall. Nature: (in press).

Turner, R. E. and D. F. Boesch. 1987. Aquatic animal production and wetland relationships: Insights gleaned following wetland loss or gain. In: D. Hooks (ed.), Ecology and Management of Wetlands. Croon Helms, LTD, Beckenham, Kent, UK. pp: 25-39.

Turner, R.E., and N.N. Rabalais. 1991. Changes in Mississippi River Water Quality this Century and Implications for Coastal Food Webs. Bioscience 43:in press).

Turner, R.E., N.N. Rabalais and Z.-N. Zhang. 1990. Phytoplankton biomass, production and growth limitations on the Huanghe (Yellow River) continental shelf. Cont. Shelf Res. 10:545-571.

Turpin, D.H., and P.J. Harrison. 1980. Cell size manipulation in natural marine, planktonic, diatom communities. Can. J. Fish. Aquat. Sci. 37:1193-1195.

Valiela, I. 1984. Marine Ecological Processes. Springer Verlag, New York.

United States Department of Agriculture (USDA), Annual Reports. Agricultural Statistics, United States Printing Office, Washington, D.C.

van Westernhagen, H., and V. Dethlefsen. 1983. North Sea oxygen deficiency in 1982 and its effect on the bottom fauna. Ambio 12:264-266.

White, F.C., J.R. Hairston, W.N. Musser, H.F. Perkins and J.F. Reed. 1981. Relationship between increased crop acreage and nonpoint-source pollution: A Georgia case study. J. Soil Water Conservation 36:172-177.

Wulff, F., and L. Rahm. 1988. Long-term, seasonal and spatial variations of nitrogen, phosphorus and silicate in the Baltic: An overview. Mar. Environmental Res. 26:19-37.

COASTAL SYSTEMS: ON THE MARGIN

Howard Levenson[1]

Abstract

Coastal systems -- bays, estuaries, wetlands, and general nearshore waters -- are located on the margins of our continental landforms and are on the "margin" in terms of threats to their health and productivity. Programs and procedures established by Clean Water Act and the Marine Protection, Research, and Sanctuaries Act have helped reduce the amounts of pollutants entering marine waters, but they are not enough to provide sufficient protection -- many estuaries and some coastal waters will continue to degrade or begin to do so during the next few decades unless additional protective measures are taken. The increasingly rapid pace of coastal development will be a major factor in future resource degradation.

One missing link is a framework for deciding where and how to provide additional management, either to reverse current degradation or to prevent degradation from occurring. A systematic framework should be developed -- one that uses a broad, environmental quality-based approach to provide additional management where needed. It will have to include difficult decisions about coastal development and land use changes. This paper reviews major legislation moving through Congress to address these issues and the status of efforts to develop international marine protection protocols.

The Problems

Estuaries and coastal waters have been used -- some say abused -- for years by coastal communities and industries for the disposal of various wastes (e.g., municipal sewage effluent and sludge, industrial effluent, dredged material). (In contrast, the open ocean exhibits few adverse impacts from waste disposal, although it is unclear whether toxic pollutants will cause long-term damages to open-ocean resources.)

Such disposal, whether via dumping or pipeline discharge, has been relatively cheap and has often been the preferred short-term approach. In part this can be attributed to the fact that many national statutes and

---

[1] Oceans and Environment Program, Office of Technology Assessment, United States Congress, Washington, D.C. 20510.

75

regulations and international treaties and conventions protect single environmental sectors (e.g., air, land, water) or address single waste management activities (e.g., ocean dumping). Consequently, disposal decisions often are based on using the least regulated environment, rather than on evaluating the risks and benefits of alternative management options. This approach has not protected the environment, either overall or components such as marine environments (IMO et al, 1990; U.S. Congress, 1987), nor has it provided for rational waste management.

This is tragic, because coastal waters are the most important of all marine environments with respect to their commercial resources, recreational uses, and ecological roles. They provide habitat during critical portions of the lifecycles of roughly two-thirds of the fish and shellfish caught in U.S. marine waters -- this fact alone distinguishes them from freshwater lakes and rivers. The value of marine commercial and sport fishing has been estimated at over $10 billion, and tourism and other recreational pursuits in the associated shoreline areas is the lifeblood of many communities. As a result, individual citizens and fishermen, towns, and county and State governments have grown increasingly concerned about the impacts of waste disposal on harvests and tourism and on the property values of real estate located near these waters.

Numerous adverse effects around the United States have been associated with such waste disposal:

o low dissolved oxygen levels leading, for example, to fish kills;
o contamination of shellfish with microorganisms, leading to restrictions on about one-third of our productive shellfish areas;
o contamination of shellfish, fish, birds, and mammals with metals and organic chemicals, in some cases leading to ulcers and tumors;
o declines in submerged aquatic vegetation;
o contamination of sediment with metals and organic chemicals;
o restrictions on fishing and consumption of certain fish species;
o restrictions on swimming; and
o increased reporting of human diseases such as viral gastroenteritis.

The degree and distribution of these impacts have varied widely among different waterbodies and organisms, but no region of the country has been immune to serious impacts. Public attention has focused mostly on the northeastern United States (e.g., the Chesapeake Bay and the New York Bight), southern California, and Puget Sound. Serious effects, however, also have occurred or are beginning to occur in the less-publicized (and relatively less-studied) Gulf of Mexico and along the southern Atlantic coast.

As a result, the Office of Technology Assessment (OTA) concluded that without additional protective measures, the next few decades will witness new or continued degradation in many estuaries and some coastal waters around the country, even in some that showed improvements in the past (U.S. Congress, 1987).

Although positive steps have been taken since OTA's report was released, its conclusions have not changed -- indeed, events during the last few years have confirmed them. For example, concerns heightened in 1987, when several incidents focused public and government attention on the quality of U.S. marine waters. These include the infamous barge of trash that wandered to three countries before returning to New York; the deaths of hundreds of dolphins along the east coast; and the washing ashore in New Jersey of illegally dumped medical wastes during the height of the summer tourist season. The summers of 1988 and 1989 saw further wash-ups of medical wastes on the beaches of numerous States, as well as continued occurrence of the other adverse effects enumerated above.

Moreover, on a global scale, a 1990 report issued by the United Nations Environment Programme concluded that "encroachment on coastal areas continues worldwide ... [i]f unchecked, this trend will lead to global deterioration in the quality and productivity of the marine environment." Eutrophication appears to be occurring at many or most of the world's river mouths, often associated with toxic algal blooms. Diversions of freshwater inputs are increasing. Questions about global climate change also loom -- e.g., how might it affect nutrient distributions and phytoplankton populations.

In the United States, at least, industrial and municipal pipeline discharges are at least as important as dumping of sewage sludge and dredged material in polluting marine resources and causing damage to them.[2] Combined (storm and sanitary) sewer overflows also are important sources of pollutants in some marine areas. Runoff (nonpoint source pollution) from agricultural lands and urban areas -- although not classed as a disposal activity -- also contains large amounts of certain pollutants (such as suspended solids, fecal coliform bacteria, nutrients, pesticides, and some metals) and enters marine waters in large quantities. In some waterbodies, runoff is considered the leading cause of degradation.

Land-based sources of marine pollution are not the only way in which marine waters can be degraded. Another potential source of pollutants is acid deposition from the air; interest in this was sparked by a 1988 Environmental Defense Fund report. The Clean Air Act Amendments of 1990 required EPA to conduct research on atmospheric deposition in the Great Lakes and coastal waters.

Moreover, we cannot forget that "non-polluting" activities also greatly affect marine waters. OTA and many others have noted that filling of wetlands, channeling or diversion of rivers, and overharvesting of fish and shellfish significantly affect marine waters and resources throughout the country. The first two of these are frequently related to the increased pace of development along our marine shorelines.

More than anything else, it is this development factor that makes the situation with estuaries and coastal waters so unique and precarious. About three-fourths of the U.S. population is expected to live within 50 miles of a coastline, including the Great Lakes, by the end of this century. As might be expected, this will mean increased development and redevelopment along shorelines. In turn, unless careful consideration is given to their ultimate environmental effects, development and redevelopment will entail both habitat modification in some cases and increased pollutant inputs into marine waters in almost all cases.

---

[2] Alternative, land-based disposal of most of these wastes is limited by economic feasibility, public opposition, and regulatory restrictions.

Among the activities that degrade our coastal waters, the issues of nonpoint pollution from agricultural areas and of coastal development patterns have received little Federal attention. Both are related to land-use planning, which traditionally has been a matter for local and State consideration. Many observers have identified land-use planning as a critical factor in determining the pace and extent of development and in controlling urban and agricultural nonpoint pollution.

U.S. Coastal Pollution Management Efforts

Two statutes form the basis of most Federal regulatory efforts to combat marine pollution: the Marine Protection, Research, and Sanctuaries Act (MPRSA) and the Clean Water Act (CWA). The MPRSA regulates the dumping of waste material in coastal and open ocean waters, while the CWA has jurisdiction over pipeline discharges in all marine waters, dumping in estuaries, and (theoretically) runoff.

The MPRSA and CWA use pollution control strategies based both on marine quality standards and on emission standards. The pollutant control programs under the CWA and the dumping criteria and permitting procedures under the MPRSA have helped reduce the amounts of pollutants entering marine waters. Under the CWA, the construction or upgrading of municipal sewage treatment plants -- towards which the Federal government has contributed almost $50 billion during the past 15 years -- has led to some important successes, namely increased oxygen levels and decreased nutrient levels in some estuaries and coastal waters. The CWA also has succeeded in reducing the levels of some metals and organic chemicals in industrial discharges into marine waters. The MPRSA, despite the controversy over the dumping of municipal sewage sludge, has been effective in regulating dumping in coastal waters and the open ocean; in particular, the dumping of industrial wastes has declined dramatically since the early 1970s.

However, these provisions have not fully protected estuaries and coastal waters from degradation associated with waste disposal. Only partial implementation of the relevant CWA provisions has been achieved to date, for example. Moreover, even with total compliance, existing regulations still will not be sufficient to maintain or improve the health of all estuaries and coastal waters for several reasons:

o current programs do not address all important toxic pollutants;
o current programs do not adequately address runoff;
o urban runoff and legal pipeline discharges will increase as population and industrial development expand in coastal areas; and
o Federal resources for sewage treatment are declining, and the ability of States and communities to fill the breach is uncertain.

The United States has been unable to develop a comprehensive pollution control strategy, one that simultaneously considers multiple environments, types and sources of pollutants, and exposure pathways. For example, the laudable reduction of "conventional" pollutants (e.g., suspended solids, fecal coliform bacteria) in municipal sewage effluent also resulted in the increased production of sewage sludge -- and we have not yet provided adequate long-term management for this voluminous, controversial byproduct.

There is consensus among public interest groups and local, State, and Federal officials that a more comprehensive strategy is needed -- in particular, one that considers the entire "system" that generates pollutants. One part of this strategy concerns avoiding the generation -- and hence the need to dispose -- of waste materials in the first place. This can involve, for example, on-site "source" reduction of hazardous and non-hazardous industrial wastes. Programs for waste reduction, and incentives to develop and use low-waste or "clean" technologies, must be a central priority and permanent feature of national strategies for environmental protection -- a principle adopted by the 1972 U.N. Conference on the Human Environment.

Another part is to avoid using as much of some products that provide important benefits but which also contribute pollutants that eventually enter coastal waters. For example, using fewer pesticides and herbicides in lawn care and agricultural production can reduce the levels of pollutants in runoff. A third part is to recycle those waste materials that are generated. For example, sewage sludge can be applied on land to enhance soil productivity or reclaim degraded lands -- providing the sludge meets appropriate standards regarding concentrations of metals, bacteria, and other constituents.

Clearly, to the extent that this strategy can be implemented, then the amounts of pollutants in runoff and the need to intentionally dispose of various wastes in different environments can be reduced. However, little attention has been given to these efforts, both in general and more particularly with respect to coastal waters. There is a clear need for greater incentives -- be they regulatory or non-regulatory -- to reduce the generation of wastes as the best means of reducing waste disposal-related risks.

Unless such incentives are in place and until they have been shown to be effective, however, large amounts of municipal and industrial wastes and dredged material will continue to require disposal for the foreseeable future. These wastes will continue to contain metals and synthetic organic chemicals, limiting possibilities for their beneficial re-use.

Protecting Coastal Waters

Various strategies might help to maintain or improve the health of our estuaries and coastal waters. One strategy to begin with is to improve the current system by increasing enforcement and by amending the Clean Water Act to expand controls over a broader range of sources and to promote reduced generation of wastes. Several bills in the 101st Congress addressed some of these issues.

For example, H.R. 2647 contained provisions that would have, for example:

o applyed existing ocean discharge criteria to all coastal water discharges;
o required EPA to develop coastal water quality criteria and States to adopt standards based on those criteria;
o extended the scope of these criteria from their traditional focus on water quality to other measures of environmental health, such as sediment quality; and

o   established an effluent fee system, to encourage reduced generation of
    waste and strengthen the industrial pretreatment program.

The effluent fee provision, while difficult to implement, could help
internalize some of the costs of marine disposal and presumably provide an
incentive for eliminating such disposal, or at least for eliminating the
substances that cause problems during disposal (e.g., heavy metals, organic
chemicals). While it has limits in that it does not address urban and
agricultural runoff, it could provide some of the economic incentives needed
to encourage direct dischargers to reduce the volume and toxicity of their
discharges. A fee on runoff might be imposed by taxing the products that
contribute pollutants to urban and agricultural runoff.

Senate bills 1178 and 1179 contained other provisions that would help
improve the Clean Water Act, including regulating combined sewer overflows
and stormwater discharges into marine waters; requiring an inventory of
releases of toxic substances and a strategy to combat subsequent problems;
and linking permit renewals for discharges with a determination that available
waste minimization practices are being fully applied.

This strategy would help bolster our existing regulatory tools and
would likely be sufficient to prevent or correct problems in some marine
waters. But it would require some combination of continued Federal, State,
and local investments in enforcement efforts for all regulations, as well as
industry investments in changing processes and technologies to comply with
these regulations. The development of additional coastal water quality criteria
would be difficult, but it would provide an important tool in future permit
evaluations.

Such a strategy also will require monitoring and research, for purposes
of enforcement and evaluating long-term trends. Many monitoring programs
unfortunately do not provide adequate information on the state of the
environment or on the effects of anthropogenic activities. Greater emphasis
must be placed on specifying goals and objectives, formulating testable
hypotheses, using better quality control procedures and statistical designs,
and conducting periodic scientific and administrative reviews (e.g., see NRC,
1990). H.R. 2647 contained a provision to set up a national task force to
systematically develop national monitoring guidelines, coordinate existing
Federal monitoring activities, and establish regional monitoring programs.
The bill would have promoted uniformity among monitoring programs, which
is critical for comparing results, but it also recognized that flexibility in
monitoring is needed to address local needs.

These provisions are similar in concept, albeit not in detail, to the
regional marine research programs proposed by the Senate. S. 587 and
other Senate bills contained provisions specifically regarding research on, for
example atmospheric deposition, algal blooms, monitoring of floatables,
monitoring protocols, and health and economic impacts. Some also would
require the use of whole effluent biological toxicity testing as an indicator of
pollution problems. In general, the regional approach -- if properly
coordinated with ongoing efforts conducted by NOAA, EPA, and other
Federal and State agencies -- could provide flexibility that standardized
national monitoring and research programs cannot conduct.

Even if such measures are implemented, the strategy of strengthening existing Clean Water Act provisions will not be sufficient to protect many areas because of the adverse effects of runoff, increasing development, and other problems. A more comprehensive approach is needed. Many areas will require additional forms of management that use a broader, ecosystem-oriented planning approach -- i.e., on a watershed, drainage basin, or waterbody basis.

The Federal government and some States have already recognized this need and developed what might be called "waterbody management" programs to attack site-specific problems that cannot be handled by existing CWA programs. The Chesapeake Bay Program and the Puget Sound Water Quality Authority are two well-known regional waterbody programs. They attempt to bring together all parties with vested interests (including ones with authority to address land-use and runoff issues), identify the most important problems and their causes, and devise a long-term management plan to alleviate those problems. In essence, they use planning on an ecosystem basis to establish site-specific goals and identify the most environmentally and economically effective means of achieving those goals.

This site-specific approach is still in its infancy. The first Federal legislation authorizing it is the Water Quality Act of 1987, which authorized several activities and programs, including the National Estuary Program. Under the National Estuary Program, EPA is authorized to convene management conferences for individual estuaries. Several commercially important but degraded waterbodies have been included in the program through various pronouncements and activities by Congress, the States, and EPA. However, the process of selecting waterbodies has not involved a systematic, scientific identification of those waterbodies that need such management now or those that are likely to become degraded in the future.

The absence of a **system** for deciding when and how to provide additional, site-specific management raises important questions about the relative roles of Federal, State, and even local governments. Will enough waterbody management programs be developed and implemented? Should a national effort to select waterbodies and spur development of plans be developed? Will existing State and Federal mechanisms be used effectively to provide such management? An expanded system of any type would have to be flexible enough to integrate existing pollution control programs into new management plans. It might be most effective if it focused on identifying "needy" waterbodies and stimulating (e.g., with technical and financial assistance) the development and implementation of basin-wide management plans (i.e., vigorously implementing relevant provisions of the 1987 Water Quality Act). Some procedure is needed to account for ongoing, effective State and Federal efforts.

H.R. 2647 attempted to address many of these questions. It emphasized basin-wide approaches to coastal resource and land use management and it attempts to make our efforts much more systematic. For example, it would have:

o   required States to develop comprehensive coastal water quality protection programs, to be submitted to EPA for approval;

o required the States, as part of the programs, to identify coastal waters now
   experiencing degradation or likely to suffer degradation in the future;
o provided for an enforceable system of tradable credits for point and
   nonpoint source pollution, with the intent of reducing overall pollutant
   loadings; and
o extended the authorities of the National Estuary Program, for example by
   requiring management conferences to be convened and requiring
   approval and implementation of conservation and management plans.

     With respect to nonpoint pollution, the Water Quality Act of 1987
authorized EPA to administer a nonpoint pollution program. In the program,
States are to identify waters not expected to meet water quality standards
because of such pollution and submit a management program to EPA.
However, Congress has not yet appropriated sufficient resources for this
program. Some States have developed nonpoint pollution programs, but in
general there has not been a serious look at the problem with respect to
marine waters.

     Several bills in the 101st Congress also attempted to link the Coastal
Zone Management Act and coastal water quality. As noted above, coastal
development is increasing rapidly and is a major factor contributing to the
degradation of some marine waters. Although land-use planning has
traditionally been a local and State matter, H.R. 2647 would have: 1) included
coastal water quality within the policy goals of the CZMA; and 2) required
State coastal management agencies to develop an approved program to
implement management measures (e.g., density restrictions, buffer strips) for
land-based sources of nonpoint source pollution.

     Senate legislation contained provisions to: 1) identify croplands that, if
enrolled in the Conservation Reserve Program, would reduce nonpoint
pollution; 2) establish a voluntary registry of coastal lands, owners of which
would receive information about practices to help protect marine waters; 3)
require the regulation of stormwater discharges; and 4) establish the eligibility
of nonpoint pollution programs for grant assistance.

     These programs would involve other State and local authorities, and
they would be developed in conjunction with existing water quality programs
of the Clean Water Act, including the National Estuary Program. Given the
revised policy goals, they would begin to address a critical issue in coastal
water quality, i.e., coastal development and associated land-use planning.
Whether this would extend to filling of wetlands and channeling or diversion of
rivers is unclear.

International Efforts

     Several activities at the international level, briefly discussed here, have
direct relevance to coastal water protection. First is the ongoing attempt to
apply the principle of sustainable development -- i.e., social and economic
development and, at the same time, protection of the long-term viability of
renewable resources. As articulated by the Brundtland Report (World
Comm., 1987), this requires that the usage of material, energy, and land
resources and the environmental implications of waste generation be
considered an integral part of every human activity or practice, whether
planned or in progress. Thus, establishing a comprehensive waste

management framework also necessitates that industrialized societies focus on how they use and dispose of materials and products in general -- from the extraction of materials (e.g., from mines, forests, oil fields, waste streams) and their incorporation into manufacturing; to the distribution of products and their purchase and use by residential, institutional, and commercial consumers; to the management of waste residuals from each one of these steps.

Second, the International Maritime Organization, U.N. Environment Program, and others have been working through Group of Experts on the Scientific Aspects of Marine Pollution (GESAMP) to develop an overall framework for the protection of marine waters. While many international conventions already exist that cover selected aspects of marine pollution -- e.g., the London Dumping Convention -- there is no global convention that provides for systematic, comprehensive control of marine pollution and other degrading activities. No binding convention, for example, covers, on a global basis, land-based sources of marine pollution; the Montreal Guidelines are a beginning, but they are only guidelines. The upcoming World Conference on Environment and Development, to be held in Brazil in 1992, may consider the adoption of a comprehensive marine protection convention.

In addition, multilateral donors are increasingly recognizing the importance of protecting our marine environments. For example, a new multilateral fund -- a Global Environmental Fund or "Green Fund" -- was established in November 1990 to provide loans at concessionary rates for projects on CFCs, greenhouse gases, biodiversity, and marine pollution (World Bank, 1990). The fund will be coordinated by the World Bank, U.N. Environment Program, and U.N. Development Program.

References

International Council for Exploration of the Sea, *Report of the Advisory Committee on Marine Pollution, 1989*, ICES Co-operative Research Report No. 167, section 2 (Copenhagen: 1989).
IMO/FAO/Unesco/WMO/IAEA/Un/UNEP Joint Group of Experts on the Scientific Aspects of Marine Pollution (GESAMP), *The State of the Marine Environment*, Reports and Studies No. 39 (London: 1990).
National Research Council, Committee on a Systems Assessment of Marine Environmental Monitoring, *Managing Troubled Waters: The Role of Marine Environmental Monitoring* (Washington, DC: National Academy Press, 1990).
U.S. Congress, Office of Technology Assessment, *Wastes in Marine Environment*, OTA-O-334 (Washington, DC: April 1987).
World Bank, "Funding for the Global Environment, 1. Framework for Program Design and Allocation Criteria, Discussion Notes" (Washington, DC: May 1990).
World Commission on Environment and Development, *Our Common Future* (Oxford: Oxford University Press, 1987).

Contaminants and Associated Biological Effects in Coastal and
Estuarine Ecosystems near Certain Urban Areas

Bruce B. McCain, Sin-Lam Chan, Donald W. Brown,
Margaret M. Krahn, Robert C. Clark, Jr., O. Paul Olson,
John T. Landahl, and Usha Varanasi[1]

Environmental Conservation Division, Northwest Fisheries Center,
National Marine Fisheries Service, National Oceanic and Atmospheric
Administration, 2725 Montlake Boulevard East, Seattle, Washington,
98112, [Phone: (206) 553-7737].

Abstract

Evidence has been obtained that indicate contaminants from urban areas
may adversely affect coastal ecosystems. Results of chemical analyses from
the first five years of a nationwide monitoring program indicate that
sediments from sites located in Elliott Bay (Seattle), San Francisco Bay,
San Pedro Bay (Los Angeles/Long Beach), San Diego Bay, Raritan Bay,
and Boston Harbor had concentrations of AHs, PCBs, lead and copper that
were among the highest found in the United States. Bottomfish from these
sites also contained high levels of the organic compounds and their
derivatives, although tissue concentrations of the metals were not
consistently reflective of sediment concentrations. Moreover,
histopathological examination showed that certain fish species from these
urban sites had prevalences of liver lesions that were significantly higher
than those detected in the same species from nonurban sites. These data
demonstrate that high levels of potentially toxic chemicals are associated
with certain urban areas and provide examples of types of adverse
biological effects connected with exposures of fish to one or more of these
chemicals.

Effects of Fresh Water Development and Water
Pollution Policies on the World's
River - Delta - Estuary - Coastal
Zone Ecosystems

Michael A. Rozengurt[1] and Irwin Haydock[2]

The coastal ecosystem is an integral link between
land-based productivity and the open ocean. Here the
micro- and macro-processes of mixing and renewal take place
at time scales from a split second to day(s), month(s) and
year(s). Man's works on land have had a substantial
influence on the spatial and temporal course of these
events and have thereby altered the outcome on both the
short- and long-term.

Freshwater runoff and, wind- and tide-induced mixing
are the major factors which control the interaction of
shelf surface, subsurface and deepwater masses of the
shelf. Some additional energy fluxes governing physical
properties of the coastal zone are associated with short-
and long-term climatic cycles of air temperature,
precipitation, and evaporation over this zone. From year
to year the hydrologic and climatic significance of these
coastal water regime-forming factors may differ. But, in
general, the variable coinfluence of these factors
determines the spatial and temporal transport of sediment
and nutrient load, oxygen enrichment, respiration,
photosynthesis, and flushing of natural or man-induced
contaminants, etc., over very large coastal regions. There
is some indication that all natural variability is
encompassed by a set limit of 25-30% for the parameter of
interest. This work examines the consequences of exceeding
this natural limit, especially through the acts of man for
the last three to four decades.

The Effects of Freshwater Discharges on the Coastal Environment

The inter- and intra-annual interaction of the
hydrophysical and biological factors in the estuaries-sea
ecosystems has its visible climax in reaches known as
frontal zones (Almazov, 1962, Bronfman and Khlebnikov,
1985, Officer, 1976, Rozengurt, 1974, Tolmazin, 1985).
The California coast is known for its pronounced
seasonal upwelling, and satellite photographs showing
varying gyres, and river plumes of mixed, brackish
estuarine waters and shallow coastal fronts. These fronts
are the most important regime-shaping phenomena and the
basis for the organic and inorganic enrichment of coastal
shallows. This, in turn, has maintained the rich stock of
commercially valuable fish and shellfish. At least 75% to
85% of the historically valuable commercial catch were
related directly or indirectly (through the food chain) to
these natural regime features (Skinner, 1962). According
to Peters and Shaaf (cited by Mann, 1982), "about
two-thirds of the total US commercial fish landings are
taken within five miles of shore." This was attributed to

[1]Michael A. Rozengurt, County Sanitation Districts of Orange
County, California, 10844 Ellis Avenue, Fountain Valley, CA 92728
[2]Irwin Haydock, County Sanitation Districts of Orange
County, California, 10844 Ellis Avenue, Fountain Valley, CA 92728

the highest density of organic matter per square meter
which, in turn, is due to the maximum of macrophytes and
phytoplankton biomass being concentrated there.

Thus, the rich bioproductivity of this area is based
on four major components: (1) volume of flow and timing of
its discharges to estuaries and bays and coastal areas with
their nutrient and sediment loads; (2) seasonal offshore
(upwelling) and alongshore transport of mixed waters; (3)
spatial   density   of   microorganisms   (phyto-   and
zoo-plankton);   and   (4)   spatial   and   temporal   food
limitations for larval growth.

Numerous detailed field observations and published
results provide ample evidences that the mixed riverine-
estuarine/riverine-inlet discharges spread far beyond the
boundaries of the estuaries themselves. Moreover, in the
coastal zones similar to California, where predominant
wind-induced surface currents move alongshore (for example,
in the spring), these water masses exert an additional
dragging pressure on the mixed outflow bodies, and
transports them in the same direction.

It has been found that freshwater outflow entrained
and   mixed estuarine and coastal waters, resulting in a
combined volume which may exceed 10 to 100 times the volume
of the direct runoff discharges. The Corioli force acting
on this plume further stimulates the vertical mixing and
development   of   a   secondary   compensatory   upwelling
(Tolmazin, 1985) which involves the succession of large
scale water and salt exchange between inland coastal basins
and deep waters of the shelf. In practice, this implies
that the estuarine-coastal zone ecosystems are experiencing
nearly instantaneous processes of refreshenment and
enrichment. Consequently, the highest nutrient inputs and
richest biological communities occur there (Ketchum, 1983;
Rozengurt and Haydock, 1981; Rozengurt and Herz, 1981;
Rozengurt and Hedgpeth, 1989; etc.).

Implication of Alteration of Freshwater Inflows
to Estuarine and Coastal Systems

According to Chambers, 1991, diversions of freshwater
inflows from estuarine ecosystems is one of the four
highest priority threats to living marine resources
nationwide. Competition for water use is recognized as a
major problem in California, the Gulf of Mexico and other
"Sunbelt" states, but it is also a significant threat to
coastal fisheries and their supporting estuarine nursery
areas in virtually all coastal states (Duke and Sullivan,
1990, Hedgpeth, 1983, McGovern, 1989; Rozengurt et al.,
1987 a, b, c Nichols et al., 1986). As is already known,

the recruitment, growth and survival of fish and shellfish
are infinitely complicated for they subjected to forces
arising at a different level in ecosystems whose hierarchy
and intrinsic interaction is more complex than we have yet
imagined (Hedgpeth, 1977, (b).

Trends in Living Resources, Habitats, and Development in the USA

U.S. commercial fish and shellfish landings have been
estimated to be comprised primarily of estuarine-dependent
species, 77% by weight and 71% by value for 1985 (Chambers,
1991). Capitalized cost of commercial landings of these
species accounts for $39 billion (in 1981) which contribute
about $5.5 billion to the GNP while 17 million marine
recreational fishermen spent an estimated $7.5 billion
largely in pursuit of these species in 1985.

Nationwide, a number of stocks of estuarine-dependent
species have suffered significant declines, due undoubtedly
to a combination of stresses; examples of these losses
nationwide are (California Department of Fish and Game
Annual Reports, 1988-1990)

SPECIES OF COMMERCIAL FISH	Residual Percentage of Fish Stock Average (%)
Salmon and Steelhead, Columbia River Basin	15-25
Striped Bass, San Francisco Bay	1-5
Shad, San Francisco Bay	5-20
Chinook Salmon, Sacramento- San Joaquin Rivers Winter Run	1-5
Smelt, San Francisco- San Joaquin Delta	1-2
Hickory Shad, Chesapeake Bay	1-4
Alewife and Blueback Herring, Chesapeake Bay	?-8
Striped Bass, Chesapeake Bay	?-30
American Shad, Chesapeake Bay	?-34
American Oyster, Chesapeake Bay	?-10

The Gulf of Mexico

Similarly, the Gulf of Mexico is presently deprived of
up to 40 to 90% of spring runoff from 44 rivers and 30
estuaries which provides 98% of all commercial fish caught
in and beyond the gulf's rim. At the same time Alabama,
Mississippi, Louisiana, and Texas discharge in summer-fall
a significant volume of returning agricultural water
saturated with toxic chemicals. This and industrial

discharges from oil refineries and six of the nation's top
seaports further aggravates the gulf's environment. Note
that fishery and shellfish in the region have experienced
a drastic decline since the late 1970s. By that time
cumulative losses in freshwater had reached a hundred
million acre feet and spring water withdrawals exceeded 40
to 80%. Today, the distortion of natural coastal dynamics
has resulted in anoxia of over 3,000 square miles of bottom
waters off the Louisiana and the southeastern Texas coast.
Nearly 3.4 million acres (60% of the gulf's shellfish-
growing areas) are off-limits to harvest because there is
no runoff to flush out natural and man-induced contaminants
accumulated in the gulf (Duke and Sullivan, 1990). As a
result, the multi-billion dollar fishery and tourism
industries sustained enormous economic losses.

In 1980, the National Symposium of Freshwater Inflow
to Estuaries concluded (Cross and Williams, 1981):

"Published results regarding water development in
rivers entering the Azov, Caspian, Black and
Mediterranean seas in Europe and Asia all point to the
conclusion that no more than 25 to 30 percent of the
historical river flow can be diverted without
disastrous ecological consequences to the receiving
estuary. Comparable studies on six estuaries by the
Texas Water Resources Department showed that a 32
percent depletion of natural freshwater inflow to
estuaries was the average maximum percentage that
could be permitted if subsistence levels of nutrient
transport, habitat maintenance, and salinity control
were to be maintained."

San Francisco Bay

One of the largest and most productive estuaries of
the Pacific Coast--the Sacramento - San Joaquin River -
Delta - San Francisco Bay ecosystem--shows signs of
inexorable deterioration. Here, since the late 1960s,
numerous water-project dams have eliminated 80 to 100% of
the migration routes and the spawning areas of several
species of salmon. Besides, the state's water conveyance
facilities have redistributed the natural seasonal runoff,
i.e. similar to that in the south of USSR (reducing spring
runoff to equal or less than summer-autumn flows). This
made possible the accumulation behind the dams of over 30%
to 85% of the natural spring runoff (10.5 km^3) regardless
of the wetness of the year (Rozengurt et al., 1987, a, b,
c). In response to such perennial water management coupled
with the drought of 1986-1990, migration and spawning of
commercially and recreationally valuable salmon, striped
bass and shad have nearly ceased to exist, and the

inexorable deterioration of water properties of entire Delta-Bay ecosystem has occurred (Hedgpeth, 1977 (a); Whipple et al., 1987). Winter run chinook salmon populations declined to 1 to 2% of normal, as the spring diversion reached up to 65 to 85% of San Francisco Bay's "lifeblood" renewal of freshwater. Striped bass egg production is down to 2% of the measured values of the 1960s (Striped Bass Working Group, 1990); salmon (Chinook) populations have also declined to 2% of their former highs. In the San Francisco Bay region, the striped bass and salmon sport catch have decreased 90 and 98%, respectively. As a result, between 1960s-1980s the cumulative economic losses for fisheries in San Francisco Bay accounted for $2.6 billion.

In California the vast network of dams and extensive water diversions for agricultural and municipal usage have withdrawn cumulatively 500 km^3 of freshwater. The effect this has had on the coastal zone has not been previously explored; current water policy assumes such out-flows are a waste!

Overall, if one assumes that the California coastal zone is one inseparable ecosystem, whose estuarine regime is controlled primarily by runoff and induced circulation patterns, than the following, partially hypothetical, calculation may shed some light on the likely cause of the present precipitous decline of commercial and recreational fishery in this area and salinization of a significant part of formerly freshwater delta body.

1.   Since 1945 the central California coastal zone has been deprived of over 400 million acre-feet (roughly 500 km^3 of freshwater) due to upstream and downstream diversions from the Sacramento-San Joaquin Delta system.

2.   This implies that at least 5000 km^3 of coastal waters has not been subjected to the former natural mixing (one part of freshwater entrains, in mixing, at least 10 parts of estuarine-coastal waters).

3.   Other sources of freshwater along the coast down to Mexico ceased to exist over this same period due to construction of flood control dams and water storage facilities; for example, on the upper reaches of the Los Angeles, San Gabriel, and Santa Ana rivers. Roughly, these losses may account for 2 km^3 per year. Hence, average losses of these small streams in coastal freshwater supply may account for approximately

90 km^3 for the period of 1945-1990. By analogy with the above-introduced entrainment computation, this implies that nearly 900 km^3 of coastal zone waters were not subjected to mixing and transport, and oxygen and organic and inorganic enrichment because of river impoundment and diversion.

4. Therefore, with some reasonable approximation (taking into account several dry and, especially, drought periods of 1976-1977 and 1986-1990), the combined freshwater deficit has amounted to 600 km^3. Correspondingly, had these losses not taken place then at least 6000 km^3 of coastal water (remember 1:10) would have participated in salt and water and gaseous exchanges between the California estuarine-tidal-inlet-marine ecosystems. In light of the above, it is worth relating these "lost" volumes to the amount of water confined along the five mile coastal zone where presumably 90% of commercial and recreational catch has taken place (Mann, 1982).

5. In this case, the volume of the most productive water masses between the Golden Gate and Newport Bay, California amounts approximately to 320 km^3 (length of coastline = 800 km, the average width toward the ocean = 10.0 km, and the average depth = 0.04 km). Now, if one compares the volume of this successful fishery zone with that of cumulative losses sustained by lack of river runoffs (600 km^3), and the induced 6000 km^3 volume which was excluded from natural mixing processes, then the following explanations may be drawn: the cumulative precipitous decline of valuable fishery for the last four decades all over California coastal zone is intimately related to inland river impoundment that has not only reduced the functionality of the estuarine-sea ecosystems but also obliterated over 90% of the rich wetland and marsh habitats (Dahl, 1990).

Unfortunately, the negative effect of freshwater losses on coastal dynamics is only part of a myriad of subtle and gross problems which follow. Some of these include the million tonnes in cumulative reduction of oxygen, sediment, organic and inorganic load lost to the system which has triggered dramatic changes in biological productivity and fishery for the last two decades.

In light of the above, we must recognize the fact that the measured depletion of pelagic and demersal fish and shellfish in estuaries and coastal zones cannot, and must not by any means, be automatically attributed to the impact of treated wastewater effluents discharged through outfalls to these waters. On the contrary, the similarity in the dramatic decline of fish and shellfish catch along the Atlantic and Pacific coast, as well as in other parts of the world provide strong support to the statement that river impoundment coupled with significant agricultural discharges, play a major role in the deterioration of vast regions of coastal zone all over the world including California Coastal Ecosystem (Coastal Waters in Jeopardy, 1988; California DFG, 1983; McGovern, 1989).

In the following examples, the scale of effects (thousands of square miles) and the speed (10-20 years) of these changes significantly exceed anything known about natural variation of the coastal ecosystem and marine life for at least the last two centuries.

The Mediterranean Sea

Before the Aswan Dam on the Nile River started to impound between 50-80% of the annual volume of flood waters (about 35 km³, mid-August to December), the fisheries of the coastal zone of the eastern Mediterranean and some of the brackish lakes of the Nile Delta produced about 120,000 tonnes of fish per year (Aleem, 1972). The annual landings of pelagic fish such as <u>Sardinella</u> ranged between 10,000 and 20,000 tonnes; and for demersal fish, the catch was about 29,000 tonnes. Such production reflected the highly successful primary biological productivity in this Mediterranean coastal zone, which depended heavily upon river-borne detritus and dissolved organic materials (Mancy, 1979).

However, since 1965 (the first year of the operation of the Aswan Dam), there has been a 96% and 99% decrease in the catch of Mediterranean <u>Sardinella</u> (Aleem, 1972) and shrimp. In general, about 20% of the Nile runoff reduction destroyed over one hundred thousand tonnes fishery in the eastern Mediterranean, including both the pelagic and demersal species. At the same time, the Nile Delta fishery ceased to exist with the delta water quality at the edge of collapse.

The USSR Southern Seas

Water development projects on the major rivers in the southern USSR--Dniester and Dnieper (Black Sea), Don and Kuban (Azov Sea), Volga, Ural, Terek, Kura and others

(Caspian Sea), Amu-Darya and Syr-Darya (Aral Sea), and
numerous small rivers were based on two erroneous
assumptions: (1) river runoff is an inexhaustible source of
water supply, and (2) fresh- water discharges to adjacent
seas are wasteful. This linear and single-minded thinking
neglected the ecologic appraisal of a limited natural
tolerance of riverine-estuarine systems and living
resources to water diversions.

This Soviet water policy since the 1950s has brought
about a dense concentration of population, water-consuming
industries (chemical, electronic, food, etc.) and water-
hungry agricultural crops, such as cotton and rice, in
known semi-arid and arid zones which were formerly self-
limited by lack of water resources.

To satisfy these new, conflicting demands for water,
the above-mentioned rivers have been subjected to the
highest water withdrawals of all the rivers of the USSR
(Rozengurt and Herz, 1981; Rozengurt and Haydock, 1981).
This loss of flushing flows (as little as 15% to 50% of
normal runoff in springtime) has triggered uncontrolled
contamination and eutrophication of entire seas' areas.

Degradation of the Black, Azov, Aral and Caspian seas
has progressed to such a state across the southern portion
of the USSR that destruction of their habitat has become
the Achilles' heel of the Soviet economic policy. In the
shelf zone of the Black Sea and Sea of Azov, about 50,000
and 60,000 km^2, respectively, the catch of valuable pelagic
and demersal fish constitutes only 0.5 to 2.5% of the pre-
project period (100,000 to 200,000 and 200,000 to 350,000
tonnes, respectively). This has proven to be a direct
result of river impoundment and regulated outflow (Volovic,
1986).

In practice, since the 1960s, entire ecosystems of the
Black, Azov and Caspian seas' have been deprived of nearly
650, 400 and 1000 km^3 of spring freshwater (40 to 85% of
spring runoff is withdrawn annually). Therefore, a hundred
million tonnes of sediment load, a dozen million tonnes of
oxygen, and inorganic and organic matter, and many other
constituents so vital to maintaining the unique diversities
of these ancient ecosystems have been lost in the Soviet
irrigation network (Rozengurt, 1991).

The dissection of rivers by dams has created an eight
to ten-fold increase in the detention time of natural and
man-induced contaminants accumulated and uncontrolled
eutrophication in coastal ecosystems. This, in turn, has
led to development of vast "red tides and, ultimately,
oxygen deficiency (hypoxia) over many thousands of square

kilometers which spelled out the demise of the shallow southern seas' coastal zones.

Unprecedented depletion has made the residual runoffs incapable of controlling the salt regimes and mixing processes which maintain the flora and fauna of coastal ecosystems. The discharge of drainage waters from a million hectares of cotton and rice fields, saturated with fertilizers and pesticides, further aggravated water quality conditions and spelled out the gradual demise of fishery.

For example, today, the northwestern part of the Black Sea (50,000 km^2) and its wetlands (10,000 km^2) is considered a zone of ecological disaster. For many fish, birds and fur-bearing animals these changes and their seasonal extremes signified the beginning of their extinction. For all practical purposes, the demise of the coastal ecosystems in this part of the Soviet Union is an unarguable fact.

Nearly 30 years of heroic efforts and hundreds of millions of dollars spent on restoration of estuarine-dependent valuable fishes, especially Russian sturgeon, at more than 150 hatcheries placed in the Black, Azov and Caspian basins have failed. The recreational and commercial catch of Russian sturgeon ceased to exist in the northwestern Black Sea and Sea of Azov, while in the Caspian Sea it constitutes only 1 to 4% of historical catch (14,000 to 38,000 tons per year). Note that the latter was the most productive sea in the world only two decades ago (Rozengurt and Hedgpeth, 1989).

The approximate estimates of economic losses for fisheries, mollusk and seaweed harvests in the Black-Azov sea basins amounts to $0.8 to $1.5 billion and $1.0 and $1.5 billion per year; in the Caspian Sea basin the loss is almost $2.0 billion. The Aral Sea ecosystem has ceased to exist (Micklin, 1988).

Several hundred kilometers of shelf zones and recreational "golden" sand beaches of the North Caspian, Black and Azov seas are contaminated by "red" tides. As a result, the approximate estimates of economic losses for local, national and international tourism account for several hundred million dollars per year.

The cost in the reduced quality of life is incalculable.

Conclusion

River-delta-estuarine-coastal zone ecosystems all over the world are very vulnerable to disturbances by man. This is primarily because of upstream withdrawals of water for agricultural, industrial and domestic purposes or due to contamination by industrial and sewage discharges, or from eutrophication by agricultural drainage carrying nutrients, as well as pesticides and herbicides from the fields, or all three. Salinization of estuarine ecosystems has also inevitably been accelerated as in-stream flows are reduced, leading to sea water intrusion and eventual destruction of natural seasonal gradients. The result is the current alarming deterioration of freshwater intakes in rivers and deltas. All of these in concert trigger drastic declines in commercial and recreational fish and shellfish catch.

This effect also carries over to the coastal zone where the cumulative losses of sediments and organic and inorganic nutrients are felt in the eventual destruction of coastal dependent fish and invertebrate species in areas even hundreds of kilometers from the estuaries.

In sum, the single-minded approach to unlimited freshwater withdrawals, which significantly exceed the natural threshold of normal runoff deviations of ± 30%, leads to a chain reaction whose feedback and interaction with ecological, economic and societal elements have started to shatter vast reaches of river - delta - estuary - coastal zone systems throughout the world.

The basic relationships are shown in Figure 1: Reduction of runoff through river impoundment ---> increase in salinity ---> decrease in organic and inorganic material and sediment load ---> increase in detention time ---> increase in pollutants load ---> significant decrease or elimination of spawning grounds and commercial and recreational valuable fish.

Similar examples of these natural dependencies being upset are becoming more frequent throughout the world, including the Gulf of Mexico, the Pacific Coast (Sacramento and Columbia Rivers), the Mediterranean (Nile River) and Soviet Union Southern Seas (Aral, Azov, Black, Caspian).

In some extent the current water pollution control policy has only exacerbated the situation by denying these same ecosystems the possibility of receiving treated waters and some small renewal of the natural sediment and nutrient loads now trapped upstream behind dams or shunted through other man-made diversions. Excessive enhancement of treatment for treatment sake alone may give the opposite

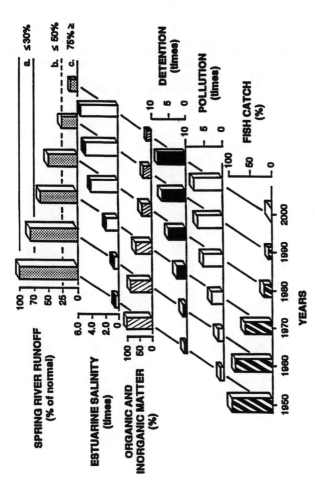

FIGURE 1. Conceptual chain reaction between spring river runoff and some major chemical, physical and economic parameters in the delta-estuary-sea economy.

a. Range of natural limitations in spring fresh water diversions (≤ 30% of normal).

b. Range of spring diversions detrimental for living and non-living resources (≤ 50% of normal).

c. Range of spring runoff diversion irrevocably damaging to environment and economics of ecosystems (75% ≥ of normal).

result of that expected. Where reduced natural flushing combines with reduced inorganic and organic nutrient replenishment, the environmental costs may outweigh the benefits to be gained from treatment. In our opinion examples discussed demonstrate how this single-minded approach has led to even greater reduction in fish and shellfish in spite of a tremendous increase in pollution control expenditures. The two alternative policies, water usage and water pollution control must be brought into balance now or we risk destroying both our water resources and their dependent living ecosystem components.

REFERENCES

Aleem, A.A, 1972. "Effect of River Outflow Management on Marine Life." Marine Biology 15: 200-208.

Almazov, A.M., 1962. Hydrochemistry of Estuaries. Academy of Science, USSR, Kuiv.

Bronfman, A.M., and E.P. Khlebnikov, 1985. The Sea of Azov (Fundamentals of Modification). Gidrometeoizdat. Leningrad. USSR.

California Department of Fish and Game (Anadromous Fisheries Branch), 1983. Salmon Management in California prepared for the Pacific Fisheries Management Council. p. 5 Calif. Dept. F & G. 1988-1990

Chambers, J.R, 1991. Coastal habitat degradation and fisheries declines. In: R.H. Stroud (editor). Proceedings of the Symposium on Coastal Fish Habitat Conservation. Baltimore, MD. (in press).

Coastal Waters in Jeopardy: Reversing the Decline and Protecting America's Coastal Resources. Oversight Report of the Committee on Merchant Marine and Fisheries, December, 1988. Washington: Congress of the USA.

Cross, R.D. and D. L. Williams (eds), 1981. Proceedings of the National Symposium of Freshwater Inflow to Estuaries. V.I and II. Washington, D.C.: Dep. of the Interior, FWS.

Dahl, T.E, 1990. Wetland Losses in the United States, 1780's to 1980s. U.S. Department of the Interior, Fish and Wildlife Service, Washington, D.C.

Duke, T.W. and E.E. Sullivan, 1990. America's Sea at Risk. Progress Report, Gulf of Mexico Program, U.S. EPA., Rockville: Technical Resources, Inc.

Hedgpeth, J.W., Seven ways to obliteration: factors of estuarine degradation, in Proc. Estuarine Pollution Control and Assessment, Environmental Protection Agency, Washington, D. C., 2, 723, 1977.

Hedgpeth, J.W., Models and Muddles: Some philosophical observations. Helgolander Meeresunterscuhungen, 1-4, 30: 92-104, 1977.

Hedgpeth, J.W., Coastal Ecosystems Marine Ecology, Vol. 5, part 2, 1983.

Ketchum, B.H. (Ed), 1983. Ecosystems of the World: Volume 26, Estuaries and Enclosed Seas, Amsterdam, Elsrvier.

98COASTAL WETLANDS

Mancy, K.H., 1979. "The Aswan High Dam and its
Environmental Implications." Socita Internationalis
Limnologiae Workshop on Limnology of African Lakes,
Nairobi, Kenya.

Mann, K.H., 1982. Ecology of Coastal Waters, A Systems
Approach. Berkeley, CA., University of California Press.

McGovern, D., May, 1989. Review DEIS of Bureau of
Reclamation Proposed Water Contracting Programs in the
Sacramento River. American River, and Delta Export Service
Area.

Micklin, Ph.P., 1988. "Desiccation of the Aral Sea: A Water
Management Disaster in the Soviet Union." Science 241:
1170-1176.

Nichols, F.H., J.E. Cloern, S.N. Luoma and D.H. Peterson,
1986. The modification of an estuary. Science. 231:567-
573.

Officer, C.B., 1976: Physical Oceanography of Estuaries
(and Associated Coastal Waters). New York, J. Wiley &
Sons.

Rozengurt, M.A., 1974. Hydrology and prospects of
reconstruction of natural resources of the northwestern
part of the Black Sea estuaries. Kiev. "Scientific
Thought," Academy of Sciences, USSR. (in Russian) Library
of Congress GB2308.B55R69.

Rozengurt, M.A., and I. Haydock, 1981. "Methods of
Computation and Ecological Regulation of the Salinity
Regime in Estuaries and Shallow Seas in Connection with
Water Regulation for Human Requirements." In R.D., Cross
and D.L., Williams (eds.). Proceedings of the National
Symposium on Freshwater Inflow to Estuaries V.II: 474-507.
Washington, DC: US Department of the Interior.

Rozengurt, M.A. and M.J.Herz, Water, water everywhere but
just so much to drink, Oceans, 5, 65, 1981.

Rozengurt M.A., M.J. Herz and M. Josselyn, 1987. The
impact of water diversions on the river-delta-estuary-sea
ecosystems of San Francisco Bay and the Sea of Azov. In:
D.L. Goodrich (editor). San Francisco Bay: issues,
resources, status, and management. NOAA Estuary-of-the-
Month Seminar Series No. 6 U.S. Department of Commerce,
Washington, D.C. pp.35-62.

Rozengurt, M.A., M.J. Herz and S. Feld, 1987 (a). Analysis
of the Influence of Water Withdrawal on Runoff to the

Delta-San Francisco Bay Ecosystem (1921-1983). Technical Report No. 87-7, Tiburon Center for Environmental Studies (TCES), San Francisco State University (SDSU).

Rozengurt, M.A., M.J. Herz and S. Feld, 1987 (b). The Role of Water Diversions in the Decline of Fisheries of the Delta-San Francisco Bay and other Estuaries. Technical Report No. 87-8, Tiburon: TCES, SFSU.

Rozengurt, M.A. and J.W. Hedgpeth, 1989. "The Impact of Altered River Flow on the Ecosystem of the Caspian Sea." Critical Review in Aquatic Sciences V.1 (2): 337-362.

Rozengurt, M.A., 1991. Strategy and Ecological and Societal Results of Extensive Resources Development in the South of the USSR. In Proceedings "The Soviet Union in the Year 2010", Symposium sponsored by USAIA and Georgetown University, 26 - 27 June 1990.

Tolmazin, D. M., Changing coastal oceanography of the Black Sea, Progr. Oceanog., 15, 217, 1985.

Skinner, J.E., 1962. An Historical Review of the Fish and Wildlife Resources of the San Francisco Bay Area. California Fish and Game Water Projects Branch Rep. No. 1. 226 p.

Volovic, S.P., the fundamental features of transformation of the Sea of Azov ecosystems in connection with industrial and agricultural development in its watershed. The Problems of Ichthyology, Scripta Technica, Wiley & Sons, 1986.

Whipple, J.A., R.B MacFarlane, M.B. Eldridge, and P. Senville, Jr., 1987. The impacts of estuarine degradation and chronic pollution on populations of anadromous striped bass in the San Francisco Bay-Delta: a summary. In: D.L. Goodrich (editor). San Francisco Bay: Issue, Resources, Status and Management. NOAA Estuary-of-the-Month Seminar Series No. 6. Washington, D.C., U.S. Department of Commerce, NOAA Estuarine Programs Office. pp. 77-106.

RECENT CHANGES IN ESTUARINE WETLANDS OF THE COTERMINOUS UNITED STATES

Ralph W. Tiner[1]

Abstract

The U.S. Fish and Wildlife Service's National Wetlands Inventory Project has recently completed an update of the national wetland trends analysis study covering the period 1974-1983. In 1983, an estimated 5.5 million acres of estuarine wetlands remained in the coterminous United States. Annual losses of estuarine vegetated wetlands averaged about 7,900 acres. Loss rates declined 42% since the mid-1970s. Louisiana and Texas experienced the heaviest losses of coastal wetlands.

Introduction

Estuarine wetlands are among our country's most valuable wetlands. These salt marshes, mangrove swamps, and tidal flats provide critical habitats for many fish and wildlife species that help support the U.S. economy. For example, over two-thirds of the U.S. recreationally and commercially important marine fishes depend on estuarine marshes and associated waters for nursery and spawning grounds (McHugh 1966). Shrimp, crabs, oysters, and other shellfish are also dependent on these wetlands. Estuarine wetlands, therefore, are the foundation of a multi-million dollar fishing and shellfishing industry as well as a major component of local tourism. These wetlands also provide other values including: (1) protection of shoreline from erosion by wave and storm action, (2) protection of private property from floods by temporary storage of water, (3) maintenance of water quality by removing sediments, nutrients, and other materials from flood waters, and (4) providing other recreational opportunities (e.g., waterfowl hunting, nature photography, and bird watching) that attract people to the coastal zone.

----

[1]Regional NWI Coordinator, U.S. Fish and Wildlife Service, Northeast Region, 1 Gateway Center, Newton Corner, MA 02158

Despite these values, estuarine wetlands have been greatly abused in the past. Prior to the passage of state and federal laws to regulate uses of coastal wetlands (in the 1960s and 1970s), there was much destruction of coastal wetlands. The Fish and Wildlife Service (FWS) estimated a loss of about 400,000 acres of estuarine wetlands between the mid-1950s and mid-1970s based on the first national wetland trends analysis (NWTA) study (Frayer, et al. 1983; Tiner 1984). Since then, protection of coastal wetlands has been strengthened by state and federal laws.

The FWS recently completed an update of the NWTA study which examined wetland changes between the mid-1970s and the mid-1980s. The purpose of this paper is to summarize the results of this study for estuarine wetlands in the United States. It focuses on coastal wetlands along the Atlantic and Gulf Coasts where these wetlands are most prominent.

Trend Analysis Techniques

Statistical sampling techniques are proven methods for making estimates of populations, land cover types, and other variables. The FWS's original NWTA study used a stratified random sampling technique where four-square mile plots were selected for sampling (Frayer, et al. 1983). The same study design was also used in the present update of the NWTA study since it was a proven sampling technique to estimate wetland acreages and changes, and sample plots had already been selected.

The sampling strata for the NWTA study were derived from state boundaries, the 35 physical subdivisions described by Hammond (1970), and coastal zone boundaries (marine and estuarine systems and the Great Lakes). This amounted to over 150 strata nationwide. Estuarine wetlands are found within the coastal zone stratum. For the present study, the coastal zone stratum was expanded to include significant portions of coastal wetlands that were unknowingly omitted from this stratum in the original study. This zone was expanded to conform to the estuarine-riverine and estuarine-palustrine breaks on existing National Wetlands Inventory maps. Consequently, acreage totals of estuarine wetlands were expected to differ from original estimates for the mid-1970s. A total of 3,629 four-square mile sampling plots were evaluated for the national study. About forty-one percent of these plots (or 1,474 plots) occurred in the coastal zone stratum allowing acreages of estuarine wetlands and their recent trends to be estimated (T. Dahl, pers. comm.).

The type and extent of wetlands within each sample plot were determined through aerial photo interpretation techniques. Aerial photographs from the mid-1970s (mean year - 1974) and early 1980s (mean year - 1983) were obtained for each plot. The approximate study interval was 9 years. Wetlands were classified using the U.S. Fish and Wildlife Service's wetland classification system (Cowardin, et al. 1979). Initial interpretations were made on 1:58,000 color infrared photographs acquired by the U.S. Geological Survey's

National High-altitude Photography Program. Recent wetlands for
each plot were then compared with the 1970s photos (1:80,000 black
and white) to detect changes in wetland boundaries and/or cover
types and to prepare an updated wetlands overlay using a Bausch &
Lomb stereo zoom transfer scope. When identifying changes, the
recent photos were examined to determine the causes of change,
either natural (i.e., changes in type) or direct human-induced
(i.e., agriculture, urbanization, or other land uses). The superior
quality of the 1980s photography allowed for improved detection of
wetlands and consequently the estimated acreages were expected to
vary from those of the earlier NWTA study. Wetland status and
trends data shown on an overlay for each plot were tabulated for
computer analysis. The data were analyzed by Dr. Edward Frayer of
Michigan Tech University.

Interpretation of Results

        The status and trend results for estuarine wetlands presented
are estimates of acreages in 1974 (mean year) and 1983 (mean year)
based on statistical sampling. For each estimate of acreage, a
standard error is given. The standard error is expressed as a
percentage of the estimated total (%SE). It gives an indication of
the reliability of the estimated number. In general, when the
standard error is 25 percent or less of the estimated number, the
estimate is considered reliable. The lower the %SE, the higher the
reliability of the estimate and vice versa. If the %SE is 50 or
more, one cannot even be 95 percent confident that the true value is
not zero. In some cases, such as changes in estuarine wetlands, a
high %SE may indicate that the underlying distribution violates the
assumption of a normality (i.e., normal distribution). The trend
figures, however, are our best estimates to date, in lieu of a
detailed examination of estuarine wetlands in each state.

National Estimates

        According to the present study, the coterminous U.S.
possessed an estimated 5,532,090 acres (3.9%SE) of estuarine
wetlands in 1974 and by 1983, only 5,472,825 acres (4.0%SE)
remained. A net loss of 59,265 acres (22.7%SE) of estuarine
wetlands occurred during the 9-year period. Estuarine vegetated
wetlands were most seriously impacted, dropping from 4,853,915 acres
(4.2%SE) in 1974 to 4,783,116 acres (4.2%SE) in 1983 for a net loss
of 70,799 acres (18.2%SE). About 1.5% of the 1974 acreage of
estuarine vegetated wetlands was lost, with annual losses averaging
about 7,900 acres. In contrast, acreage of estuarine nonvegetated
wetland (e.g., intertidal flats) increased from 678,175 acres
(11.8%SE) in 1974 to 689,709 acres (11.6%SE) in 1983 for a gain of
11,534 acres (36.3%SE). Much of this 1.7% gain came from estuarine
deepwater habitats, while a small amount came from estuarine
vegetated wetlands.

        The majority (or 98%) of the estuarine vegetated wetland
acres present in 1974 survived into 1983 (Table 1). Those that did

change, however, were mostly converted to estuarine deepwater
habitat presumably due largely to sea level rise and resultant
coastal subsidence, erosion, and to a lesser extent by dredging.
Nearly 20,000 of the original estuarine vegetated acres changed to a
different wetland type - mostly palustrine wetland. About 4,000
acres were estimated to be converted to agriculture, while an
estimated 4,500 acres were lost to urban development. Estuarine
wetlands are dynamic as evidenced by the losses and gains from
various land and water types shown in Table 1 below.

Table 1. Recent trends (1974-1983) in estuarine vegetated
wetlands of the coterminous United States.

Nature of Change	Estimated Acreage (%SE)	
Losses to:		
Marine Intertidal Wetland	1,614	(39.3)
Estuarine Nonvegetated Wetland	6,077	(21.2)
Palustrine Wetland	11,577	(48.0)
Deepwater Habitat	57,004	(12.7)
Agricultural Land	3,980	(79.9)
Urban Development	4,502	(34.5)
Other Land*	7,126	(18.9)
Gains from:		
Marine Intertidal Wetland	1,428	(88.2)
Estuarine Nonvegetated Wetland	5,271	(23.8)
Palustrine Wetland	148	(58.1)
Deepwater Habitat	13,095	(20.1)
Agricultural Land	276	(98.2)
Other Land (non-urban)*	863	(35.2)
No Change	4,762,035	(4.2)

----
*Other Land includes rangeland, forests, strip mines, quarries, and
land being converted to an, as yet, unknown use, for example.

State Estimates

        Estuarine wetlands are found in 22 states of the coterminous
United States. Estimated acreages in 1983 are presented for 18 of
these states in Table 2. The other four states (California, Oregon,
Pennsylvania, and Washington) did not have any sampling plots within
the coastal zone. In future updates of the NWTA study, additional
stratification to include the coastal zones of these states should
be done. Louisiana and Florida combined contained over half of the
estuarine wetlands with 35.2% and 24.8%, respectively. Other states
with extensive acreages were Texas (8.0%), South Carolina (7.9%),
and Georgia (6.7%). These six states accounted for over 80% of the
estuarine wetlands in the coterminous United States.

Table 2. Estimated acreage of 1983 estuarine wetlands by state.

State	Estimated Acreage (%SE)		State	Estimated Acreage (%SE)	
Alabama	6,049	(52.0)	Mississippi	9,934	(44.8)
Connecticut	12,325	(35.3)	New Hampshire	2,200	(92.4)
Delaware	54,800	(34.2)	New Jersey	188,241	(14.7)
Florida	1,359,482	(9.0)	New York	170,058	(46.2)
Georgia	366,700	(6.4)	North Carolina	162,479	(18.6)
Louisiana	1,925,413	(6.3)	Rhode Island	16,422	(74.6)
Maine	81,494	(39.5)	South Carolina	432,982	(12.3)
Massachusetts	73,885	(31.2)	Texas	438,119	(12.5)
Maryland	73,716	(23.5)	Virginia	98,525	(29.6)

Table 3 summarizes the recent status and trends data for
estuarine vegetated wetlands in each state sampled. The largest
loss of estuarine vegetated wetland occurred in Louisiana where an
estimated 57,000 acres were lost. Other states with substantial
acreages lost included Texas, New Jersey, South Carolina, and
Florida. Nearly 10,000 acres of these wetlands were lost in Texas,
while about 1,000 acres disappeared in each of the other states.
Consistent with the national trends, most of the losses in these
states were to estuarine open water and to estuarine nonvegetated
wetland (e.g., intertidal flats). Other states where similar
changes have dominated the cause of estuarine vegetated wetland loss
included Delaware, Georgia, Massachusetts, Maryland, North Carolina,
and Rhode Island. Conversion of these wetlands to urban land still
took place. States where such losses were detected included:
Alabama (17 acres lost, 94.1%SE), Connecticut (8 acres, 87.5%SE),
Delaware (68 acres, 92.6%SE), Florida (2,691 acres, 52.7%SE),
Georgia (28 acres, 85.7%SE), Louisiana (1,103 acres, 51.5%SE), New
York (104 acres, 97.1%SE), and Texas (501 acres, 50.7%SE).

Other study findings of special interest involved Florida,
Louisiana, and Texas. In Florida, 4,145 acres (34.0%SE) of
estuarine emergent wetlands (salt marsh) changed to estuarine scrub-
shrub wetlands (presumably mangroves), while 2,748 acres (90.5%SE)
of mangroves changed to salt marsh. Also 6,208 acres (101.8%SE) of
Florida's estuarine aquatic beds disappeared; mostly they became
nonvegetated intertidal areas. Urban development had its greatest
coastal wetland impact on Florida's estuarine forested and scrub-
shrub wetlands (mangrove swamps) with an estimated 2,454 acres
(57.3%SE) filled. Louisiana lost 3,370 acres (93.7%SE) of estuarine
emergent wetlands to agriculture and another 4,569 acres (19.6%SE)
to other (non-urban) land uses. An estimated total of 2,776 acres
(61.9%SE) of estuarine emergent wetlands were converted to
lacustrine deepwater habitats (e.g., impoundments and reservoirs) in
Texas, while an additional 1,321 acres (55.4%SE) were altered for
other uses (excluding agriculture and urban development).

Table 3.   Recent status and trends in estuarine vegetated wetlands
           by state.

State	# Plots Sampled	Estimated 1974 Acreage (%SE)		Estimated 1983 Acreage (%SE)		Estimated Change in Acreage* (%SE)	
Alabama	26	4,711	(65.8)	4,757	(66.0)	+46	(84.8)
Connecticut	24	10,929	(36.5)	10,921	(36.6)	-8	(87.5)
Delaware	19	53,145	(35.0)	52,996	(35.0)	-149	(63.8)
Florida	276	956,600	(11.5)	955,757	(11.5)	-843	(312.0)
Georgia	40	349,870	(6.8)	349,529	(6.8)	-341	(268.3)
Louisiana	348	1,970,264	(6.2)	1,913,167	(6.3)	-57,097	(20.9)
Maine	4	21,895	(9.3)	21,895	(9.3)	0	(0)
Massachusetts	27	37,352	(32.1)	37,190	(32.1)	-162	(102.5)
Maryland	92	72,669	(23.8)	72,413	(23.9)	-255	(60.0)
Mississippi	31	6,072	(62.5)	6,059	(62.6)	-13	(169.2)
New Hampshire	5	284	(91.9)	284	(91.9)	0	(0)
New Jersey	34	185,552	(15.2)	184,404	(15.2)	-1,149	(40.2)
New York	30	134,579	(47.4)	134,476	(47.4)	-104	(97.1)
North Carolina	148	158,977	(18.9)	158,886	(18.9)	-91	(422.0)
Rhode Island	6	11,682	(81.5)	11,657	(81.5)	-24	(112.5)
South Carolina	72	422,474	(12.6)	421,342	(12.6)	-1,132	(224.5)
Texas	142	360,636	(14.1)	351,170	(14.3)	-9,466	(30.8)
Virginia	92	96,224	(30.0)	96,214	(30.0)	-10	(490.0)

----

* Acreage number may vary by 1 from difference between 1970s and
1980s estimates due to round-off.

Comparison with Other Studies

In the absence of detailed inventories, statistical estimates
generated by the NWTA study provide reasonable estimates of the
current status and recent trends in wetlands. Of course, the most
accurate assessment would be a complete inventory of two-time
periods, but this is much too labor-intensive and costly, especially
on a nationwide basis. Such studies are perhaps best suited for
site-specific analysis of regional and local concerns, such as
learning more about losses of Louisiana's coastal wetlands or losses
of wetlands in a particular county. Nonetheless, statistically
designed studies do provide a cost-effective means of assessing
wetland trends in larger areas and may help pinpoint problem areas
that require more detailed analysis.

Comparison of estimated losses from the NWTA study with other
studies (site-specific) such as have been conducted in Louisiana's
coastal zone is not really meaningful, since any well-designed site-
specific inventory should generate more reliable trend data for that
area than the NWTA study. Even a statistical study for a specific
area could produce improved results, provided the study area was
stratified appropriately and sampled sufficiently.

Comparison of the current estimates with previous estimates
of estuarine wetland loss produced by the NWTA study is useful for
identifying general trends, although the actual estimates of wetland
acreage differ due to an expansion of the coastal zone stratum and
to the use of higher resolution aerial photography that improved
detection of estuarine and other wetlands. From the mid-1950s to
the mid-1970s, the coterminous U.S. lost 372,300 acres (8.4%SE) of
its estuarine vegetated wetland acreage (Frayer, et al. 1983). This
represented a 7.6% loss. This time period included the period
during which alterations of coastal wetlands were largely
unregulated, so many acres were filled or dredged for alternative
uses. In contrast to the mid-50s to mid-70s, only 70,799 acres
(18.2%SE) of estuarine vegetated wetlands disappeared. This
represented only a 1.5% loss and suggests that losses of these
wetlands have declined substantially. Urban development in these
wetlands dropped from 106,500 acres (16.4%SE) between the mid-50s to
mid-70s to 4,502 acres (34.5%SE) between 1974 and 1983. Overall,
the annual loss of estuarine vegetated wetlands decreased from about
18,600 acres to about 7,900 acres for a 42% decline in the annual
loss rate. This reduction in the loss rate may be largely
attributed to protection afforded by numerous state laws and
increased federal wetland regulations in the 1970s and 1980s.

Although nationally, the status of estuarine vegetated
wetlands has dramatically improved since the 1970s, Louisiana's
coastal marshes remain threatened. According to Frayer and others
(1983), Louisiana lost about 34,000 acres. The present study
estimated a loss of 57,097 acres (20.9%SE). This amounts to a 2.9%
loss in acreage between 1974 and 1983. These findings were not
unanticipated, since Louisiana's coastal marshes were previously

designated as a national wetland loss problem area by the U.S. Fish and Wildlife Service based on the first NWTA study (Tiner 1984). Moreover, much work at national and state levels has been initiated to better understand the specific causes of these losses and to develop management options to reduce the losses of these valuable wetlands (e.g., Boesch 1982; Duffy and Clark 1989; Turner and Cahoon 1987).

Besides Louisiana, the only other state with heavy losses of estuarine vegetated wetlands was Texas which lost an estimated 9,466 acres (30.8%SE) or about 2.6% of these wetlands between 1974 and 1983. During this period, 8,737 acres (28.5%SE) changed to estuarine deepwater habitat and 3,547 acres (23.6%SE) of new estuarine vegetated wetlands were established in former deepwater habitat for a net loss of 5,190 acres. This loss represented about 55% of the net loss of estuarine vegetated wetlands. Additional work should be conducted to learn more about the causes of this loss. This will apparently be accomplished under the recently passed Coastal Wetlands Planning, Protection, and Restoration Act (1990) which includes a provision to assess the current status and recent trends in Texas' wetlands.

The effects of rising sea level on coastal wetlands have recently received much attention (e.g., Titus 1988, Park et al. 1989a, Titus, et al. 1991). Global warming due to the "greenhouse effect" may cause sea level to rise as much as six feet or more by the year 2100 (Thomas 1986). Accelerated sea level rise could permanently inundate vast acreages of estuarine wetlands, since vertical accretion of these wetlands is not likely to keep pace with the amount of rise (Park, et al. 1989b). While the latest NWTA study data seem to support this view, a closer examination of individual plot data would be advisable before drawing any further conclusions as there may be other explanations. This new data set does provide some opportunities for a real-time assessment of the effects of recent sea level rise on estuarine wetlands.

Summary

In 1983, approximately 5.5 million acres of estuarine wetlands existed in the coterminous United States. From 1979 to 1983, the U.S. experienced a 1.5% loss of estuarine vegetated wetlands and a 1.7% gain in estuarine nonvegetated wetlands. Nationwide, annual losses of estuarine vegetated wetlands dropped from about 18,600 acres (mid-1950s to mid-1970s) to 7,900 acres (1974-1983) for a 42% decline in the loss rate. Extensive losses of these wetlands occurred in Louisiana and Texas. Currently, steps are being taken to more closely examine the causes of these losses with the intent to develop management strategies to help minimize future losses. In addition, much attention has focused on the potential impacts of global warming on sea level rise and coastal wetlands. Given the value of these wetlands, this matter must continue to be one of the major wetland issues of the 1990s and beyond.

Acknowledgements

The study was conducted by the U.S. Fish and Wildlife
Service's National Wetlands Inventory Project (NWI) at St.
Petersburg, Florida. The following individuals provided information
useful to this paper: Dr. Bill Wilen, Thomas Dahl, and Mary Bates
of the NWI Project and Dr. Edward Frayer of Michigan Tech University
at Houghton. Ms. Joan Gilbert typed the manuscript. Their
contributions to this paper are gratefully appreciated.

References

Boesch, D.F. (editor). 1981. Proceedings of the Conference on
Coastal Erosion and Wetland Modification in Louisiana: Causes,
Consequences, and Options. U.S. Fish and Wildlife Service,
Washington, DC. FWS/OBS-82/59. 259 pp.

Cowardin, L.M., V. Carter, F.C. Golet, and E.T. LaRoe. 1979.
Classification of Wetlands and Deepwater Habitats of the United
States. U.S. Fish and Wildlife Service, Washington, DC. FWS/OBS-
79/31. 103 pp.

Duffy, W.G. and D. Clark (editors). 1989. Marsh Management in
Coastal Louisiana: Effects and Issues - Proceedings of a Symposium.
U.S. Fish and Wildlife Service, Washington, DC and Louisiana
Department of Natural Resources, Baton Rouge. U.S. Fish. Wildl.
Serv. Biol. Rep. 89(22). 378 pp.

Frayer, W.E., T.J. Monahan, D.C. Bowden, and F.A. Graybill. 1983.
Status and Trends of Wetlands and Deepwater Habitats in the
Conterminous United States, 1950s to 1970s. Dept. of Forest and
Wood Sciences, Colorado State University, Ft. Collins. 32 pp.

McHugh, J.L. 1966. Management of estuarine fishes. Amer. Fish
Soc., Spec. Pub. No. 3: 133-154.

Park, R.A., M.S. Trehan, P.W. Mausel, and R.C. Howe. 1989a. The
effects of see level rise on U.S. coastal wetlands. In: J.B. Smith
and D.A. Tirpak (editors). The Potential Effects of Global Climate
Change on the United States. U.S. Environmental Protection Agency,
Washington, DC. EPA-230-05-89-052. Appendix B - Sea Level Rise.
pp. 1-1 - 1-55.

Park, R.A., M.S. Trehan, P.W. Mausel, and R.C. Howe. 1989b.
Coastal wetlands in the twenty-first century: profound alterations
due to rising sea level. In: D.W. Fisk (editor). Wetlands:
Concerns and Successes. American Water Resources Association,
Bethesda, MD. pp. 71-80.

Thomas, R. 1986. Future sea level rise and its early detection by
satellite remote sensing. In: J.G. Titus (editor). United Nations
Environment Programme and U.S. Environmental Protection Agency,
Washington, DC. 19-36 pp.

Tiner, R.W., Jr. 1984. Wetlands of the United States: Current
Status and Recent Trends. U.S. Fish and Wildlife Service, National
Wetlands Inventory, Washington, DC. 59 pp.

Titus, J.G. (editor). 1988. Greenhouse Effect, Sea Level Rise and
Coastal Wetlands. U.S. Environmental Protection Agency, Washington,
DC. EPA-230-05-86-013. 152 pp.

Titus, J.G., R.A. Park, S.P. Leatherman, J.R. Weggel, M.S. Green,
P.W. Mausel, M.S. Trehan, S. Brown, C. Grant, and G.W. Yohe. 1991.
Greenhouse Effect and Sea Level Rise: Loss of Land and the Cost of
Holding Back the Sea. Coastal Management 19(2).

Turner, R.E., and D.R. Cahoon (editors). 1987. Causes of Wetland
Loss on the Coastal Central Gulf of Mexico. Volume 2. Technical
narrative. Mineral Management Service, New Orleans, LA. OCS
Study/MMS 87-0120. 400 pp.

EFFECTS OF ACCELERATED SEA-LEVEL RISE
ON COASTAL SECONDARY PRODUCTION

Roger J. Zimmerman[1], Thomas J. Minello, Edward F. Klima
and James M. Nance

Abstract

Sea-level rise alters biochemical processes and the geomorphology of coastal habitats when saltwater submerges marshes and uplands. This coastal inundation can change the way marshes function as estuarine habitats. One change can occur through increased abundance of marsh algae which stimulates production of primary consumers. When grazers in drowning marshes become more accessible to estuarine predators, greater production may result at higher trophic levels. But such changes are transitional, and the benefits can disappear when drowning marshes convert to open-water habitats without plants.

Our evidence indicates that marsh submergence on the order of 1 cm per year enhances utilization of food resources by shrimp and crab predators. The apparent near-term effect is that production of secondary consumers is increased, but the long-term effect is not clear. Since estuarine-dependent shrimp and blue crab fisheries are among the largest in the U.S., the consequences of sea-level rise affecting their productivity is a concern. Moreover, subsiding marshes may simulate effects of submergence by the sea, providing a model for predicting changes in fisheries attributable to sea-level rise from global warming.

Introduction

Recent investigations emphasize that access to marsh surfaces may greatly influence productivity of estuarine consumers. Previously, investigators noted that fishery

---

[1]National Marine Fisheries Service, Galveston Laboratory, 4700 Avenue U, Galveston, Texas 77551

species use marsh creeks (Weinstein 1979), marsh infauna
are impacted by estuarine predators (Bell and Coull
1978), and some offshore fishery yields can be related
to the amount of marsh area inshore (Turner 1977). But
until development of drop trap (Zimmerman et al.1984) and
flume (McIvor and Odum 1986) sampling methods, the extent
of utilization of marsh surfaces by transient estuarine
consumers was unknown. These studies revealed that large
numbers of estuarine animals often directly invade tidal
marshes (Zimmerman and Minello 1984; Rozas and Odum 1987;
McIvor and Odum 1988; Hettler 1989; Mense and Wenner
1989) in densities exceeding those of subtidal habitat.

However, utilization of marsh surfaces by invading
nekton differs regionally, with comparatively less
exploitation (fewer animals per unit area) in East Coast
marshes (Hettler 1989; Mense and Wenner 1989) than in
Gulf Coast salt marshes (Zimmerman and Minello 1984;
Thomas et al. 1990). The reasons behind regional
differences are unclear, but tides, inundation patterns
and marsh morphology are likely involved. Differences
in tidal scaling, geophysical effects and marsh surface
accretion rates can modify amplitude, frequency, and
duration of tidal inundation patterns regionally (Provost
1976). These factors affect the time available to
estuarine predators for exploitation of marshes and the
quantity and quality of marsh prey.

Regional differences have been expressed in trophic
dynamic terms, through indentification of carbon sources
and food chain pathways. East Coast marshes are valued
for outwelling of organic materials which fuel estuarine
food chains (Odum 1980). Tracing of this organic carbon
from marshes has been elusive and, to date, evidence that
large energetic contributions of marsh detritus control
estuarine food chains is not convincing (Pomeroy 1989).
Stable isotope ratios reveal that algal carbon is at
least equal to the vascular detritus carbon in food
chains associated with East Coast salt marshes (Haines
1976; Peterson and Howarth 1987). In the Gulf of Mexico,
Sullivan and Moncrief (1988a) demonstrated that edaphic
algal production in marshes is higher than on the East
Coast and propose (Sullivan and Moncrief 1988b) that more
algal carbon is incorporated into food chains associated
with Gulf estuaries. The algae and their grazers are
common foods of juvenile shrimps and crabs (Gleason 1986;
Thomas 1989; Stoner and Zimmerman, 1988; McTigue and
Zimmerman, in press) and small fishes (Minello et al.
1989) which exploit marshes.

Thus, secondary production of estuaries can be

modified through production of algae in marshes and the
accessibility of primary consumers. Primary consumers,
such as amphipods and tanaidaceans (peracarid
crustaceans) and annelid worms, are among the most
abundant macroinvertebrate components of salt marshes
(Thomas 1976; Kneib 1982; Rader 1984) and, as prey, these
organisms may transfer energy to higher trophic levels.
Accordingly, mechanisms which control availability of
primary consumers to estuarine predators may greatly
determine the degree of coupling between the marsh and
secondary productivity.

It has been postulated that greater productivity
occurs in juveniles of fishery species through increasing
their access to high abundances of foods (mainly primary
consumers) in marshes (Boesch and Turner 1984; Zimmerman
and Minello 1984; Minello and Zimmerman in press).
Childers et al .(1990) and Morris et al (1990) have
related annual fluctuations in sea level to secondary
productivity of fisheries, and Zimmerman and Minello
(1984) have shown that fishery juveniles increase their
utilization of marsh surfaces during periods of
seasonally high water. However, specific investigations
relating utilization of marsh surfaces by transient
predators to abundances of resident marsh prey have not
been conducted.

In this paper, we show predator-prey relationships
between three transient predators and their prey on a
marsh surface undergoing submergence. Marsh submergence
is the mechanism which in this case increases prey
availability and elicits a response in predation. The
submergence rate of the marsh simulates a moderate
increase in eustatic sea-level rise. The estuarine
predators are brown shrimp (Penaeus aztecus), white
shrimp (P. setiferus) and blue crab (Callinectes
sapidus). The prey are macro-invertebrate peracarid
crustaceans and annelid worms which are common temperate
marsh infauna and epifauna.

Methods

The salt marsh described in this study, located on
Galveston Island in the northwestern Gulf of Mexico, is
the site of an investigation continuing since 1981 (the
site is described in Zimmerman et al. 1984). The marsh
is undergoing submergence at a rate which exceeds its
ability to accrete and maintain its elevation above sea
level. The result is a high degree of reticulation
between marsh and open-water habitat, including a large
amount of marsh to open-water edge, and inundation by

flood tides for a relatively large percentage of time. The duration of flooding at the marsh edge ranges between 10 and 75 % of each month depending upon season and year (unpublished water level record data).

Densities of shrimp, crab and fish fauna, on the marsh surface and in open-water habitat adjoining the marsh, were measured using drop trap sampling (Zimmerman et al. 1984). Habitat along the outer marsh edge was subtidal nonvegetated mud bottom. Each habitat was sampled within 2 m of the marsh edge, and 8 samples were taken from each habitat monthly. The area enclosed by each sample was 2.6 m^2. Decapods and fishes larger than 5 mm length were removed, identified and enumerated. For predators in this paper, only brown shrimp, white shrimp and blue crab are considered.

To quantify densities of infauna and epifauna as food organisms (prey), substrate cores, 10 cm dia. x 5 cm deep, were taken from within each drop trap. Samples from the marsh surface included 6 to 8 culms of marsh (<u>Spartina</u> <u>alterniflora</u>) vegetation. Cores were sieved through a 500-micron screen to retain peracarids, molluscs and annelids. These small macrofauna were removed, identified and enumerated.

Marsh submergence was estimated using subsidence measured by land elevation changes at four USGS bench marks in the area of the Galveston Island State Park (data from the Harris-Galveston County Subsidence District in Friendswood, Texas). Seasonal, annual, and long-term changes in sea level for the greater area were obtained from a permanent NOAA tide station (No. 877-1450) in Galveston (data from Lyles et al. 1988).

This paper covers the period in 1985 beginning with the spring immigration of predators and ending with their fall emigration.

<u>Animal Densities</u>

Predaceous brown shrimp, white shrimp, and blue crab occurred in comparatively high densities on the marsh surface, and densities were usually significantly greater in the marsh than in nearby open water (Figure 1). Variation in monthly abundances followed predictable patterns of seasonal immigration. Brown shrimp occurred in highest abundance during spring recruitment, and white shrimp abundance peaked in the summer (described by Baxter and Renfro 1967).

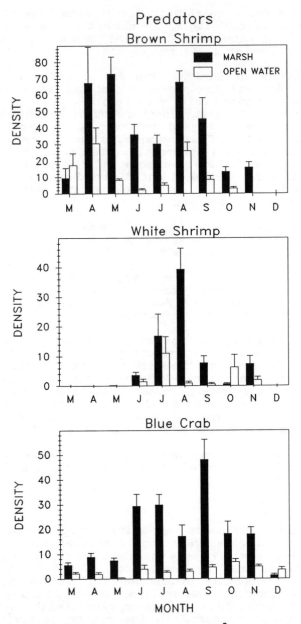

Figure 1. Densities (mean # per 2.6 m^2 ± 1 SE) of decapod crustacea, as predators, on marsh surface and adjacent mud bottom, in a subsiding area of Galveston Bay.

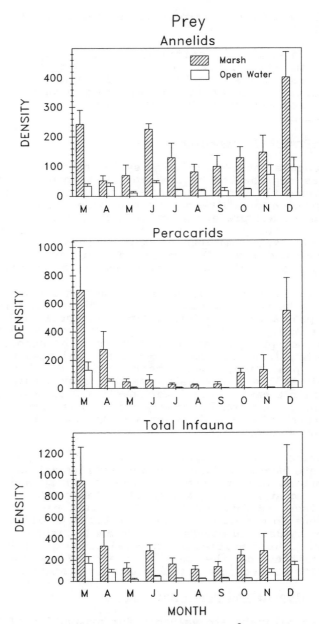

Figure 2. Densities (mean # per 78 cm^2 ±1 SE) of infauna, as prey, on the marsh surface and adjacent mud bottom, in a subsiding area of Galveston Bay.

Blue crabs recruited in large numbers in the fall (Thomas et al. 1989).

The infauna and epifauna, as prey, were most abundant in the winter and early spring and least abundant from late spring to late fall (Figure 2). Mean prey densities were often several times higher in the marsh, in association with marsh plants, as compared to open-water habitat.

Predator-Prey Relationships

Predators entered the marsh system beginning in the spring, as represented by immigration of brown shrimp, and continued through the summer and fall, as represented by immigration of white shrimp and blue crabs. Peak abundances of all predators combined occurred during the summer and early fall before the onset of fall emigration. Thus, maximum predation pressure occurred during the warm months (Figure 1). The winter season was virtually free of effects of large numbers of predators.

The effects of predation are evident in the seasonal changes in prey populations. Populations of infauna build to peak abundances during March and April, and then decline sharply by May (Figure 2). The rapid decline suggests a cropping effect by predators. Throughout the warm months, infaunal populations were subjected to the heaviest predation, but their densities remained relatively stable. This indicates that prey populations can maintain themselves at lower levels and suggests that consequent production is largely going to the predators. The pattern of predation appears to be similar in peracarids and annelids.

Previous field caging studies and manipulative experiments have established that peracarids and annelids are important prey to juvenile penaeid shrimp and blue crabs (Young et al. 1976; Nelson 1981; Leber 1985; Thomas 1989; Nelson and Capone 1990). Thus, responses of prey populations in our study can be attributed to effects of predation by decapod crustaceans and small fishes. Since shrimp and crab densities are several times greater than fish densities on the marsh surface (Zimmerman and Minello 1984), decapods may dominate as predators.

Habitat Value of the Marsh Surface

Marsh habitats are valuable for production of high densities of infauna and epifauna which are at least potentially available to aquatic consumers (Kneib 1982,

1984; Rader 1984; Fleeger 1985). These prey may be exported to subtidal habitats near the marsh edge or fed upon directly on the marsh surface. Our data suggest that predators may preferentially exploit marsh habitat because of the high prey abundances.

Marsh habitat also is valuable as refuge for the juveniles of transient estuarine species. The structure of marsh plants provides protection for juvenile shrimp and crabs against predation by fishes (Minello and Zimmerman 1983; Minello et al. 1989; Thomas 1989). Thus, survivorship of these decapods, as well as their impacts as predators, may be increased by greater access to marsh habitat (Minello et al. 1989; Minello and Zimmerman in press).

Sea-Level Change

Between 1960 and 1985, the cumulative increase in relative sea level at the Galveston Pier 21 NOAA tide gage was approximately 20 cm (Figure 3). The rate was about 8 mm per year (Paine 1990). Eustatic change in sea level over that period is assumed to be about 1 mm per year (Etkins and Epstein 1982; Gornitz et al. 1982) with the remainder due to land elevation changes at the tide gage. The record depicts occurrence of large annual fluctuations in relative sea-level, including some periods of rapid rise and other periods of sharp decline (Figure 3).

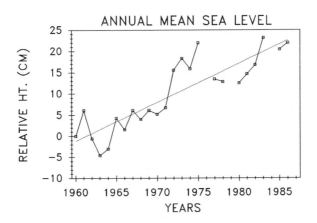

Figure 3. Annual mean changes in relative sea level at NOAA tide station No. 887-1450 on Galveston Island.

Marsh Submergence

Subsidence rates at the marsh site have been increasing at least since 1958 (Table 1; data from the Harris-Galveston County Subsidence District, Friendswood, Texas). Accordingly, land surfaces became lower by rates of 3.6, 7.4 and 13.7 mm per year between the intervals 1958-1964, 1964-1978 and 1978-1987, respectively. The subsidence rates plus eustatic sea-level rise provide an estimate of the marsh submergence and drowning rate. Since the marsh surface appears to be undergoing erosion rather than accretion, the estimated submergence rate may be understated. We note that submergence at the site is greater than that approximated by the Galveston NOAA tide gage, located about 20 km east of the marsh site.

Table 1. National Geodetic Survey bench marks (elevation in ft.) in the area of a salt marsh on Galveston Island.

Designation	1958	1964	1978	1987
J 1186	----	9.842	9.516	9.18
K 1186	----	4.091	3.775	3.42
Park R M 2	----	----	7.774	7.25
D 460	9.567	9.495	9.119	----

Flooding of the marsh surface varied monthly, following usual seasonal tide patterns for the northern Gulf of Mexico (Lyles et al. 1988). Seasonal tides cause prolonged inundation in the spring and fall and less inundation in the winter (Figure 4).

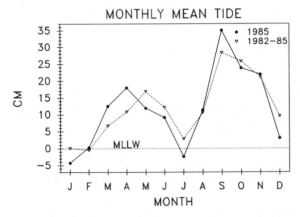

Figure 4. Monthly mean tide level changes at NOAA Station No. 887-1450 on Galveston Island.

Relationships Between Marsh Submergence and Habitat Value

The value of habitat, in terms of food and refuge, increases for consumers when marshes undergo submergence and prolonged inundation, extending the time and area available for exploitation. In the process of becoming more submerged, marsh plants also begin to die as they drown (Mendelssohn and McKee 1988), and thus eventually the marsh converts into open water. During the transition, a greater amount of marsh to open-water edge is created which further improves the accessibility of marshes to consumers. Decreasing plant density in drowning marshes may further accommodate ease of access.

Regional Differences

Compared to other regions, the number of estuarine animals using northwestern Gulf marshes is relatively high. The few density measurements from the East Coast of shrimp and blue crabs utilizing marshes (Hettler 1989; Mense and Wenner 1989) are orders of magnitude lower than those of the northwestern Gulf (Zimmerman and Minello 1984; Thomas et al. 1989). We propose that such differences reflect dissimilarity in the regional levels of secondary production derived directly from marshes.

Since land surface movements and scales of tides differ regionally, processes and functions of marshes may vary on a regional basis. East Coast marshes tend to have larger tidal amplitudes and less subsidence than northwestern Gulf marshes. Meso-scale tides promote exportation of marsh materials and importation of sediments, but may limit use of marsh surfaces to specialized organisms. Micro-scale tides can lessen transport of materials, but may promote the use of marsh surfaces by organisms. Marsh submergence, through subsidence or water-level rise, also can promote direct utilization by consumers. Increased utilization, such as is occurring in northwestern Gulf marshes, translates into more secondary production from marsh habitat. Less direct utilization of East Coast marshes by consumers may result in lower secondary production.

Implications of Marsh Submergence to Sea-Level Rise

Biological response patterns, including productivity and trophic dynamic responses, in submerging marshes provide a model for predicting effects of accelerated sea-level rise. For example, our evidence suggests that changes in coastal fisheries production can occur with accelerated sea-level rise. Productivity of estuarine-dependent fisheries may increase or decline because of the effects of sea-level rise on marsh function. The most immediate changes will occur in marsh areas with micro-tides and high subsidence rates. In particular, modifications to fishery productivity may be already underway in the northwestern Gulf of Mexico, where water-level changes are similar to those predicted by global warming/sea-level rise models.

Secondary production may increase, at least temporarily, in marshes undergoing submergence due to sea-level rise at rates of 5 to 10 mm per year. In this case, habitat function and utilization is improved for consumers due to greater inundation of marsh surfaces and to the relatively slow die-back of plants at the marsh edge. If salt marsh plants invade upland at the same rate as the outer edge decays (due to drowning), the overall area of the marsh is maintained.

Secondary production may decline sharply, however, when eustatic sea-level rise rates exceed 1 cm per year because rapidly drowning marshes may not be able to maintain habitat area. At some point in sea-level rise, inland re-establishment rates will not keep pace with the outer edge die-back. The loss of habitat in this case outweighs the improvement to habitat function for consumers. This scenario will almost assuredly occur in areas presently undergoing high rates of subsidence.

Acknowledgements

The authors gratefully acknowledge Robert E. Thompson, of the Harris-Galveston Coastal Subsidence District, and Jeffery G. Paine, of the University of Texas Bureau of Economic Geology, for assistance in providing data on subsidence and sea-level rise trends. We also thank Joan Kostera, Janet Thomas-Martinez, Marie Castiglione-Patillo, David Smith, Timothy Baumer and others, of the NMFS Galveston Laboratory, for assistance in the field, processing samples, analyzing data and preparing figures. The marsh study site was provided by the Texas Parks and Wildlife Department.

References

Baxter, K. N., and W. C. Renfro 1967. Seasonal occurrence and size distribution of postlarval brown and white shrimp near Galveston, Texas, with notes on species identification. Fish. Bull. 66:149-158.

Bell, S.S., and B. C. Coull 1978. Field evidence that

shrimp predation regulates meiofauna. Oecologia 35:141-148.

Boesch, D. F., and R. E. Turner 1984. dependence of fishery species on salt marshes: the role of food and refuge. Estuaries 7:460-468.

Childers, D. L., J. W. Day, Jr. and R. A. Muller 1990. Relating climatological forcing to coastal water levels in Louisiana estuaries and the potential importance of El Nino-Southern Ocillation events. Climate Research 1:31-42..

Etkins, R., and E. Epstein 1982. The rise of global mean sea level as an indication of climate change. Science 215:287-289.

Fleeger, J. W. 1985. Meiofaunal densities and copepod species composition in a Louisiana, U.S.A., estuary. Trans. Am. Microsc. Soc. 104:321-332.

Gleason, D. F. 1986. Utilization of salt marsh plants by postlarval brown shrimp: carbon assimilation rates and food preferences. Mar. Ecol. Prog. Ser. 31:151-158.

Gornitz, V., S. Lebedeff and J. Hansen 1982. Global sea level trend in the past century. Science 215:1611-1614.

Haines, E. B. 1976. Stable carbon isotope ratios in the biota, soils and tidal water of a Georgia salt marsh. Estuar. Coast. Mar. Sci. 4:609-616.

Hettler, W. F., Jr. 1989. Newton use of regularly-flooded saltmarsh cordgrass habitat in North Carolina, USA. Mar. Ecol. Prog. Ser. 56:111-118.

Kneib, R. T. 1982. Habitat preference, predation, and the intertidal distribution of gammaridean amphipods in a North Carolina salt marsh. J. Exp. Mar. Biol. ecol. 59:219-230.

Kneib, R. T. 1984. Patterns of invertebrate distribution and abundance in the salt marsh: causes and questions. Estuaries 7:392-412.

Kneib, R. T. 1986. The role of *Fundulus* *heteroclitus* in salt marsh trophic dynamics. Amer. Zool. 26:259-269.

Leber, K. M. 1985. The influence of predarory decapods, refuge, microhabitat selection on seagrass communities. Ecology 66:1951-1964.

Lyles, S. D., L. E. Hickman and H. A. Debaugh Jr. 1988. Sea level variations for the United States 1855-1980. NOAA/NOS Rep., National Ocean Survey, Tides and Water Levels Branch, Rockville, MD., 182 pp.

McIvor, C. C., and W. E. Odum 1986. The flume net: a quantative method for sampling fishes and macrocrustaceans on tidal marsh surfaces. Estuaries 9:21-224.

McIvor, C. C., and W. E. Odum 1988. Food, predation risk, and microhabitat selection in a marsh fish assemblage. Ecology 69:1341-1351.

Nelson, W. G. 1981. Experimental studies of decapod and fish predation on seagrass macrobenthos. Mar. Ecol. Prog. Ser. 5:141-149.

Nelson, W. G., and M. A. Capone 1990. Experimental studies of predation on polychaetes associated with seagrass beds. Estuaries 13:51-58

McTigue, T., and R. J. Zimmerman (in press). Carnivory versus herbivory in juvenile Penaeus setiferus (Linnaeus) and P. aztecus (Ives). J. Exp. Mar. Biol. Ecol.

Mendelssohn, I. A., and K. L. McKee 1988. Spartina alterniflora dieback in Louisiana: Time course investigation of soil waterlogging. J. Ecol. 76:509-521.

Mense, D. J., and E. L. Wenner 1989. Distribution and abundance of early life history stages of the blue crab, Callinectes sapidus, in tidal marsh creeks near Charleston, South Carolina. Estuaries 12:157-168.

Minello, T. J., and R. J. Zimmerman 1983. Fish predation on juvenile brown shrimp, Penaeus aztecus Ives: the effect of simulated Spartina structure on predation rates. J. Exp. Mar. Biol. Ecol. 72:211-231.

Minello, R. J., and R. J. Zimmerman (in press). The role of estuarine habitats in regulating growth and survival of juvenile penaeid shrimp. In: Dougherty, W.J. and M.A. Davidson (eds.), Frontiers of shrimp research. Elsevier Science Publishers.

Minello, T. J., R. J. Zimmerman and T. E. Czapla 1989. Habitat-related differences in diets of small fishes in Lavaca Bay, Texas, 1985-1986. NOAA Tech. Memo. NMFS-SEFC-236, 16 pp.

Minello, T. J., R. J. Zimmerman, and E. X. Martinez 1989. Mortality of young brown shrimp Penaeus aztecus in estuarine nurseries. Tran. Am. Fish. Soc. 118:693-708.

Morris, J. T., B. Kjerfve and J. M. Dean 1990. Dependence of estuarine productivity on anomalies in mean sea level. Limnol. Oceanogr. 35:926-930.

Nance, M. J., E. F. Klima and T. E. Czapla 1989. Gulf shrimp stock assessment workshop. NOAA Tech. Memo. NMFS-SEFC-239, 41 pp.

Odum, E. P. 1980. The status of three ecosystem-level hypotheses regarding salt marsh estuaries: tidal subsidy, outwelling, and detritus-based food chains. pp. 485-495. In: V. S. Kennedy (ed.), Estuarine Perspectives. Academic Press, New York, N.Y.

Paine, J. G. 1990. Recent vertical movement and sea-level changes, Texas coastal zone. Eos, Am. Geophy. Union Trans. 71:479.

Peterson, B.J., R. W. Howarth and R. H. Garrit 1986. Sulfur and carbon isotopes as tracers of salt-marsh organic matter flow. Ecology 67:865-874.

Pomeroy, L. R. 1989. Estuaries of southeastern U.S. an ecological overview. pp. 7-15. In: Barrier Island/Salt Marsh Estuaries, Southeast Atlantic Coast:Issues, Resources, Status, and Management. NOAA Estuarine Programs Office, Estuary-of-the-Month Seminar Series No. 12, Proc. Feb. 17, 1988, Wash. D.C.

Provost, M. W. 1976. Tidal datum planes circumscribing salt marshes. Bull. Mar. Sci. 26:558-563.

Rader, D. N. 1984. Salt-marsh benthic invertrbrates: small-scale patterns of distribution and abundance. Estuaries 7:413-420.

Rozas, L. P., and W. E. Odum 1987. Use of tidal freshwater marshes by fishes and macrocrustaceans along a marsh stream-order gradient. Estuaries 10:36-43.

Stoner, A. W., and R. J. Zimmerman 1988. Food pathways associated with penaeid shrimps in a mangrove-fringed estuary. Fish. Bull. 86:543-551.

Sullivan, M. J., and C. A. Moncreiff 1988a. Primary production of edaphic algal communities in a Mississippi salt marsh. J. Phycol. 24:49-58.

124 COASTAL WETLANDS

Sullivan, M. J., and C. A. Moncreiff 1988. An evaluation
of the importance of algae and vascular plants in salt
marsh food webs using stable isotope analyses.
Mississippi-Alabama Sea Grant Consortium Rep. MASGP-88-
042, 62 pp.

Thomas, J. D. 1976. A survey of gammarid amphipods of
the Barataria Bay, Louisiana region. Contrib. Mar. Sci.
20:87-100.

Thomas, J.L. 1989. A comparative evaluation of Halodule
wrightii Aschers, Spartina alterniflora Loisel and bare
sand as nursery habitats for juvenile Callinectes
sapidus. M.S. Thesis, Texas A&M University, 119 pp.

Thomas, J. T., R. J. Zimmerman and T. J. Minello 1990.
Abundance patterns of juvenile blue crabs (Callinectes
sapidus) in nursery habitats of two Texas bays. Bull.
Mar. Sci. 46(1):115-125.

Titus, J. G. (ed.) 1988. Greenhouse effect, sea level
rise and coastal wetlands. Environmental Protection
Agency, Office of Policy, Planning and Evaluation, Wash.
D.C. EPA-230-05-86-013, 151 pp.

Turner, R. E. 1977. Intertidal vegetation and commercial
yields of penaeid shrimp. Trans. Am. Fish. Soc. 106:411-
416.

Weinstein, M. P. 1979. Shallow marsh habitats as primary
nurseries for fishes and shellfish, Cape Fear River,
North Carolina. Fish. Bull. 77:339-357.,

Young, D. K., M. A. Buzas and M. W. Young 1976. Species
densities of macrobenthos associated with seagrass: a
field experimental study of predation. J. Mar. Res.
34:577-592.

Zimmerman, R. J., and T. J. Minello 1984. Densities of
Penaeus aztecus, Penaeus setiferus, and other natant
macrofauna in a Texas salt marsh. Estuaries 7:421-433.

Zimmerman, R. J., T. J. Minello and G. Zamora 1984.
Selection of vegetated habitat by Penaeus aztecus in a
Galveston Bay salt marsh. Fish. Bull. 82:325-336.

Trends in Eutrophication in Five Major U.S. Estuaries: Management Implications

Donald W. Stanley[1], Virginia Lee[2], and Alan Desbonnet[2]

Estuarine managers continue to be hampered by a lack of understanding of the quantitative relationships between nutrient loading and eutrophication. The problem has been approached by means of experimentation and various modelling efforts. The NOAA-sponsored study that we undertook employed a third technique; i.e., using historical data on nutrient loading, estuarine nutrient concentrations, etc. as evidence for changes in eutrophication status in five major U.S. estuaries: Narragansett Bay, Delaware Bay, Albemarle-Pamlico Sounds, Galveston Bay, and San Francisco Bay. Our primary objective here is to explore the management implications of trends observed in nutrients in the bays. Also, we will summarize what we have learned about the adequacy of past and current water quality monitoring in our nation's estuaries.

There is strong evidence that improved municipal and industrial wastewater treatment have led to substantial reductions in point source nutrient loading to these estuaries during the past two decades. However, future population growth near some of the bays could offset these gains. There is very little direct evidence of change in non-point source loading. The use of one major nonpoint nutrient source -- chemical fertilizers -- has peaked and may be declining. Potential increases from atmospheric loading and farm animals are threats, but there is scant evidence concerning the fractions of these inputs that actually reach the bays.

There is no evidence of increasing nutrient concentrations within these bays during the past two decades. Rather, the N and P concentrations appear to have either declined (near point source outfalls), or not to have changed significantly (in the more open waters of the bays). In light of this finding, the current emphasis on further reducing loading to "improve" water quality and enhance fisheries could be questioned. Clearly, more quantitative knowledge about the loading-concentration relationship is needed before costly treatment improvements are undertaken.

The largest and "most important" estuaries are not necessarily those with the most complete historical water quality data bases. Initiation and maintenance of monitoring programs have often depended on fortuitous circumstances involving collaboration between state agencies, industries, and university scientists. In some cases, there is an overly optimistic misconception regarding the quantity and quality of data available. The data collected by state and federal agencies is often placed in large computerized databases which are relatively easily to access; but

[1]Institute for Coastal and Marine Resources, East Carolina University, Greenville, North Carolina 27858-4353; [2]Coastal Resources Center, Graduate School of Oceanography, University of Rhode Island, Narragansett, Rhode Island 02882

documentation concerning methodology and other aspects of quality control may be very poor. Data generated by individual researchers on the other hand is sometimes difficult to access. These deficiencies indicate the need for a national concerted effort on long-term monitoring of our most important estuaries.

Compliance and Enforcement in Pretreatment Programs:
A Comparison of Three Programs

Jon G. Sutinen[1]

Abstract
    This paper characterizes the extent and nature of
noncompliance with pretreatment regulations in Rhode Island. The three pretreatment enforcement programs described and compared are the Narragansett Bay Commission
(NBC), Blackstone Valley District Commission (BVDC), and
the City of East Providence.

Introduction
    Wastewater discharged by industry often contains a
variety of toxic and other harmful substances, including
heavy metals and cyanide from electroplating and metal
finishing shops, and lead from battery manufacturers.
Most sewage treatment system's were not designed to properly treat industrial wastes of this type. Such industrial wastes can interfere with the effective operation
of sewage treatment plants, pass through untreated and
contaminate local bodies of water, and increase the
health and environmental risks of sewage sludge disposal.
    In its report to Congress, the Office of Technology
Assessment (Levenson and Barnard, 1988) identified compliance and enforcement as one of four key issues in the
management of industrial effluents being discharged into
the marine environment. In particular, OTA identifies
the need for (1) more and better information on the extent of noncompliance, and (2) for assessments of enforcement effectiveness in order to design improvements in
the framework that regulates industrial discharges. This
study is an attempt to satisfy those two demands. Using
data from three pretreatment programs in Rhode Island,
our purpose is to fully characterize the nature and ex-

1 Professor, Department of Resource Economics, University
of Rhode Island, Kingston, RI 02881. This is an abridged
version of a larger and more complete report, which can
be obtained from the author.

tent of noncompliance with pretreatment regulations, and
to assess the effectiveness of the enforcement efforts in
each program.
     Enforcement is an essential link in the effort to
manage the discharge of industrial effluents in the nations lakes, streams and marine waters. Levenson and
Barnard (1988) review most of the literature concerning
compliance and enforcement under the National Pollution
Discharge Elimination System (NPDES) and the National
Pretreatment Program. The existing evidence exposes
weaknesses in three areas. First, the evidence shows a
low level of response to permit violations, with less
than half of significant violations receiving enforcement
responses in the cases studied. Second, both formal and
informal enforcement actions were ineffective in
eliminating violations in a timely manner. And third,
the ability to impose meaningful penalties is absent,
collection of penalties is infrequent, and assessed
penalties are usually small and inconsequential to the
violator. Their review also concludes that it is difficult to reliably measure the extent of industry compliance with pretreatment regulations. This is due, they
suggest, to the heavy reliance on self-monitoring by industries which is often not independently verified by
pretreatment program monitoring.
     The present study of three pretreatment programs in
Rhode Island will investigate many of these same issues.
Noncompliance with the regulations is measured and patterns of noncompliance are characterized. The period under study is June, 1985, through June, 1988, and changes
and developments since then will not be discussed unless
they are relevant to the analysis.

Three Pretreatment Programs in Rhode Island
     This section describes the principal elements of the
industrial wastewater pretreatment programs operated by
the Narragansett Bay Commission, the Blackstone Valley
District Commission, and the City of East Providence.
     Narragansett Bay Commission NBC operates the
Field's Point Wastewater Treatment Facility in Providence. The POTW processes sewage of Providence, North
Providence, Johnston, and sections of Cranston and Lincoln, serving a population of about 200,000. It is the
largest POTW in Rhode Island. The plant, which discharges near the mouth of the Providence River, is a
secondary activated sludge facility and has a design capacity of 64 million gallons per day (mgd). About 20
percent of the plant's wastewater influent comes from industrial sources, most originating with electroplaters
and metal finishers. (EPA, 1987)
     NBC's pretreatment program was approved by the En-
vironmental Protection Agency (EPA) in September, 1984.

NBC implements and enforces its pretreatment program
through its Rules and Regulations, which provide for industrial wastewater discharge permits to be used as the
IU control mechanism (NBC, 1985, 1987). Each permit
usually requires that the user meet the limits on discharges at all times and monitor its discharges, among
other things (NBC, 1988).

Blackstone Valley District Commission The Blackstone Valley District Commission (BVDC) operates the
Bucklin Point Sewage Treatment Plant in East Providence.
The plant, discharging directly into the Seekonk River,
serves a population of about 100,000 in the cities of
East Providence, Central Falls and Pawtucket, and the
towns of Cumberland, Lincoln and part of Smithfield. It
is the second largest POTW in the State. The treatment
plant, a secondary activated sludge facility, has a design capacity of 31 mgd and a peak capacity of 46 mgd.
(EPA, 1987) The plant's wastewater influent is 14 percent industrial, largely from metal finishing and textile
industries (Ferreira, 1990) .

BVDC's pretreatment program was approved by EPA in
March 1983. Its Rules and Regulations provide for industrial wastewater discharge permits to be used as the IU
control mechanism (EPA, 1987).

East Providence The City of East Providence opera-
tes the Ponham Terrace treatment plant in Riverside, RI.
The plant, which discharges through an outfall into the
Providence River - Narragansett Bay, serves about two-
thirds of the City and the entire Town of Barrington.
The plant is an aerated activated sludge wastewater
treatment facility with a design capacity of 10.5 mgd.
The plant receives industrial wastewater from about 70
IUs. (EPA, 1985)

East Providence's pretreatment program was approved
by EPA in September 1983. The East Providence pretreatment program is enforceable through sewer ordinances of
the City of East Providence and the Town of Barrington,
and a formal agreement between the City and the Town
(EPA, 1985). The City of East Providence uses permits as
its mechanism to establish program requirements and discharge conditions.

DISCHARGE LIMITATIONS

Narragansett Bay Commission Of the more than 200
permits issued by mid 1988, 131 were issued to electroplaters, metal finishers, and other IUs subject to
categorical standards. During the early part of the
study period NBC applied federal categorical standards to
the electroplaters and metal finishers in its pretreatment program. In response to operational and pass-
through problems attributed to industrial discharges, NBC
developed a set of local discharge limits. The more
stringent local limits became effective in September,

1987, and apply to all IUs discharging industrial
wastewater into NBC's sewers regardless of the amount of
flow (NBC, 1987d).
   Blackstone Valley District Commission Of the 94
permits issued by June, 1988, 48 were issued to IUs subject to categorical pretreatment standards (DEM, 1988).
During the study period, electroplaters and metal finishers were regulated by a set of interim limits (DEM,
1988). Most of the standards are the same as the monthly
average federal categorical standards.
   East Providence The City has two industrial
category groups. Forty-nine industrial users had been
permitted by early 1988, including 11 metal finishers,
one textile mill, and one fastener manufacturer as
Priority I users, and 35 as Priority II users. The City
replaced the categorical electroplating and metal finishing standards with local limits in December 1985.
                        MONITORING
   Under the pretreatment program, sampling and analysis of industrial wastewater for the purpose of regular
compliance determinations is done through industry self-
monitoring and compliance monitoring done by the CA.
   Narragansett Bay Commission The required frequency
of self-monitoring reports depends on the type of IU and
the volume of wastewater discharges by the IU. The frequency of self-monitoring may be increased to weekly when
an IU consistently fails to meet the terms of its permit.
IUs subject to federal categorical standards are required
to resample their wastewater for any parameters violating
the standards within 30 days of detecting the violation.
Samples by IUs may be either composite or grab samples,
but most sampling consists of composite samples (NBC,
1989). Except for a 12 month period ending in mid 1988,
NBC has satisfied the RI Department of Environmental Management's (DEM) frequency of compliance monitoring (EPA,
1987; DEM, 1988).
   Blackstone Valley District Commission All categorical IUs are required to submit self-monitoring reports at
least twice a year which indicate the nature and concentration of pollutants in their effluent which are
limited by the discharge limitations. Some IUs are required to submit self-monitoring reports as often as
every month. The reports contain measures of the average
and maximum daily flow of the pollutants (BVDC, 1986).
Both grab and composite samples are required for self-
monitoring. An audit by EPA and inspections by DEM indicate that several IUs have been delinquent in submitting
their self-monitoring reports (DEM, 1988, EPA, 1987).
   The compliance monitoring activities of the BVCD
pretreatment staff were minimal for most of the study period. As of mid-1988, the end of the study period, the
BVCD Pretreatment Program had not consistently fulfilled

its obligation to monitor each significant IU at least
once a year. This failure was reportedly due to lack of
trained staff and equipment (DEM, 1988).
East Providence All Priority I users are required
to conduct monthly self-monitoring. The categorical industries are required to sample monthly for specific pollutants of concern and bi-annually for all regulated pollutants. Other Priority I users conduct monthly sampling
for pollutants likely to be found in their discharge.
Compliance monitoring, conducted by the City, consists of
compliance inspections along with announced and un-
announced sampling events at each permitted industry.
There is no evidence that the POTW has had difficulty
meeting any of its monitoring and inspection obligations.

Enforcement
        Each POTW, as the Control Authority for its
pretreatment program, is the principal enforcement entity
governing industrial users' discharges into the sewer
system. In addition, enforcement actions may be taken by
EPA and DEM, and citizens may file suit against
violators. Enforcement actions are classified as minor
and formal. The minor enforcement actions include telephone calls, letters, and informal meetings to discuss
violations and measures for coming into compliance. Formal enforcement actions include the Notices of Failure,
of Deficiency, of Violations, Civil Penalties (fines),
and Publication of the names of significant violators in
the local newspaper. The volume of minor enforcement actions (telephone calls, meetings, etc.) is the largest
and roughly proportional to the size of each pretreatment
program. There is not much informational content in
these statistics. The statistics on the formal enforcement actions, however, are informative.
        Narragansett Bay Commission NBC employed all of its
formal enforcement actions. Notices of Failure were the
most common, the numbers steadily increasing over the
study period. The overall pattern shows very little formal enforcement activity during the first half of the
study period, followed by a significant increase in the
numbers of all formal actions in the second half of the
study period. NBC increased significantly its use of
Notices of Failure in 1987 and 1988. From mid 1985 to
the end of 1986, NBC issued an average of 4.5 NOFs per
month. For all of 1987 and through mid 1988, it issued
an average of 27 NOFs a month. The use of Notices of
Deficiency, on the other hand, remained steady at between
4 and 5 NODs issued per month on average. Notices of
Violation were issued less frequently, and their use increased from the end of 1986 forward. Only one NOV was
issued during the first half of the study period;
whereas 7 NOVs were issued during the second half of the
study period.

The assessment of Civil Penalties occurred with roughly the same frequency as NOVs were issued. The amount of one of these penalties was initially assessed at nearly $220,000 (the Ronci case). However, this penalty was later negated by the RI Supreme Court. Four other civil penalties were successfully assessed in 1987 and 1988, averaging just under $40,000 a case. All of the penalties were assessed after 1986.

Publication of the names of significant violators in the Providence Journal occurs annually. NBC published the names of 53 IUs in 1986, 37 in 1987, and 74 in 1988. Of these, 51, 28, and 50 IUs, respectively, were in significance noncompliance for exceeding discharge limits in each of the three years.

Blackstone Valley District Commission BVDC employed Notices of Failure and Violation, and Civil Penalties during the study period. Like NBC, the overall pattern for BVDC shows very little formal enforcement activity during the first half of the study period, followed by an increase in the numbers of all formal actions in the second half of the study period. During the first half of the study period, mid 1985 through 1986, only a few Notices of Failure were issued (7 in total), and no Notices of Violation or Civil Penalties were used. During the second half, the incidence of enforcement actions increased somewhat. Notices of Failure were issued an average of one (1) a month, Notices of Violation nearly two (2) a month, and two Civil Penalties were assessed, one for $1000 and another for $18,000 (both in 1987).

East Providence, with its compliant set of IUs, had less enforcement activity than NBC and BVDC, and what activity existed remained somewhat steady over the entire study period. East Providence issued 17 Notices of Deficiency, one (1) Notice of Violation, and no Civil Penalties during the entire study period.

The coverage of these enforcement actions is now examined. Specifically, we calculate the extent to which significant violators have had formal enforcement actions taken against them. This calculation results in a measure of the percent of IUs in Significant Noncompliance which have had at least one formal enforcement action taken against them[2]. **NBC's** coverage was poor in the early months, modest in the middle months, and good in the latter months of the study period. **BVDC's** coverage is poor until late 1987, and good in the last four months of the study period. **East Providence's**

---

2 SNC is being used here only as an indicator of major or serious violations. EPA's definition of SNC did not become effective until 1987, and prior to 1987 each program had its own criteria for initiating enforcement actions.

coverage varies widely reflecting the small numbers of
enforcement actions.

An important feature of enforcement is its timeli-
ness, the amount of time from detection to when a formal
enforcement action is taken. NBC averaged 39 days to an
enforcement action for the entire study period; and for
the second half the average is 34 days[3]. BVDC's average
was slightly better at 36 and 33, respectively. The
averages for the early part of the period are deceptive,
however, since there were so few enforcement actions by
BVDC. Interestingly, East Providence's formal enforcement response time was higher than the other two
pretreatment programs. East Providence averaged 58 days
to a formal enforcement action over the entire period,
but this is also deceptive because so few enforcement actions were taken. This higher average may also reflect
the POTW's policy of "working with its IUs" on compliance
issues rather than sanctioning them (i.e., emphasizing
the carrot more than the stick in their approach to noncompliance).

Approval Authority Enforcement EPA and DEM, as Approval Authories, oversee the implementation and enforcement of the pretreatment programs in the state. As part
of their oversight role, EPA and DEM conduct evaluations
and can take direct enforcement actions in cases of national or regional significance.

DEM took enforcement actions directly against four
IUs in NBC's pretreatment program. Two of the IUs closed
their businesses, one case was taken over by NBC, and the
fourth case resulted in the IU coming into compliance.
Two additional enforcement actions were joint with EPA.
The joint enforcement actions by DEM and EPA resulted in
two civil penalties, one for $50,000 and the other for
$125,000. EPA initiated two actions against IUs in the
BVDC pretreatment program. The actions consisted of letters requesting information, and resulted in enforcement
actions being taken by BVDC's pretreatment program.

Citizen Enforcement The Clean Water Act, through
section 505, also provides citizens with the authority to
bring civil action against violators of the pretreatment
regulations. In Rhode Island, the Conservation Law Foundation of New England and Save the Bay, Inc. have filed
citizens suits against IUs in violation of pretreatment
regulations. In 1987, the Conservation Law Foundation of
New England and Save the Bay, Inc., joined together to
file legal suits against two IUs in NBC's pretreatment
program. In one case (Rolo Manufacturing), the suit
resulted in a $25,000 civil penalty. The other case

---

3 It should be noted that this data is only for those
violations where a formal enforcement action was taken.

(RIBCO) was subsequently taken over by EPA and DEM, and
resulted in a $125,000 civil penalty.
    In summary, the investigation of enforcement reveals
low levels of enforcement activity during the first half
of the study period. Enforcement activity increased significantly in the NBC and BVDC programs during the second
half of the study period. Even in the second half, however, low and modest levels of serious major enforcement
actions were taken. Relatively few notices of violation
were issued, and even fewer civil penalty actions were
taken.

Noncompliance with Wastewater Discharge Limits
    The following description and assessment of non-
compliance is not intended to indicate the success or effectiveness of the three pretreatment programs. Here, we
measure noncompliance, which is not necessarily related
to the effectiveness of a pretreatment program. The appropriate measure of effectiveness of a pretreatment program is the extent to which it reduces loadings of pollutants to the influent of the POTW. Successful and effective pretreatment programs, as measured by influent
loadings, may exhibit either high or low levels of noncompliance. For example, a successful pretreatment program may exhibit high levels of noncompliance because of
stringent standards and intensive monitoring. Conversely, a less effective pretreatment program may exhibit low
levels of noncompliance because of liberal discharge
standards and/or inadequate monitoring.
             MEASURING NONCOMPLIANCE
    There are numerous ways to measure noncompliance
and, unfortunately, no one way is fully satisfactory for
our purposes. Next, we discuss the five measures used in
this study, including their advantages and disadvantages,
followed by a presentation of the evidence characterizing
noncompliance in the three pretreatment programs[4]. The
five basic means of quantifying and characterizing noncompliance used in this study are Significant Noncompliance, Noncompliance, Compliance Style, and Time to
Compliance.
    Significant Noncompliance (SNC) Significant Non-
compliance is where either or both of the following conditions hold: (a) Sixty-six percent or more of the
wastewater samples analyzed in a six-month period exceed
the same daily maximum, or average, limit. (b) Thirty-

---

[4] Only noncompliance with established discharge limits is
measured here. Violations of compliance schedule mile-
stones, failures to submit reports on schedule, failures
to report noncompliance, and any other violation are not
considered.

three percent or more of the wastewater samples analyzed
in a six-month period exceed the same daily maximum, or
average, limit by more than 20 percent. An advantage of
SNC as a measure is that it distinguishes minor from
major violations, and chronic from occasional violations.
Some disadvantages of this measure is that the conditions
are arbitrary and, for some purposes, six months and 20
percent are too liberal.

Noncompliance (NC) Noncompliance is any violation
of the discharge limits, regardless of its size or frequency. This measure does not distinguish between large
and small, chronic and occasional violations. Two
measures of NC are calculated, one for IUs and one for
samples. An IU is in Noncompliance if one or more of the
samples analyzed during the period violates a discharge
limit for any parameter. The percent of IUs in NC is
calculated for monthly intervals. The percent of samples
in NC is the proportion of all IUs samples analyzed during the period for which one or more discharge limit is
violated. The sample NC measure also is presented for
selected parameters subject to discharge limits. Each NC
rate is calculated for monthly intervals. The NC and SNC
measures are useful complements since NC captures the
violations missed by the SNC measure.

Percent Exceedance (%EX) Prior to 1987, when EPA's
definition of significant noncompliance became effective,
each pretreatment program used different criteria to
trigger an enforcement action. Each POTW had its own
policy regarding the length, magnitude, and severity of
violation that would trigger enforcement actions. This
measure is an alternative to SNC for quantifying the magnitude and severity of a given violation.

The extent to which sampled discharges exceed the
discharge limits is measured by the Percent Exceedance
measure of noncompliance. It is defined as the ratio of
the actual discharge quantity, for only those samples exceeding the limit, divided by the standard. The measure
presented here is an average of averages. First an average is calculated for each parameter of each IU's monthly
samples. Finally, an average is calculted for each parameter across all sampled IUs in the month. The principal advantage of this measure is that it reflects the
seriousness of noncompliance across parameters and over
time. A disadvantage is that it is an average and, as
such, masks the extreme values of the more serious violations.

Compliance Styles This measure, adapted from
Brubaker and Byrne (1989), summarizes the various styles
of compliance behavior exhibited by IUs during the study
period. IUs are grouped into the following categories:
(a) Regular Compliance. IUs with less than five violations over the 36 month period.

(b) Improved Compliance over time. IUs with a decreasing incidence of violations over the 36 month period, and compliance in the later months.

(c) Erratic Compliance. IUs in and out of compliance, with no trend over the 36 month period.

(d) Deteriorated Compliance over time. IUs which exhibited compliance in the early months, and an increasing incidence of violations in the later months of the 36 month period.

(e) Consistent Noncompliance. IUs consistently out of compliance over the entire 36 month period.

Any and all violations are used in this calculation, so there is no distinction between minor and major violations.

Time to Compliance (TVC) Time to Compliance is simply the number of days between the time a violation is detected (reported on a self-monitoring or a compliance-monitoring report) and the time compliance is next reported. This measure is calculated as an average over all IUs against which formal enforcement actions were taken.

PATTERNS OF NONCOMPLIANCE

Significant Noncompliance (SNC) NBC's SNC rate improves through September, 1987, and then increases, reflecting the imposition of the more stringent local limits. But for most of the study period, NBC's SNC varies between 30 and 40 percent, higher than the SNCs of the other two programs. BVDC's SNC rate varied widely in 1986, reflecting the small number of IUs monitored. However, for the second half of the study period, BVDC's SNC rate stabilized at between 20 and 30 percent. East Providence's SNC rate began high in 1986 and dropped to zero by the end of 1986.[5] For most of the second half of the study period, East Providence's SNC rate remained below 10 percent.

For most parameters, NBC's discharge standards are more stringent than those for the other two programs. When NBC's standards are imposed on BVDC and East Providence, the rates of SNC are somewhat higher. With NBC standards, BVDC's SNC rate is between 30 and 40 percent during the last 24 months of the study period. East Providence's SNC rate with NBC standards ranges between 10 and 20 percent during the last 12 months of the study period. The picture when common discharge standards are use remains roughly the same: East Providence exhibits the lowest rates of SNC, and NBC and BVDC exhibit nearly comparable high rates of SNC for much of the period.

---

5 East Providence was monitoring all of its categorical IUs in 1986, therefore, this pattern accurately reflects the situation in the pretreatment program.

Noncompliance (NC) Since the NC measures all, including the less serious, violations, its values are
higher than the values of SNC. As expected, the NC patterns are similar to the SNC patterns. **NBC's** NC pattern
reflects again the more stringent local standards invoked
in September 1987. When allowing for the change from the
federal categorical standards to local limits, the data
show an oscillating but improving trend up to September
1987. In October 1987 there is a shift up in NC followed
by a gradual improvement. By the July 1988 the percent
of IUs in NC is down to about where it was in August of
1987.
    BVDC's NC exhibits an erratic pattern of NC in the
early months. This erratic pattern is due to the small
number of samples being taken at that time. The pattern
from mid-1986 to the end is more reliable, showing little
improvement with the NC rate oscillating between 60 and
40 percent. **East Providence** presents a picture of low NC
over the entire period, with most values between zero and
20 percent. Since the number of all firms being
monitored is small (13), the large variation in NC is not
surprising.
    NBC's sample NC rate shows substantial variation
during the first half of the study period. The measure
improves steadily during the first nine months of 1987,
reaching a low of about 15 percent in August and September. As the more stringent local limits were imposed,
the sample NC rate rose dramatically to over 50 percent
by the end of 1987. By mid 1988, however, the sample NC
rate improved to less than 40 percent.
    BVDC's sample NC rate exhibits wide variation in the
first half of the study period, reflecting the small number of samples taken. Considerable variation remains in
the second half, with sample NC rate values regularly
above 40 percent. In 1988 the sample NC rate was near or
above 50 percent in four of the six months. **East Providence's** sample NC rate varies between zero and 20 percent for most of the study period. The monthly average
for the entire study period is just above 10 percent.
    NBC's most common problem pollutants during the
study period were with copper, nickel, zinc, and cyanide.
Copper, nickel, and cyanide reflect the overall pattern
of NBC's sample NC rate with improvements in 1987 through
September. After the local limits were imposed, the NC
rates for these three pollutants rose in late 1987, followed by significant improvement during 1988. Zinc exhibits a NC rate consistently between one and 15 percent,
with an average of less than 10 percent for the entire
study period.
    BVDC's most common noncompliance problem was with
cyanide, which nearly mirrors the overall sample NC rate.
In the second half of the study period, cyanide's NC rate

was regularly above 20 percent, once at 45 percent. The
most frequent NC rates for other pollutants is zero, with
occasional high values occurring for copper, nickel, and
zinc. **East Providence**'s most problematic pollutant is
nickel, but the incidence of noncompliance is much lower
than for the problematic pollutants of NBC and BVDC. The
only other pollutant to record a violation in the last
six months of the study period was zinc.

Percent Exceedance (%EX)   This evidence is similar
to the NC measures for individual pollutants. For NBC,
cyanide was more than 1000 percent above the limit briefly in the early part of the study period; since mid-1986
it has varied between 100 and 500 percent above the
limit. Zinc, nickel, and copper exhibit considerable
variation over the entire period, with occasional peaks
above 500 percent.

For BVDC, cyanide was more than 1000 percent above
the limit in most of 1987 and early 1988. Zinc, nickel
and copper were frequently above 100 percent of the
limits during the study period. For East Providence, the
amount of exceedance by nickel is relatively low over the
study period.

Compliance Styles   Less than half (46 percent) of
NBC's IUs regularly comply or have improved their compliance over the study period. Over 20 percent of NBC's
IUs were in consistent noncompliance, and over 25 percent
exhibited erratic noncompliance.

The pattern of Compliance Styles for **BVDC** is less
encouraging. Only 30 percent of the IUs regularly comply
or have improved their compliance. A hefty 70 percent of
the IUs are either in consistent noncompliance or display
erratic noncompliance. As expected by now, the picture
for **East Providence** is the brightest. All of the 13 IUs
regularly comply, with only an occasional violation.

Time to Compliance (TVC)   **NBC** realized a substantial
improvement in TVC from over 400 days in the first two
semesters to 90 days during the last semester of the
study period. **BVDC** experienced short TVCs during the
first half of the study period, but TVC rose to around
100 days in the second half. **East Providence** took very
few formal enforcement actions; but on average the TVC
was less than 100 days.

In summary, **NBC** exhibits an overall pattern of having serious noncompliance problems early in the study period, followed by significant improvements in all areas.
When local limits were introduced in September 1987, noncompliance escalated, but by the end of the study period,
noncompliance had returned to its low level once again.
NBC's noncompliance problems center around at least three
pollutants (copper, nickel, and cyanide), a group of
about 25 IUs that are consistently in noncompliance, and
another 32 IUs in erratic noncompliance. **BVDC** exhibits

an overall pattern consistent with the slow start in implementing its pretreatment program. By the second half of the study period the noncompliance situation is comparable in many respects to NBC's. BVDC, however, has fewer IUs to monitor and has noncompliance problems with only one pollutant (cyanide). BVDC, with a much smaller program than NBC, has consistent noncompliance problems with 6 IUs, and erratic noncompliance problems with another 12 IUs. East Providence's small pretreatment program exhibits an overall pattern of consistent compliance. What noncompliance problems existed occurred early in the study period.

Summary and Conclusions
     This study has investigated enforcement and compliance in three pretreatment programs in Rhode Island. Noncompliance with pretreatment regulations is measured and patterns of noncompliance are characterized.
     Regarding the patterns of noncompliance, we found that NBC exhibits an overall pattern of having serious noncompliance problems early in the study period, followed by significant improvements in all areas. When local limits were introduced in September 1987, noncompliance escalated, but by the end of the study period, noncompliance had returned to its low level once again. NBC's noncompliance problems center around at least three pollutants (copper, nickel, and cyanide), a group of about 25 IUs that are consistently in noncompliance, and another 32 IUs in erratic noncompliance.
     BVDC exhibits an overall pattern consistent with the slow start in implementing its pretreatment program. By the second half of the study period the noncompliance situation is comparable in many respects to NBC's. BVDC, however, has fewer IUs to monitor and has noncompliance problems with only one pollutant (cyanide). BVDC, with a much smaller program than NBC, has consistent noncompliance problems with 6 IUs, and erratic noncompliance problems with another 12 IUs. And East Providence's small pretreatment program exhibits an overall pattern of consistent compliance. What noncompliance problems existed occurred early in the study period.
     The investigation of enforcement reveals low levels of enforcement activity during the first half of the study period. Enforcement activity increased significantly in the NBC and BVDC programs during the second half of the study period. Even in the second half, however, low and modest levels of serious major enforcement actions were taken. Relatively few notices of violation were issued, and even fewer civil penalty actions were taken.

References
Narragansett Bay Commission:
EPA. 1987. Audit of Pretreatment Program, Narragansett Bay Commission, Rhode Island. U.S. Environmental Protection Agency, Region I, Boston MA. (October)
DEM. 1988. Narragansett Bay Commission Pretreatment Compliance Inspection. RI Department of Environmental Management, Division of Water Resources. (October)
NBC. 1983. Sewer Use Ordinance. Narragansett Bay Commission, Providence, RI. (September)
NBC. 1985. Rules and Regulations for the use of Wastewater Facilities within the Narragansett Bay Water Quality Management District. Narragansett Bay Commission, Providence, RI. (August)
NBC. 1986. Industrial Pretreatment Program Annual Report, October 1985 - September 1986. Narragansett Bay Commission, Providence, RI.
NBC. 1987a. Rules and Regulations for the use of Wastewater Facilities within the Narragansett Bay Water Quality Management District. Narragansett Bay Commission, Providence, RI. (August)
NBC. 1987b. Industrial Pretreatment Program Annual Report, October 1986 - September 1987. Narragansett Bay Commission, Providence, RI.
NBC. 1987c. Annual Report. Narragansett Bay Commission, Providence, RI.
NBC. 1988. Industrial Pretreatment Program Annual Report, October 1987 - September 1988. Narragansett Bay Commission, Providence, RI.
NBC. 1989. Industrial Pretreatment Program Annual Report, October 1988 - September 1989. Narragansett Bay Commission, Providence, RI.
Blackstone Valley District Commission:
BVDC. 1986. Rules and Regulations of the Blackstone Valley District Commission. Blackstone Valley District Commission, Rumford, RI. (February)
BVDC. 1987. Annual Report, Blackstone Valley District Commission Pretreatment Program. Blackstone Valley District Commission, Rumford, RI.
BVDC. 1987. Notice of Decision, R.I.G.L. 46-21-30. Blackstone Valley District Commission, Rumford, RI.
DEM. 1988. Blackstone Valley District Commission Pretreatment Compliance Inspection. RI Department of Environmental Management, Division of Water Resources. (August)
EPA. 1987. Audit of Pretreatment Program, Blackstone Valley District Commission, Rhode Island. U.S. Environmental Protection Agency, Region I, Boston MA. (June)
Ferreira, Susan. 1990. Personal communication. Telephone call to Jenny Martin, Narrangansett Bay Project, April 9.
Gleber, E. 1987. Memo to M. Paul Sams re Guidelines for the Administration of Enforcement Action. (September)

Volkay-Hilditch, C. 1987. Letter to A. Malloy re BVDC
Draft Pretreatment Audit Report. (July)
East Providence:
Booth, K. 1987. Letter to C. Volkay-Hilditch re Implemen-
tation of PCI Recommendations. (December)
Booth, K. 1988. Letter to C. Volkay-Hilditch re PCI
Recommendations. (April)
East Providence. 1986. 1985 - 1986 Status Report. In-
dustrial Wastewater Pretreatment Program, Water Pollu-
tion Control Division, City of East Providence. (July)
East Providence. 1987. 1986 - 1987 Status Report. In-
dustrial Wastewater Pretreatment Program, Water Pollu-
tion Control Division, City of East Providence. (July)
EPA. 1985. Pretreatment Program Audit of the City of East
Providence, Rhode Island. U.S. Environmental Pro-
tection Agency, Region I, Boston, MA. (May)
DEM. 1987. East Providence Pretreatment Compliance In-
spection. RI Department of Environmental Management,
Division of Water Resources. (March)
DEM. 1988. East Providence Pretreatment Compliance In-
spection. RI Department of Environmental Management,
Division of Water Resources. (March)
Volkay-Hilditch, C. 1987. Letter to K. Booth re Pretreat-
ment Annual Report. (January)
General:
Brubaker, K. 1989. Down the Drain: Toxic Pollution and
the Status of Pretreatment in Rhode Island. Save The
Bay, Inc., Providence, RI. (September)
Brubaker, K. and J. Byrne. 1989. Zero Tolerance: Reducing
Toxic Pollution in Narragansett Bay. Special Report,
Save The Bay, Inc., Providence, RI. (March)
Dyer, J., A. Vernick, and H. Feiler. 1981. Handbook of
Industrial Wastes Pretreatment. New York: Garland
STPM Press.
EPA. 1986. Pretreatment Compliance Monitoring and Enfor-
cement Guidance. Office of Water Enforcment and
Permits ,Environmental Protection Agency, Washington,
DC. (July)
EPA. 1988a. "General Pretreatment Regulations for Exist-
ing and New Sources, 40 CFR Parts 122 and 403." Feder-
al Register, October 17.
EPA. 1988b. "Environmental Protection Agency Pretreatment
Standards," Environment Reporter (November 4):115-142.
EPA. 1989. A Primer on the Office of Water Enforcment and
Permits and Its Programs. Office of Water Enforcment
and Permits ,Environmental Protection Agency, Washing-
ton, DC. (February)
Levenson, H. and W. Barnard. 1988. Wastes in Marine En-
vironments. New York: Hemisphere Publishing Corp.

Approaches to Environmental Management Planning: The CZMA and the National Estuary Program

Mark T. Imperial[1], Donald Robadue, Jr.[2], and Dr. Tim Hennessey[3]

Introduction

The ecologic problems that the estuaries of this nation face are too multifaceted and complex for any individual administrative agency, community, or interest group to address alone. They involve complex issues of habitat protection, multimedia and nonpoint sources of pollution, land-use planning, and resource management. These problems are not adequately addressed by the existing conventional pollution control and land management programs which use regulation and enforcement.(EPA, 1990a)

In order to deal with coastal environmental problems such as water quality degradation, Congress has enacted various programs. These enactments have been progressive over time with each subsequent enactment building on the success and experience of the prior programs. Experiences of past efforts are valuable to those planning as well as undertaking new efforts in environmental management.(Albert,1988)

The birth of the "new federalism" has fostered a process of devolving federal responsibilities in a wide range of social and environmental program areas to the states which will likely continue well into the 1990s.

---

[1]Graduate Student, Department of Marine Affairs, University of Rhode Island, Kingston, RI 02881.
[2]Coordinator, Policy Development and Evaluation, Coastal Resources Center, University of Rhode Island, Kingston, RI 02881.
[3]Department of Political Science, University of Rhode Island, Kingston, RI 02881.

Budgetary shortfalls at all levels of government have
shaped responses to the current environmental problems
that this nation faces. Accordingly, Congress and the
United States Environmental Protection Agency (EPA) have
shifted their attention to more cost effective and issue
oriented approaches to environmental cleanup. There is
also increased attention to further coordination of
environmental regulatory responsibilities and the
elimination of duplicative efforts in environmental
management. This shift has placed the EPA more
frequently in the role of a facilitator rather than
manager or administrator of water pollution control
policies.

     One of the new programs enacted by Congress to deal
with the integrity of the nation's coastal waters is the
National Estuary Program(NEP). The NEP is administered
by the EPA and currently is funding the preparation of
Comprehensive Conservation and Management Plans (CCMPs).
These estuaries include both heavily urbanized and rural
water sheds. Thus, the NEP's management efforts tend to
deal with a wide assortment of environmental issues.

     The National Estuary Program is in many ways both a
product of and important step beyond these past
approaches to water pollution control and coastal
environmental protection. This paper focuses on the
environmental management planning process of the NEP as
it relates to the process employed by the Coastal Zone
Management Program.

     This comparison is particularly relevant because
the recent Coastal Zone Management Act Reauthorization
requires the formulation of nonpoint source pollution
control programs and places improved water quality as a
concern of state coastal zone management programs. (P.L.
101-508) Additionally, the Coastal Zone Management
Program(CZMP) is a critical implementation mechanism for
the the CCMPs. Thus, a comparison is particularly
relevant with respect to the compatibility of a CCMP
with individual coastal zone management programs. This
paper will examine this compatibility and assess the
future of the relationship between the CZMP and the NEP.

Environmental Management Planning

     Environmental planning involves the identification
of policies and programs to ameliorate and address
environmental problems. Environmental management
incorporates the administration and implementation of
these policies and programs. Environmental management

planning links environmental planning and environmental
management together in order to improve program
design.(Dean,1979)

The 1972 Coastal Zone Management Act

The Coastal Zone Management Program is an excellent
example of early environmental management planning.
This federal grant-in-aid program extends beyond a
single functional area of concern and confronts issues
as diverse as economic development, land use, fisheries
management, coastal resource protection, and pollutants
from land use activities.(Dean,1979) The 1972 Coastal
Zone Management Act(CZMA), subsequent amendments, and
Congressional funding were the impetus for voluntary
state-level planning for the wise use and protection of
the coastal zone. The incentives for state
participation included federal funding for both the
planning and implementation stages as well as having the
approved plan subject to the federal consistency
provisions of the CZMA. The federal Coastal Zone
Management(CZM) program is administered by the National
Oceanographic and Atmospheric Administration(NOAA).

The intent of the CZM program was to develop a
comprehensive program which could recognize and respond
to a variety of resource management problems. It was
intended as an integrated approach to management of the
coastal zone; sensitive to resources and their
uses(Knecht,1979). Indeed a prescribed goal of the CZMA
is the preservation, protection, development, and where
possible, the enhancement of the nation's coastal
zone.(16 U.S.C. 1452, 1982) The CZMA is intended to
encourage the states to exercise full authority over
their coastal lands and waters. (Sec. 302(h), 1972 CZMA)

The CZMA established only broad national goals and
a process by which the states, in cooperation with
federal officials, could develop specific objectives for
their given coastal areas This emphasis on process
anticipated an adaptive, experimental approach to
individual state program design.(Lowry,1985) These
state-managed multi objective resource management
programs were to coordinate resource use activities of
public agencies and private actors. The federal role in
the CZM program was to provide funding during the
planning stage of program development, coordinate the
development of these plans, and provide technical
expertise.(Lowry,1985; Matuszeski,1985)

The development of state CZM plans was a multi-
stage process with separate phases of planning and
implementation. An interesting factor in the
development of the national CZM program is the diffusion
of innovations and learning that took place in program
development. The national office served as a conduit
for information among the various programs such that
states could learn from on another's experiences in this
new area of environmental management.(Matuszeski,1985)

It was not the original intention of the federal
CZM agency to become involved in this collaborative and
experimental approach to program development. This
practice emerged from the the "practical requirements of
developing a new program around general goals and
particular state problems, conditions, and political
traditions."(Lowry,1985:295) This approach only lasted
as long as political support remained.

As pressure for approved state programs increased,
the federal program shifted direction and decreased its
emphasis on a collaborative approach and became more
directive. The federal government determined more than
ever what it saw as the key issues and elements that had
to be addressed, the content of the policies to address
these issues, and the laws and regulations that were
required to implement these issues. Thus, the CZM
program itself adapted at the federal level as it
learned from its own experiences.

Three basic elements emerged as critical to the
development of workable state programs. The first
element is the need for state programs to address the
major coastal issues in that state. This gave the
federal government the flexibility to define and
negotiate program requirements with the different states
reducing the scope and political vulnerability of a
given state program. The final CZM plans of many states
frequently contain exhortative rather than directive
language. This is attributed to the fact that the
breadth of coastal concerns and the intensity of
conflicting interests in the uses of coastal resources
makes it difficult to obtain consensus about the program
except in broad policy generalities.

The second element was the establishment of
policies to deal with major coastal issues in each
state. Once an issue was identified as a part of the
state program, then the state had to delimit its
position on the issue and elaborate on how it would
reach decisions with regards to that issue. This gives

the federal government some control over the specificity
of a state's plan.

The third crucial element to a workable state
program was the adequacy of a state's statutes,
regulations, and other authorities to enforce the
enumerated policies in the CZM plan. The burden was
placed upon the individual states to prove that they had
adequate laws, executive orders, and regulations to
carry out the plan's
policies.(Lowry,1985;Matuszeski,1985)

There were a large number of permutations with
respect to the management tools utilized. The principal
tools that have come to be used by state programs
include regulatory permit systems, comprehensive
planning, land use designations, land acquisition,
negotiation, promotion of desirable coastal development,
and federal/state consistency.(Healy and Zinn,1985)

States adopted many different approaches to the
implementation of their CZM programs. These
implementation methods range from inter-agency
agreements and the coordination of policy and decision
making procedures to changes in permitting procedures.
The definition and clarification of administrative and
regulatory responsibilities and the streamlining and/or
creation of permit procedures are additional methods of
program implementation. The use of the federal
consistency provisions is another potentially strong
implementation tool that can be utilized by the states
to implement their CZM programs.(Dean,1979;Healy and
Zinn,1985)

The CZM program has had many impacts on the
nation's environmental management planning process. One
of the greatest impacts has been the improved planning
capabilities of local governments and improved
communication between state and local governments with
respect to coastal zone issues.(Sorensen,1979)

This experiment in environmental planning yielded
many lessons that have carried over into other
environmental management and planning initiatives.
First, incentive-based, voluntary programs with broad
goals must stress collaborative relationships and
implementation stages that account for participant
learning and program development.(Godschalk and
Cousins,1985) The collaborative and learning aspects of
program design help to control factors which regulate
the political vulnerability of management plans.

A second lesson learned was the necessity to recognize the political vulnerability of coastal issues. It was learned that the designers of new environmental management planning programs must account for the nature of legislatures(Godschalk and Cousins,1985). Thus, institution and constituency building is an important part of this program.

Another general lesson learned from the CZM experience is that there is a time to plan and a time to act. This requirement is illustrated in the separation of the planning and management stages of the CZM program.(Godschalk and Cousins,1985) Another important factor was having clear goals and time limits involved in these stages. This experience has shown that a highly structured management process surrounding broad goals can be useful in dealing with complex environmental issues.

Perhaps the single most important lesson learned was simple flexibility. The negotiation which occurred between the federal and state authorities prevented confrontation, polarization, and/or political escalation with respect to program requirements. The federal-state efforts had to retain this level of flexibility to ensure that the program retain its ability to diffuse innovations throughout the management system. This diffusion of innovations is important to environmental management planning efforts.

The CZM program has certainly seen its share of success. It has continued to develop with time. For example, the recent reauthorization amendments include requirements for new nonpoint source pollution control plans, coastal enhancement grants, and expand the limits of the federal consistency provisions. Indeed, the status of the CZM program appears to be improving as its relationship with the EPA grows. The Section 319 nonpoint pollution control plans required by the 1987 Water Quality Act will likely see state coastal zone management programs as their prime method for implementation in order to satisfy the new nonpoint source pollution plan requirements. In addition, the NEP, pursuant to a Memorandum of Understanding between the EPA and NOAA, will utilize state coastal zone management programs as the method to implement their CCMPs.

The National Estuary Program

        Despite years of effort by federal and state
governments, many of the original problems addressed by
the early 1970s environmental legislation remain. In
order to address issues surrounding coastal water
quality and living resources, Congress established the
National Estuary Program with the passage of the 1987
amendments to the Clean Water Act and embarked on
another environmental management planning
initiative.(1987 Water Quality Act; P.L. 100-4; Section
320, Clean Water Act as Amended; 33 U.S.C. 1330)
According to the 1987 WQA, the NEP is managed by the
Environmental Protection Agency (EPA). The EPA is to
identify nationally significant estuaries that may be
threatened by pollution, development, or overuse and to
facilitate the preparation of a comprehensive
conservation and management plan (CCMP) through the use
of a management conference. The CCMP is to ensure and
enhance the ecologic integrity of the estuary(EPA,
1989b). The NEP is intimately related to the CZM
program because state CZM programs are to serve as the
prime implementation mechanisms for an approved CCMP.

        The NEP consists of seventeen estuaries in fourteen
state jurisdictions. These estuaries have been added to
the program in an incremental fashion. This incremental
growth in the NEP has fostered a learning capacity
within the program to an even greater extent than the
CZM program. Essentially, it allows the newer estuaries
to learn directly from the experiences of other
estuaries that have already completed stages in the
management conference process.

        Recent and subsequent additions to the NEP must
submit nominations from the State(s) governor to the
administrator of the EPA for an estuaries inclusion in
the program. Estuaries are selected by the EPA on the
basis of their national significance and the degree to
which they enlarge the pool of management and governance
experience for the Nation's estuaries.(EPA,1990b)
Therefore, both the NEP and the CZM program share the
similar characteristic of being voluntary with the NEP
adding a selective aspect(EPA, 1990b).

        The NEP approach to creating a Comprehensive
Conservation and Management Plan for each accepted
estuary contains two important elements: a specific
series of steps for identifying and solving problems and
a collaborative decision making process called the

"management conference".(EPA, 1989b). These elements
combine to give the NEP some characteristics which
distinguish it from past environmental management
planning efforts.

The first characteristic is that the NEP emphasizes
partnerships. This is a result of the fact that the
problems to be addressed are inter-jurisdictional in
nature and are often too complex to be addressed by a
single agency, community or interest group. The
management conference is the mechanism for the creation
of these partnerships. These partnerships are the heart
and soul of the NEP's collaborative decision making
process. While the CZM program employed a collaborative
approach to decision making, it has done so to a lesser
degree than the NEP.

The second distinguishing characteristic is that
the NEP provides strong management and regulatory tools
to help carry out these plans. These tools range from
traditional technical assistance in monitoring, to
helping communities find innovative financing mechanisms
for their environmental protection programs to the NEP's
ability to coordinate and leverage other programs such
as the National Oceanic and Atmospheric Administration
(NOAA) and the Department of Agriculture's Soil
Conservation Service (SCS). By leveraging other
environmental and resource based programs, members of
the estuary's "management conference" gain access to the
resources and expertise of these other agencies.

The third and most critical characteristic is that
these partnerships and tools emphasize action through
systematic problem solving. In other words, the NEP is
intended to produce results and is not intended to be
just another planning exercise or procedural requirement
in order to get federal funds. The Comprehensive
Conservation and Management Plan developed by the
management conference is designed to bind its
participants to specific financial, institutional and
political commitments in order to address priority
problems (EPA, 1990a).

The NEP's primary goals are the protection and
improvement of water quality and the enhancement of
living resources(Section 320 of 1987 WQA). These goals
are broad and allow states flexibility with respect to
the issues considered for management. The use of broad
goals around which a structured management process is
developed bears resemblance to the CZM program. To
accomplish these goals involves numerous distinct

federal and state agencies and resource users who
collectively share portions of the jurisdiction and
power to implement management objectives. The
management conference is composed of these groups. The
shared authority and efforts of the management
conference is the means to develop the CCMP for an
individual estuary program rather than the creation of a
new management authority. This management conference
then becomes the audience for program findings and
recommendations such that it can take action.(Demoss,
1986) Its members represent water quality and resource
managers, members of the regulatory community, and
others representing community and environmental
interests(EPA,1989b). For these management committees
to be successful, there has to be effective
collaboration and communication among its members.(EPA,
1989b).

    Because several of the estuaries span state
boundaries, a number of the individual programs in the
NEP are multi-jurisdictional in nature or involve some
assistance from neighboring states to implement their
CCMPs. These management committees may therefore
contain members from more than one state. This is a
tremendous difference between the NEP and the CZM
programs. It is rare when coastal environmental issues
are addressed on an inter-state basis. This is
primarily due to the complexities which surround land
use controls and the sensitivity of states to home rule
issues. In the case of the NEP, management plans
affecting both states are jointly formulated.

    One of the characteristics that differentiates the
NEP from the CZM approach is its highly structured
management process. This process has three initial
steps conducted concurrently with activities to satisfy
the seven principal purposes of a management conference.
Due to the individuality of the programs, as well as
that varying degrees to which states practice
environmental management, there are differences in the
structured process across all seventeen programs. What
follows is a brief explanation of the general process
which is followed by the original estuaries in the
program.

    The first step in the management conference is to
elucidate the environmental concerns that the NEP is
going to address. The identification of environmental
concerns and the reevaluation of these concerns is one
of the primary focuses of the program. A second step
that the management conference follows is to develop a

Data Information and Monitoring System(DIMS). The DIMS
is to facilitate the collection and organization of the
essential data to be utilized by the management
conference and assist long term evaluation of management
activities. The third step which has been carried out
primarily by the more recently accepted estuary projects
is the development of "action now agendas". These are
priority actions that can be achieved early in the
management conference process to build participation and
enthusiasm at the state and local levels for the
planning effort.

The remaining seven activities of the management
conference are directed towards satisfying the seven
statutory purposes of the 1987 law. The first three
purposes are: the development of a status and trends
report; the identification of the probable causes of the
environmental concerns that have been identified; and a
report documenting proposed relationships between
pollution loads and potential uses. Frequently, the
reports from the first three purposes are combined and
distributed as one inclusive status and trends report.
The NEP emphasizes using limited applied research to
link resource use impairments to pollutants and then
develop corrective actions to alleviate these resource
use conflicts.

The fourth purpose is the development of the
Comprehensive Conservation Management Plan. The CCMP is
intended to address critical problems and outline
corrective actions that will help attain/maintain uses
of the resource that are threatened due to current human
interactions with the environment. In addition, it is
designed much like the CZM plans to coordinate and
improve the regulation and administration of resource
use activities. The last three purposes are: an
implementation plan demonstrating the institutional and
financial commitments necessary; a monitoring program to
keep track of implementation effectiveness; and a review
of federal financial assistance programs for their
consistency with the CCMP. These generally become
chapters of the CCMP. Once all seven of these purposes
are completed, the CCMP is ready for acceptance by the
EPA and implementation.

The management system employed by the NEP is highly
structured and synoptic. At the same time, the
management conference encourages, and indeed requires,
participation by diverse and varied constituencies at
the state and local levels thereby assuring a wide range
of views. The NEP management system also includes

numerous outlets for public participation and review of
NEP work products. This participatory feature is
buttressed by a consensus building element so that
agreement about priorities among issues is possible.
The management conference provides flexibility and
adaptability into what would otherwise be a synoptic
planning process mandated by the federal government.

The management system employed by the NEP has also
displayed a capacity to learn and adapt just as the CZM
program did during its development. The NEP as a whole
is very amenable to the diffusion of innovations across
the different programs. The inclusion of estuaries of
various types also allows estuaries and coastal zone
programs that are not directly involved with the NEP to
learn from its experiences.

To date, the NEP has dealt with a number of
problems similar to those addressed by the CZM program.
The main difference in the nature of the issues that
these programs deal with is that the CZM program is more
"land" oriented while the NEP focuses on the "wet" side
of the coastal environmental issues. This distinction
continues to blur with time. New requirements in the
recent CZMA reauthoriztion will force the CZM programs
to deal with the "wet" side of coastal zone management
addressed by the NEP.

Flexibility with respect to the issues addressed by
the NEP is important just as it was for the CZM
programs. Even when the issues appear to be the same,
they are not always identical in importance, scope, or
complexity. The varying importance that these concerns
receive at the state level is the result of differing
public perceptions, differing causes, and differing
jurisdictional complexity. The NEP does not attempt to
solve the whole universe of concerns that the estuary
faces. Instead, a CCMP selects several of the primary
concerns and seeks appropriate action to address these
concerns. For example, the Long Island Sound Study is
primarily concerned with toxics and hypoxia. Thus, it
is similar in nature to the special area management
aspects of the CZM program. The interesting thing about
the NEP is that it appears flexible enough to handle the
diversity of issues and remedial actions that are
required to address these concerns. This uniform
management structure yields outputs which can differ
significantly from one NEP program to another.

The NEP is quite different from the CZM program in
its method of implementation which involves the

incorporation of approved CCMPs into state CZM plans.
The implementation stage is separate and distinct from
the planning phase in that it is intended to occur under
the jurisdiction of a separate agency. Unlike the CZM
program, there is no significant promise of
implementation funding other than those available from
the CZM program. This approach removes a large portion
of the implementation responsibility from the federal
government, and places it with the state governments.
In addition, the federal government has embarked on a
new course of environmental planning by having one
agency of the federal government design and administer
the development of a program that is to be implemented
in another agency.

Just as with many portions of state CZM plans, the
CCMP's frequently are required to use exhortative rather
than directive language. This is a direct result of the
consensus based decision making approach that has been
employed and the political vulnerability of many of
these sensitive issues. Thus, the challenge faced by
the CZM program is a familiar one. The CZM program will
have to work to safeguard potential pitfalls such as
failing to adopt the CCMP's as state policy and
coordinate the administrative, regulatory, and financial
support neccessary to carry out the policies set forth
in the CCMP, and.

The NEP and CZM Program Compatibility: Outlook Towards the Future

Despite the recent reauthorization of the CZMA and
the added and much needed vitality that it has received,
questions remain with regards to the role of the CZM
program in the future. Despite these recent expansions
in the program, it seems as if the CZM program is
growing into an implementation mechanism for innovative
new environmental planning programs such as the NEP and
Section 319 nonpoint source planning program. This
raises questions. First, how successful will the
implementation of programs developed by one agency be
when implemented under the aegis of another program.
For example, the increased flexibility of the NEP when
compared to the CZM program yields drastically different
end products across different NEP programs. This may
pose a problem. Some of the end products will look like
modified Section 208 plans prepared under the 1972
Federal Water Pollution Control Act Amendments. These
208 plans saw little coordination with state CZM
programs during there development and
implementation.(Dean,1979)

Second, how does the approach taken to
environmental management planning by different agencies
effect the implementation of these programs by another
agency. It seems that for implementation to be
successful, the environmental management planning
approaches should be compatible. This comparison
illustrates the fact that these programs are indeed
compatible with respect to the approaches taken to CCMP
development and CZM plan development. However, as
mentioned above, the end products of these approaches
may differ drastically.

Third, if the CZM program is growing into an
implementation mechanism so that innovative programs
designed and administered by the EPA can be subject to
federal consistency provisions, then one is left to ask
what states without CZM programs are to do with respect
to implementation. It is conceivable that the
development of the CZM program as an implementation
mechanism for other programs could incite those states
without CZM programs to develop and implement them in
order that they receive additional federal
implementation funding and have them subject to the
newly expanded federal consistency provisions.

One thing is certain from this comparison. The
NEP, as the nation's newest innovative approach to
environmental management planning, is very compatible
with the state CZM programs. They employ similar
techniques with the NEP representing the amalgamation of
the nation's past effort in environmental management
planning. The NEP is an innovative new approach to
environmental management planning which.has quickly
refined and adjusted its approach based on its own, as
well as past experiences. It creates a workable
framework to identify problems and build consensus for a
plan of action with public involvement incorporated
throughout the process. In the years to come, CZM
programs must face the challenge of developing and
incorporating appropriate implementation mechanisms such
that the CCMPs can become nested in state CZM programs.

Acknowledgement

The preparation of this paper was financed in part
by the NOAA Office of Sea Grant, U.S. Department of
Commerce, under grant #NA 89 AA-D-SG-082

References

Albert, Richard C.. June 1988. "The Historical Context of Water Quality Management for the Delaware Estuary". Estuaries. Vol. 11, No. 2. pp. 99-107.

Dean, Lillian F.. 1979. "Planning for Environmental Management: New Directions and Initiatives". Coastal Zone Management Journal. Vol. 5, No. 4. pp. 285-306.

Demoss, Thomas B..(1986). "Management Principles For Estuaries", Estuarine and Coastal Management-tools of the trade, Proceedings of the Tenth National Conference of the Coastal Society, Oct. 12-15,1986, pp. 24.

Environmental Protection Agency.(February, 1990a). Progress in the National Estuary Program: Report to Congress,.

Environmental Protection Agency.(January 1990b). The National Estuary Program: Final Guidance on the Contents of a Governors Nomination.

Environmental Protection Agency. (August 1989b). Saving the Bays and Estuaries: A Primer for Establishing and Managing Estuary Projects.

Godschalk, David R. and Kathryn Cousins. Summer 1985. "Coastal Management: Planning on the Edge". American Planning Association Journal. pp. 263-265.

Healy, Robert G. and Jeffrey A. Zinn. Summer 1985. "Environment and Development Conflicts in Coastal Zone Management". American Planning Association Journal. pp. 299-311.

Kitsos, Thomas R..Summer 1985. "Coastal Management Politics". American Planning Association Journal. pp. 275-287.

Knecht, Robert W..1979. "Coastal Zone Management: The First Five Years and Beyond". Coastal Zone Management Journal. Vol. 6, No. 4. pp. 259-272.

Lowry, Kem. Summer 1985. "Assessing the Implementation of Federal Coastal Policy". American Planning Association Journal. pp. 288-298.

Matuszeski, William. Summer 1985. "Managing the Federal
    Coastal Program". American Planning Association
    Journal. pp. 266-274.

Sorensen, Jens. 1979. "State-Local Relations in Coastal
    Zone Management: Implications for Change". Coastal
    Zone Management Journal. Vol. 6, No. 4. pp. 295-
    302.

SEDIMENT DIVERSION AS A FORM OF WETLAND CREATION IN THE MISSISSIPPI DELTA

Jack Moger [1]
Kenneth J. Faust [2]

ABSTRACT

The Mississippi River drains parts of 32 states and brings sediment-laden freshwater to the Mississippi Deltaic Plain located in southeast Louisiana. Controlled channelization of the Mississippi has formed defined passes in the delta consisting of natural banks formed by the deposition of sediments. This controlled low has starved the surrounding wetlands of both freshwater and sediment and has allowed saltwater intrusion and erosion of marshes. Over time, natural cuts that have been made through the banks have created significant areas of new land for vegetation to spread.

This concept was utilized in designing crevasses in the Pass a Loutre Wildlife Management area located in the Mississippi Delta. Six sediment diversions were designed - three in South Pass and three in Pass a Loutre and were constructed in early 1990. These diversions were designed hydraulically as well as using historical information collected from quadrangle maps and infrared photos over the last ten years. Past experience indicates each diversion will rebuild 100 acres of marsh.

Hydraulically, water that could be diverted through the designed diversions could not exceed 10% of the flow in the pass. This percentage was set by the U. S. Army Corps of Engineers to accommodate navigational needs. Using this criteria, different sizes were selected for each of the six diversions and a capacity analysis was

---

[1] Coastal Manager, Brown & Root, USA, Inc, 1112 Engineers Road, Belle Chasse, LA 70037.

[2] Coastal Engineer, Brown & Root, USA, Inc., 1112 Engineers Road, Belle Chasse, LA 70037.

completed.

The widths and depths of the cuts were determined
using historical information collected and by studying
the capacities of the outfalls. The angles of the cuts
relating to the pass were then designed to achieve the
best land growth using information developed by the U.S.
Army Corps of Engineers and other references on sediment
transport.

At two of the six sites, fencing techniques were
used in the outfall to develop different bifurcation
plans. These sites will then be compared to sites
without fencing plans to predict what method optimizes
land growth.

Monitoring systems for each site were established,
and data will be collected every year. The life of the
diversions is predicted to be about five years. After
this time the channel will choke itself off allowing new
crevasses to be dredged in the passes at other
locations.

INTRODUCTION

The Mississippi River drains parts of 32 interior
states and southern Canada. It brings freshwater laden
with nutrients and sediment to the Mississippi Deltaic
plain. Approximately 14,000 square miles of new land
was created in Louisiana by the Mississippi River making
four major course changes over the past 7,000 years
(LeBlanc, 1988). This process resulted in the
formation of four delta systems: the Teche Delta, the
St. Bernard Delta, the Lafourche Delta and the Modern
Delta.

The Modern Delta is situated between the old St.
Bernard and Lafourche Deltas. Unlike the older delta
formations that were constructed within broad,
relatively shallow waters, the Modern Delta is building
near the outer margin of the Continental Shelf. This
position results in sediments being washed into the deep
waters of the Gulf of Mexico rather than contributing to
further delta development. This loss of sediment
together with saltwater intrusion, controlled
channelization of the Mississippi for navigational
needs, and subsidence of previous laid sediments has
resulted in erosion of the surrounding marshes.

Historical infrared and aerial photos show that
natural and manmade cuts along the banks of channels in
the modern delta have created significant areas of new
land. Through surveys taken by the United States Army

Corps of Engineers, it has been estimated that the
Mississippi River carries an average of 186 million
tons of sediment each year, equating to enough material
to fill 30 Louisiana Superdomes (U. S. Army Corps of
Engineers, New Orleans District, pers. comm.).

METHODS & RESULTS

The concept of diverting sediments for marsh growth
in Louisiana has been used in designing crevasses in
the Pass a Loutre Wildlife Management Area located in
the Mississippi Delta. Six sediment diversions were
constructed in early 1990 (Figure 1), including three
diversions in South Pass and three diversions in Pass a
Loutre.

To prevent shoaling in each of the navigable
passes, the total flow diverted was limited to ten
percent of the flow in each pass. The Mississippi
River has an average flow of 1.1 million cubic feet per
second. Of this flow, 17 percent is allocated to South
Pass, 31 percent to Pass a Loutre and 52 percent to
Southwest Pass below Head of Passes (U. S. ARmy Corps
of Engineers, New Orleans District, pers. comm.).
These limits were used to size the crevasses in each
pass.

Each diversion is a box cut, with a flat bottom and
sloped sides, through the natural banks along the
passes. The width of the cuts were determined so that
the amount of diverted water would not disrupt the
allocated distribution of flow.

South Pass has the most favorable hydraulic
gradient of the two passes where diversions were to be
placed allowing for more sediment to be transferred.
Splays previously constructed along this pass developed
up to 100 acres of new land after just one year
(Howard, et. al, 1987).

Hydraulic analyses were completed for each of the
six proposed diversions using river elevations
collected from the United States Army Corps of
Engineers, New Orleans District for each month.
Hydraulic gradients were calculated for both passes
using river elevations at Head of Passes and assuming
elevations at the end of the passes to be the same as
the recorded elevations at the end of Southwest Pass.
Water elevations were then interpolated along the
gradient for each site.

The maximum amount of water that could be diverted
from South Pass was divided equally between the three

proposed diversion sites. For Pass a Loutre, the maximum diversion limits were divided so that one of the sites (#2) diverted twice as much flow as the other two sites, sites #3A and #3B, feed into the same outfall area. The maximum flow distribution in each of the passes was determined as follows:

South Pass
Average flow in Mississippi River = 1.1 X 10 cfs
Allocation = 17%
Average flow in South Pass = 17% X 1.1 X 10 cfs = 187,000 cfs
Diversion Limits = 10% X 187,000 cfs = 18,700 cfs

Flow distribution at each site:

Diversion	Maximum flow limits
South Pass #2 (SP2)	6,250 cfs
South Pass #3 (SP3)	6,250 cfs
South Pass #4 (SP4)	6,250 cfs

Pass a Loutre

Allocation = 31%
Average flow in Mississippi = 1.1 X 10 cfs
Average flow in Pass a Loutre = 31% X 1.1 X 10 cfs = 341,000 cfs
Diversion limits = 10% X 341,000 cfs = 34,100

Flow distribution at each site:

Diversion	Maximum flow limits
Pass a Loutre #2 (PAL2)	17,000
Pass a Loutre #3A (PAL3A)	8,500
Pass a Loutre #3B (PAL3B)	8,500

Once maximum flows were calculated, the diversions were designed and checked to see that they did not exceed allocated flow limits. The length of each cut was scaled from 1983 infrared photos, with the width being the variable determining the limits of how much water could be diverted at each site. Using a conservative estimate of 200 parts per million of sediment concentration in the main channel (U. S. Army Corps of Engineers, 1990), the Corps recommended dredging to depths of 15 feet on the pass side for both South Pass and Pass a Loutre to obtain appreciable accretion in the outfall areas. The cuts were sloped from -15 feet on the pass to an average depth of -2.0 feet National Geodetic Vertical Datum on the receiving side. These shallow receiving areas will allow sediment to drop out creating additional land and increasing emergent and submerged aquatic vegetation production.

With the length and depth of the box cut
configuration set, a width for each crevasse was
assumed.   River flow entering the channel simulated a
broad crested weir and friction loss was calculated by
Mannings formula.   Flows were calculated by adjusting
the width through iteration to achieve the average
water gradient in the pass at the site.   The
controlling tailwater elevation of the crevasse was
assumed to be equal to the average water elevation at
the end of the pass.

Monthly flow rates were determined, and a twelve
month average flow was computed and checked to see
whether it exceeded the maximum limits at each site.
In checking the 10 percent diversion limit of each
pass, existing manmade and natural crevasses were also
taken into consideration.   This process produced the
following design widths and flows.

Location	Width of Crevasse (ft)	Flow Rate (cubic ft per second)
SOUTH PASS		
2	300	3,807
3	350	3,872
4	350	3,710
PASS A LOUTRE		
2	750	9,541
3A	750	5,437
3B	750	5,437

The total flow in South Pass using these design
widths would divert 11,389 cfs.   This in conjunction
with existing water being diverted from the pass was
estimated to be close to the 10 percent limit of the
pass.   The total flow for Pass a Loutre (20,415 cfs),
added to existing flows in the pass, was estimated to
be lower then the 10 percent limit, allowing additional
water to be diverted at a future date.

DISCUSSION

Although hydraulic calculations show that cuts of
up to 750 feet wide could be made in Pass a Loutre, a
width of this size does not reflect the optimal width
for splay development.   Both manmade and natural
crevasse widths versus splay development was measured
from infrared aerial photographs and quadrangle maps
dating over a ten year period.   These estimates
together with potential outfall growth at each site,
were used to check the design widths.   For Pass a
Loutre, all three diversions were scaled down.   The
width of #2 was scaled down to 200 feet, and #3A and

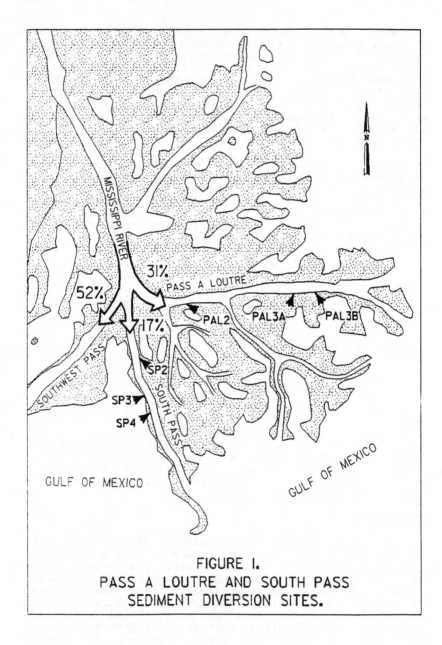

FIGURE I.
PASS A LOUTRE AND SOUTH PASS
SEDIMENT DIVERSION SITES.

#3B were scaled to 150 feet each. The design widths in South Pass, were closer to the desired values. In this pass, #2 remained at 300 feet, with #3 and #4 were scaled down to 300 feet.

Historical data as well as reference on sediment transport was also used in designing the angles of each cut (A.S.C.E., 1975). Analysis of this information resulted in a 70 degree angle for cuts measured from the centerline of the pass in a downstream direction would produce optimal splay development at each site.

In selection of the sites to be constructed, two pairs of channels (one pair in each pass) were established as study areas. These pairs include diversions #3 and #4 in South Pass and #3A and #3B in Pass a Loutre. One site in each of the designated pairs was designed with sediment fences strategically placed in the outfall to develop different bifurcation plans. Their placement relative to the upstream current was reversed between the two pairs for comparison. Each pair will be intensively monitored to analyze the relative effectiveness of the fencing technique. This information will then be used in designing future crevasses in the Mississippi Delta.

Within five years, it is estimated that the efficiency of the crevasse will decrease and eventually the process will choke itself off. This will allow new crevasses to be constructed at other locations.

REFERENCES

American Society of Civil Engineers. 1975 Sedimentation Engineering ASCE Manual & Reports on Engineering Practice #54.

Howard, Needles, Tammen, & Bergendoff, and Coastal Studies Institute, Louisiana State University. 1987. Final project report for Pass a Loutre marsh creation project. Prepared for the Lousisiana Department of Natural Resources, Geological Survey Division, Coastal Protection Section. 36 pp.

LeBlanc, R. J. 1988. The geological history of the marshes of coastal Louisiana p. 1 - 13. In: Marsh Management in Coastal Louisiana: Effects and Issues. Louisiana Department of Natural Resources and the U. S. Fish & Wildlife Service.

U. S. Army Corps of Engineers, New Orleans District. 1990. Unpublished report. 6 pp.

An Innovative Approach to Wetlands Restoration in the San Francisco Bay Estuary: the Montezuma Wetlands Project

Stuart W. Siegel[1] and James D. Levine, P.E.[2]

ABSTRACT

The Montezuma Wetlands Project is the privately-sponsored restoration of 2,000 acres of tidal brackish marsh in the San Francisco Bay estuary. The project provides relief to two major regional environmental problems: the loss of wetlands resources and the need to find environmentally sound dredge disposal options. The purpose of the project is to restore diverse, complex wetlands resources on a diked subsided historical marshland that has been in use for several decades as low-intensity agricultural and grazing land. This restoration will allow the evolutionary development of a wide variety of endemic, threatened, rare, and endangered wetlands plant and animal species.

The project purpose will be achieved in a profitable manner through the engineered placement of 15 to 20 million cubic yards of nonhazardous dredged sediment. Planned to begin construction in early 1992, the project is currently undergoing regulatory review.

This paper focuses on two specific areas of innovation associated with this project: private sponsorship of natural resource restoration and its environmental, policy and economic benefits, and application of the latest scientific and engineering knowledge in large-scale wetlands restoration and sediment management. The paper also describes an unique ecological feature of the project: the restoration of rare transitional ponds found at the wetlands/uplands boundary.

---

[1] Project Environmental Scientist, Levine·Fricke, Inc., 1900 Powell St., Emeryville, CA 94608

[2] President, Levine·Fricke, Inc., 1900 Powell St., Emeryville, CA 94608

INTRODUCTION

The restoration of tidal marshlands and the development
of options for the disposal and management of dredged
sediments in ways that protect and enhance the region
economically and ecologically is now needed in the San
Francisco Bay estuary. The development of
environmentally sound strategies to solve these
problems, however, is often constrained by limited
economic resources and the lengthy administrative
procedures inherent in government- and non-profit-
sponsored actions. The Montezuma Wetlands Project (MWP)
presents a new strategy that avoids these constraints.

The MWP is a privately-sponsored, 2,000-acre wetlands
restoration project. The project will be operated for
profit by assessing fees on the 15 to 20 million cubic
yards of sediment needed to restore the topography of
the site to elevations suitable for establishment of
tidal wetlands. The MWP does not stem from agency
imposed mitigation requirements. Instead, it has been
initiated as an environmentally and economically
beneficial use of a large tract of shoreline property.
The environmental benefits arise from addressing the
regional problems of wetlands loss and dredged sediment
disposal. The economic benefit arises from fees
assessed on the dredged sediment.

As a result of the project sponsorship and financial
mechanism that make the project possible, the MWP makes
several important contributions to wetlands restoration
and dredged sediment management. It demonstrates that
natural resource restoration and conservation can be
initiated and undertaken by the private sector without a
regulatory mandate and that the beneficiary includes the
public. It provides the funding necessary to apply the
latest scientific and engineering knowledge in wetlands
restoration and monitoring and dredged sediment
management. The MWP site also offers the opportunity to
restore transitional ponds, the rare habitats along a
wetland's natural topographic upland boundary.

Because of the complexity, scope and importance of the
activities proposed by the MWP, cooperation is essential
between the project sponsor, the regulatory and resource
agencies and the environmental conservation community.
This cooperation will best fulfill the environmental and
policy needs identified for the region and the needs of
these interested parties. Achieving this kind of
cooperation will aid not only the outcome of the MWP but
also the climate within which the issues of wetlands

conservation and dredged sediment management are
discussed.

BACKGROUND

The estuary of the San Francisco Bay and the Sacramento
and San Joaquin Rivers sustains a region of significant
economic and ecological resources. The operations of
our cities and agriculture, however, have impaired the
estuary's natural functions to the detriment of the
region's ecological and economic health.

The San Francisco Bay regional economy and ecology
continue to be threatened from the loss of tidal
wetlands resulting from human activity. The region's
historical marshlands supported a rich endemic flora and
fauna as well as commercial species of fish and
wildlife. Many of these species are now rare or absent
in the region, and others are in danger of extinction,
mainly because of the loss of tidal marsh habitats.
These historical marshes removed sediments from the
rivers, streams and tides and enriched the adjacent bays
and coastline with nutrients.

The San Francisco Bay estuary also suffers from reduced
tidal flows and increased sediment loading caused by the
loss of vast amounts of marshland, urbanization,
agriculture, mining, and water development projects.
Maintenance of shipping lanes and harbors has required
extensive dredging, and the dredged sediment was
routinely disposed in the estuary. Many of these past
dredging practices and disposal sites have had
deleterious effects on the environment by diminishing
the estuary's tidal capacity and increasing its exposure
to pollutants through direct discharge into the water
column.

The current lack of viable, large-capacity dredge
disposal options also poses a threat to the estuary's
ecology and economy. This dredge disposal problem has
recently become acute for the San Francisco Bay estuary.
Several million cubic yards of sediment are planned to
be dredged in the next few years for port and shipping
channel expansions. Several million additional cubic
yards of maintenance dredging occurs annually. By
accepting 15 to 20 million cubic yards of dredged
sediment, the MWP will utilize this sediment as a
resource rather than allowing it to become a waste. The
MWP should thus make a significant dent towards solving
the San Francisco Bay dredge disposal problem.

THE SITE

The Montezuma Wetlands Project site is located on the
eastern perimeter of the Suisun Marsh in Solano County,
California, near the towns of Collinsville and Birds
Landing (Figure 1). The site comprises about 2,000
acres of diked subsided historical marshland below
approximate local mean high water, adjacent to an
additional 2,000 acres of open upland grassland under
the same ownership (Figure 2). Direct barge access is
via a tidal inlet channel in the southeast corner of the
site that extends approximately 1 mile inland (north)
from the Sacramento River.

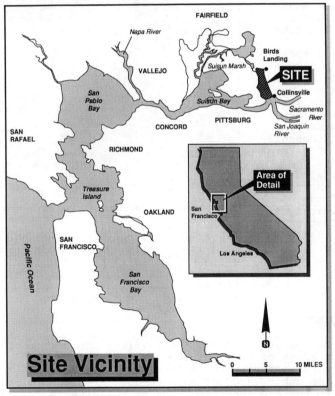

Figure 1.   Montezuma Wetlands Project Site Vicinity.

The historical marshland was diked around the turn of
the century, and has since been used primarily as cattle
pasture. A sand and oyster shell reclamation facility

operates adjacent to the tidal inlet channel. Seasonal pickle-weed wetlands have developed in the southeast portion of the site, where the historical marsh-land has subsided the most. Remnant wetland characteristics are found scattered elsewhere around the site.

The large sediment capacity at the site arose from extensive subsidence caused by soil oxidation, wind erosion and consolidation of natural sediments. Current site elevations now range from approximately 5 to 15 feet below local tidal maximums.

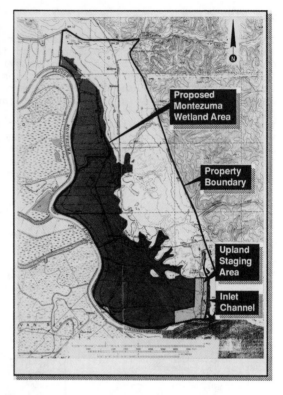

Figure 2. Project Site. Shaded area is the proposed wetlands area and barred area is the upland staging area. Barges would enter the inlet channel on the eastern perimeter of the staging area.

THE PROJECT

The Montezuma Wetlands Project involves the restoration of 2,000 acres of tidal marsh. Restoration is accomplished through the engineered placement of 15 to 20 million cubic yards of nonhazardous dredged sediments needed to restore the topography of the site to elevations suitable for establishment of tidal wetlands, and the return of tidal action to the site. Five overall components will turn this concept into functioning, high-value wetlands:

1. *Determine Sediment Quality:* prior to the beginning
    of each dredging project, a determination will be
    made of which sediments can be accepted for use at
    the site. Only materials designated by the
    regulatory agencies as non-hazardous are planned to
    be accepted at the site.

2. *Construct Upland Staging Area:* a 1,000-foot-wide
    upland staging area will be constructed on dredged
    sediment in the southeast corner of the site,
    adjacent to the existing tidal inlet channel and
    the Sacramento River (see Figure 2). The existing
    inlet will require some dredging at the onset of
    the project to facilitate barge access.

3. *Develop Site in Phased Approach:* filling, habitat
    development and return of tidal action at the site
    will occur in "cells" that will allow for
    flexibility in sediment containment and treatment
    options, and phased construction of wetlands Unit
    Landscapes.

4. *Monitoring and Maintenance:* a long-term monitoring
    and maintenance program will be developed and
    funded.

5. *Donate Habitat to Wildlife Trustee:* approximately
    2,000 acres of created wetland will be donated to a
    wildlife trustee following habitat completion.

The funding for the MWP will come from fees assessed for
dredged sediment. The fact that the project will be
operated for profit is an important element that nearly
guarantees the project's success. This aspect will be
discussed in greater detail below.

THE BENEFITS OF PRIVATE SPONSORSHIP

The Montezuma Wetlands Project will demonstrate how the
private sector can benefit from large-scale investments
in restoration and conservation. It conversely will
demonstrate how the estuary can benefit from privately-
sponsored natural resource restoration. Finally, it
will demonstrate that public and non-profit sector
involvement are not the only forces that can move
society toward natural resource restoration.

Private sector involvement in natural resource
restoration and conservation has historically been
limited to either satisfaction of agency-imposed
mitigation requirements or philanthropic actions such as

land donations or grants. The realm of seeking and
promoting natural resource restoration projects has
traditionally been left to the government or non-profit
conservation organizations. Given the limited financial
resources of these entities, excellent opportunities
often slip away.

The MWP breaks this pattern. Levine·Fricke, an
Emeryville, California-based private environmental
science and engineering firm that provides the technical
expertise, has teamed with Catellus Development
Corporation, a San Francisco-based real estate developer
that provides its property, to undertake the MWP. This
team formed out of the recognition that its property
could be used profitably in addressing the region's
wetlands loss problem and, subsequently, the dredged
sediment disposal problem.

The fact that the project will be operated for profit
generates several benefits. The for-profit nature of
the project also means that the restoration efforts are
necessarily teamed with other activities (accepting and
managing dredged sediment) that have real regional
environmental and economic benefit. These benefits
include the following.

First, the economic mechanism encourages the private
sector to engage in natural resource restoration without
a regulatory mandate, thereby benefitting the public
with increased natural resource values. Without the
financial incentives, such restoration likely would not
take place. This aspect is particularly important for
the MWP because the project is occurring on privately
owned land on which other less environmentally
beneficial options for generating economic benefit could
otherwise be considered.

Second, it allows for participation of the private
sector and private property owners in addressing short-
and long-term regional dredged sediment management
needs. In particular, it opens up a large tract of land
for short-term sediment disposal that otherwise would
not be available to dredging project sponsors. By
demonstrating the viability of the MWP's environmental
and economic concepts, other privately-owned sites in
the San Francisco Bay Area could become available for
similar use. Further, although not a part of the MWP,
the adjacent uplands could be used as a sediment
rehandling facility, thereby providing long-term relief
to dredged sediment management concerns while providing
dry material for needs such as levee maintenance.

Finally, it makes it economically feasible to involve
the extensive technical expertise necessary to design,
implement, and monitor this complex project in an
environmentally sound and beneficial manner.

APPLICATION OF THE LATEST WETLANDS RESTORATION IDEAS

Wetlands restoration on the scale and with the approach
proposed by the Montezuma Wetlands Project has not been
undertaken before in the United States. The size of the
wetlands, the goals for the restored wetlands and the
extent of the engineering proposed therefore require
application of the best and most up-to-date knowledge of
wetlands functions, values, and restoration.

The project design will reflect the nature of tidal
marshlands within the estuary, with an emphasis on rare,
endangered, and threatened species habitats. The
overall strategy is to create a physical structure for
the desired ecological community, using conventional
earth-moving equipment as necessary to create
naturalistic ponds and tidal channels. Once returned to
tidal action, this system will develop naturally into
equilibrium wetlands that will sustain themselves over
time with a minimum of maintenance.

To design such a wetlands system, the MWP will apply
recent research results for the first time. These
results provide the scientific rationale for the form,
tidal elevation, and arrangement of the landscape
features of the restored marsh as a physical system.
These results were obtained from continuing research on
the geomorphology of tidal wetlands in the San Francisco
Bay and Delta system (Collins et al, 1987; Collins,
1989; Collins, 1990; Haltiner and Williams, 1988). This
research examined the equilibrium relationships between
wetlands hydrology, geomorphology and ecology that exist
in large, natural estuarine wetlands of the region. The
results have demonstrated that establishment of
appropriate tidal marsh geomorphology is the most
critical element in the success of habitat creation, and
that success is more readily achieved with larger
physical systems such as those proposed by the MWP.

The Project will proceed with successive creation of
tidal marsh "Unit Landscapes" that will be comprised of
several separate "cells" created for sediment filling.
These units are large dendritic tidal drainage systems
that accommodate regional changes in water or sediment
supply and adjust to biotic influences on internal
channel capacity. The tidal prism and ecological

diversity of such units are conserved over time. The
completed project will be a tidal marsh ecosystem that
achieves a dynamic equilibrium such that its ecological
functions are sustainable over time.

SEDIMENT QUALITY MANAGEMENT TECHNIQUES

Sediment dredged from the ports, shipping channels and
other areas of the San Francisco Bay estuary poses
several challenges for its safe reuse. Six aspects of
the Montezuma Wetlands Project promote the
environmentally sound long-term management of the
dredged sediment.

First, evaluating sediment quality will take place
separately within the context of each dredging project.
Only non-hazardous sediments falling into one of the two
following categories will be accepted at the MWP:
sediment suitable as an ecological substrate (i.e.,
toxicological effects are not significant) and therefore
usable as "top" sediment, and chemically stable sediment
that is of relatively poorer quality (with respect to
ecological substrate) that can be contained as "bottom"
sediment without adverse environmental impacts. Thus,
ecologically stable sediment will be placed above
chemically stable sediment that is not suitable for
ecological interaction. Sediment determined to cause
potential adverse environmental effects at the site that
cannot be mitigated will not be accepted.

Second, the large vertical filling capacity, up to 15
feet in some areas of the site, yields the ability to
place poorer quality sediments safely below local mean
lower low water (MLLW). Under these conditions,
sediment will remain saturated and thus metals will
remain in immobile, reduced states. Ground-water
movement through the site into surface water has been
determined to be negligible, and a clay subsurface layer
has been identified that separates the shallow, brackish
ground water from deeper, drinking water resources.

Third, soil characteristics at the site promote
immobilization of metals and organic compounds that may
be present in the imported sediment. The high soil
organic carbon content (more than 50% wet weight in some
locations) provides excellent sorptive capacity for
retaining metals and organics. The high clay content
(more than 40%) provides low permeability and excellent
sorptive capacity for metals and typically contains high
levels of soil organic matter for increased sorptive
capacity. Finally, the high sulfate soil content will

transform into sulfides when soils are flooded. The
sulfides will bind most metals very strongly, thereby
further curtailing their leachability.

Fourth, the horizontal positioning of poorer quality
sediments will be designed to be a sufficient lateral
distance from the larger channels. Research has shown
that only larger channels (i.e., fourth order and
larger) of mature tidal marshland will cut below MLLW.
Since channels of this size have been shown to be
stationary in natural wetlands (e.g., Ahnert, 1960;
Redfield, 1965; Van Eertdt, 1985), they will be
constructed at the outset in their equilibrium
geomorphic configurations. Smaller channels, which
continually wax and wane across the marsh landscape, do
not extend to the depth of MLLW. Thus, the dynamics of
these small channels will not involve the poorer quality
sediments that will be contained below MLLW.

Fifth, flexibility in sediment containment has been
designed into the MWP through use of "cells" for
sediment filling. By filling a sufficient number of
cells simultaneously, incoming sediment can be
segregated. Thus, sediment can be placed in whichever
cell and at whatever elevation is deemed appropriate for
the identified sediment type and quality. For example,
poor quality sediment can be placed in the bottom of a
recently opened cell while high quality sediment can be
placed on the top of a cell nearly filled.

Finally, flexibility in sediment treatment has also been
designed into the sediment transport technique. Once
removed from the barges, sediment will be hydraulically
pumped to awaiting cells. Additives such as gypsum or
ferrous sulfate can readily be added to the sediment
slurry in this continuous-stream transport process.

RESTORATION OF TRANSITIONAL PONDS

A special feature of the Montezuma Wetlands Project site
is about 12 miles of the north-south trending historical
boundary between upland terrain and tidal marshland.
The historical transitional ponds along this boundary
can be discerned in the 1866 map of the area presented
in Figure 3. The habitats in this natural boundary,
among the estuary's most ecologically diverse settings,
are very rare today as they were the first lost to
agriculture and development. The transitional ponds of
these boundary areas represented extreme gradients of
environmental moisture and salinity that formed where

sheet runoff from the surrounding uplands met the marsh plain headward of the tidal drainage networks. A rich flora and fauna reflected the diverse mosaic of these habitat conditions.

These rare habitats can be included in the wetlands restoration for two reasons. First, the restored wetlands boundary will for the most part be at the natural topographic upland boundary rather than at an artificial levee. Second, because the adjacent 2,000 acres of upland is also owned by the project sponsors, buffer zone land uses can be managed to accommodate these transitional ponds.

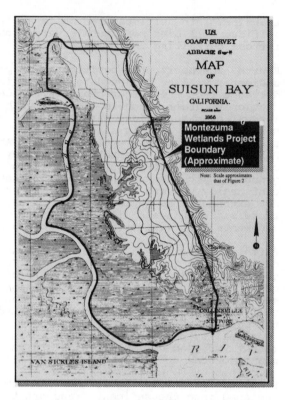

Figure 3. Historical marshlands at the site in 1866. Note the transitional ponds along much of the perimeter uplands boundary.

CONCLUSIONS

Solving the important environmental problems of wetlands loss and dredged sediment disposal facing the San Francisco Bay estuary will necessitate an expansion of involvement and resources beyond the public and nonprofit sectors. The Montezuma Wetlands Project presents an example of how the private sector can help solve these environmental problems while at the same time profiting from the endeavor. This approach can help not only to solve these problems, but also to contribute to the advancement of the policy, science, engineering, and

economics of natural resource restoration. The success
of the Montezuma Wetlands Project will thus provide the
motivation for other similar ventures in natural
resource restoration and conservation.

ACKNOWLEDGEMENTS

The contents of this paper could not have been possible
without the excellent assistance of Dr. Joshua Collins
of the University of California at Berkeley on wetlands
matters and Mr. Doug Lipton of Levine·Fricke, Inc. on
sediment quality matters.

REFERENCES

Ahnert, F., 1960. Estuarine meanders in the Chesapeake
Bay area. Geogr. Rev. 50: 390-401.

Collins, L.M., J.N. Collins, and L.B. Leopold, 1987.
Geomorphic processes of an estuarine marsh. Internat.
Geomorph. I: 1049-1071.

Collins, J.N., 1989. The concept of tidal prism
conservation: natural maintenance of tidal marshes in
relation to marsh size. Proc. First Nat. Conf., Soc.
for Ecological Rest. and Man., Oakland, CA.

_____, 1990. What is a tidal marsh? Proc. First Nat.
Conf. on Landscape Ecology: the Landscape Dimension.
Sacramento, CA.

Haltiner, J. and P.B. Williams, 1988. Hydraulic design
in salt marsh restoration. Proc. Nat. Wetland Sym.
Wetland Hydrology, Chicago, IL: 293-297.

Redfield, A.C., 1965. Ontogeny of a salt marsh.
Science 147(3653): 50-55.

Van Eertdt, M.M., 1985. The influence of vegetation on
erosion and accretion in salt marshes of the
Oosterschelde, The Netherlands. Vegetatio 62: 367-
373.

Maintaining Resources Through Re-Establishing
Physical Processes at Pescadero Marsh

Thomas L. Taylor[1]

Abstract

        Analysis of historical maps and aerial photographs
confirm that Pescadero Marsh has experienced gross changes
to physical features of the wetland, primarily a result of
sedimentation, but also due to direct human alteration for
agricultural use and flood control. Levees were
constructed along stream channels in the marsh and the
channel bed elevations raised by sedimentation. The
lagoon, an important aquatic habitat in the marsh, also
appears to have aggraded. Sedimentation has reduced the
extent and duration of tidal flooding in the lagoon and
channels. Levees greatly restrict both tidal and flood
flows from much of the marsh plain and interferes with
sediment transport through the marsh. The water quality
and biological consequences of the physical changes are
discussed. Recommended restoration actions include
restoring channel volumes and removing levees to allow
physical processes of flooding and tidal exchange to recur.
The cost of restoring suspected historic lagoon depths is
presently prohibitive.

Introduction

        Sedimentation is a serious problem for California
coastal wetlands. Loss of wetland acreage has be
attributed to fluvial sedimentation filling former wetland
areas to elevations that become uplands. A common feature
of many of these wetlands is the lagoon. The lagoon is a
permanently flooded body of water exposed to tidal flows
during the winter when it tends to be open, but after
closing in late spring or early summer, it can either
remain brackish and stratified, or become a freshwater body

---

[1] Senior Aquatic Biologist, California Department of Parks
and Recreation, P.O. Box 942896, Sacramento, CA 94296-
0001.

depending on the amount of freshwater inflow. Sedimentation affects the physical quality of aquatic habitat in the inflowing stream channels and lagoon areas of these coastal wetlands. Water quality conditions in coastal lagoons are strongly influenced by water depth. Shallow channels operate differently than deep ones. When channels become shallow, the extent of saltwater influence is restricted.

In openly tidal or closed coastal lagoons, saltwater will remain stratified from freshwater due to its greater density. In deep channels, saltwater can move upstream, along the bottom of the channel to elevations approximating the mean high water. When channels fill to above this elevation, saltwater movement is greatly restricted to areas near the beach, unless extreme tides or events occur. Shallow lagoons are also more prone to thermal heating when compared to deeper ones. Typical water quality conditions in shallow central coast lagoons include stratified conditions with warm, anoxic bottom waters. Deeper systems do not appear to develop the poor water quality bottom conditions seen in shallower lagoons.

Sedimentation of coastal lagoons is not readily apparent from studying historic maps because depths are usually not included on these early documents unless surveys had been done specifically for water depths. Aerial photographs also do not normally reveal water depths.

Study Site

The California Department of Parks and Recreation owns and operates Pescadero Marsh Natural Preserve, a 588 acre wetland/upland complex located on the central California coast in San Mateo County, about 35 miles south of San Francisco. The Natural Preserve contains a 320 acre brackish-freshwater wetland that is seasonally exposed to tidal influence. Pescadero Creek watershed is approximately 67 square miles in size. Butano Creek, a major tributary that joins Pescadero Creek in the marsh, drains a 21 square mile watershed. The wetland is confined between hills on the northeast and southwest sides of the marsh, and is a typical "drowned valley" wetland of this section of the coast. Adjacent land uses currently are confined to coastal-dependent agriculture, primarily artichokes. The wetland includes about 18 acres of permanently flooded habitat composed of a lagoon and the channels of Pescadero and Butano Creeks, and about 300 acres of marsh plain.

Pescadero Marsh is the largest wetland on the open

coast between Bolinas Lagoon, north of San Francisco and
Elkhorn Slough, on Monterey Bay. The wetland provides
habitat for the federally endangered San Francisco garter
snake,   and   numerous   additional   sensitive   species.
Steelhead, a species of special management concern within
California, are dependent upon aquatic habitat found within
the Natural Preserve.

Materials and Methods

        The plants and animals in the marsh area were
generally known (Elliott 1974). Land use history of the
marsh and the area was documented (Viollis 1979). A
history of land use in the watershed and hydrologic and
sediment information was also developed (Curry et. al.
1985). Water quality and biological studies were began in
1984 and resulted in obtaining important information on
aquatic habitat conditions and fisheries (Smith 1987).

        A topographic survey was completed in 1987. The
survey included 34 cross sections through the lagoon and
stream channels. Other evaluations were done on the upper
watershed streams. A report entitled "Pescadero Marsh
Hydrological Enhancement Plan was prepared from this
information (Williams, 1990).

Results

        Biological studies identified several sensitive
species using the marsh. The marsh supports one of the
largest remaining populations of California red-legged
frogs known in the State; the lagoon and channels support
one of the three remaining populations of the tidewater
goby in San Mateo County, the most northerly of only three
remaining populations of the California marsh snail left in
California, and is of critical importance to the
maintenance of the steelhead run in the watershed. The
wetland also supports a large population of the San
Francisco garter snake. Other sensitive species known to
be using the marsh include the saltmarsh yellow throat, and
the western pond turtle.

        Analysis of historical aerial photographs indicate
that 74% of the historic wetland is presently behind levees
and 85% of the length of the two major creek channels in
the marsh are confined by levees with 38% leveed on one
side and 47% leveed on both sides. Sedimentation has
affected 100% of the deep water aquatic habitat of the
lagoon and channels in the marsh. The lagoon is presently
a maximum of 6 to 7 feet deep during the highest annual
water elevations (about +6 feet MSL). Historic accounts by
local residents have estimated lagoon depths at 20 to 50

feet. Other evidence in the marsh indicate that the actual depth was probably somewhere in between. Whole clam and abalone shells can be found in the marsh at elevations from five to ten feet below the marsh plain along Butano Creek. Stream channels were much wider in early maps and old aerial photographs than they appear today.

In the early 1940's, there was a major phase of levee construction in the marsh. Additional levees were added in the 1950's and the final period of levee construction ended around the late 1960's or early 1970's. These levees were constructed for the purpose of converting wetlands to agricultural use. Subsequently, the marsh was eventually acquired by the State and agricultural practices ceased by the early 1980's. Only one farmer continues cultivating privately-owned historic wetland acreage.

Pescadero Marsh is open to tidal influence during the winter and closed during summer. Water quality and fishery studies have documented varying aquatic habitat quality conditions in the lagoon (Smith 1987). The shallow lagoon and channel areas can provide good quality aquatic habitat when conditions are favorable. The lagoon provides good habitat when it is openly tidal but, once the lagoon closes, a trapped saltwater lens becomes established. Aquatic habitat conditions can become poor after the lagoon closes. Because the lagoon is shallow, sunlight penetrates to the bottom, warming the sediments and the overlying saltwater layer. Temperature in this water mass can become critical, and oxygen levels can fluctuate from anoxic to supersaturated. This region can become lethal for fish and aquatic invertebrates. The trapped layer of salt water must be expelled to avoid persistent poor water quality conditions. If inflowing freshwater is adequate to fill the lagoon, the salt lens is eliminated by processes not yet fully understood. Once the water column becomes destratified, favorable water quality conditions become re-established and aquatic organisms can re-inhabit the bottom sediments and overlying water column.

Discussion

Sedimentation has caused serious degradation of wetland values at Pescadero Marsh. The problems are apparent in the reduction of depth and narrowing of the inflowing stream channels. The shallow channels are preventing the transport of salt water to a majority of the marsh plain in the wetland. Levees restrict exchange of water and nutrients between the channels and the marsh plain during tidal and flood flows. This probably influences nutrient cycling in the marsh plain and lagoon, soil salinity and silt deposition, and, ultimately,

vegetation patterns, but the actual mechanisms at work here
are unknown. Sedimentation of the lagoon and channels has
affected aquatic habitat conditions and seriously threatens
the production of steelhead and other species of fish and
invertebrates in the marsh.

The biological and physical studies were used to
develop an hydrologic enhancement plan for the marsh
(Williams 1990). Management objectives were developed in
concert with the local community to make their flood
control needs compatible with wetland restoration goals.
The overriding objective was to maintain a functional
wetland ecosystem in as natural a setting as possible,
given land use constraints and existing site
characteristics. Removing the majority of the levees from
the marsh and excavating the channels will restore the
physical processes critical to maintaining the biological
resources dependent upon the marsh.

References

Curry, R., R. Houghton, T. Kidwell and P.Tang. 1985.
Pescadero Marsh, A Plan for Persistence and Productivity.
Contract Report Prepared for the California Department of
Parks and Recreation, Resource Protection Division.
January 28, 1985. 145 pp.

Elliott, B.G., 1975. The Natural Resources of Pescadero
Marsh and Environs. California Department of Fish and Game
Coastal Wetland Series #13, Draft Report, July 1975. 70
pp.

Smith J.J., 1987. Aquatic Habitat and Fish Utilization of
Pescadero, San Gregorio, Waddell and Pomponio Creek
Estuary/Lagoon Systems. Report Prepared for the California
Department of Parks and Recreation, Resource Protection
Division. 31 May 1987. 35 pp.

Viollis, F.S., 1979. The Evolution of Pescadero Marsh.
M.A. thesis, San Francisco State University. 167 pp.

Williams, J., 1990. Pescadero Marsh Natural Preserve
Hydrological Enhancement Plan. Report Prepared for the
Department of Parks and Recreation, Resource Protection
Division. Philip Williams and Associates Report #504,
August 31, 1990. 69 pp.

NEW WETLANDS IN SOUTHERN CALIFORNIA

Andrea Bertolotti[1], P.E.
James R. Crumpley[2], P.E.

ABSTRACT

New wetland areas were created in the Seal Beach Wildlife Refuge, near Anaheim Bay, California. The wetland areas were part of a mitigation project for the Port of Long Beach. Design criteria and constraints were determined at the onset of the project with the assistance of regulatory agencies. The application of a hydraulic numerical model proved very useful in the analysis of the new tidal channel network. The project consisted of soil excavation, culvert construction, channel dredging, and utility relocation.

1.0 INTRODUCTION

1.1 Purpose of Project

The Port of Long Beach has recently expanded its facilities at Pier J with a 147-acre landfill in San Pedro Harbor, California. As part of the construction permits for the new landfill, the Port was required to mitigate the environmental impacts of the landfill by creating new wetland areas. A background and description of this process can be found in "Anaheim Bay Wetland Restoration and Enhancement - An Overview" by Larry Purcell, Port of Long Beach, presented at this conference.

---

[1] Project Engineer, Moffatt & Nichol, Engineers, 3000 Citrus Circle No. 230, Walnut Creek, CA 94598.

[2] Project Manager, Moffatt & Nichol, Engineers, 250 West Wardlow Rd., Long Beach, CA 90807.

1.2   Project Location

A total of four sites were selected for the mitigation project, as
shown in Figure 1. These sites are located in upper Anaheim Bay,
Orange County, California and are about 6 miles east of the Port of
Long Beach. The sites are bounded by the Seal Beach Wildlife Refuge
wetlands and by the Seal Beach Naval Weapons Station.

The location and size of the four sites were limited to areas without
man-made structures. Although the sites are not currently tidal
wetlands, they a located within the Seal Beach Wildlife Refuge.

1.3   Existing Conditions

The wetlands in Anaheim Bay originally extended over a much greater
area than exists today. After the development of the Naval base in
1944 and the Huntington Harbour residential area in the mid 1960's,
the wetlands were reduced in size. The National Wildlife Refuge was
created by an act of Congress in 1972 to protect the unaffected areas
of the wetlands. In 1982, the U.S. Fish and Wildlife Service and the
Navy completed a restoration program to increase the tidal circulation
by removing some of the roads built in the wetlands.

The areas selected for the mitigation project are the Forrestal and
Case Road sites, shown in Figure 2, and Seventh Street and Perimeter
Road sites, shown in Figure 3. Prior to the restoration project, the
existing habitat in the four sites was different from that of fully
tidal wetlands. The tide flow was blocked by the construction of
roads and rail lines. Without the beneficial influence of the ocean,
pickleweed, cordgrass, and other vegetation typical of wetlands was
replaced by upland vegetation.

2.0   WETLAND DESIGN

2.1   Criteria and Constraints

The features of the new wetland areas were outlined in a Memorandum of
Understanding (MOU) between the Port of Long Beach (POLB), the U.S.
Fish and Wildlife Service (FWS), the National Marine Fisheries Service
(NMFS) and the California Department of Fish and Game (CDFG). The MOU
identified the location of the new wetlands as well as a specific
habitat distribution within the wetlands. The MOU specified the
following:

   o   The perimeter of the boundary of each site was set at the
          contour line of 2.5 feet Mean Lower Low Water (MLLW).

o   Restoration work of the new wetlands should result in not
        less than 50 percent of the restored area at an elevation
        of -3.0 feet MLLW.

o   Not more than 35 percent of the area will form slopes
        between -3.0 and 2.5 feet MLLW.

o   Not more than 15 percent of the area will remain as
        islands with elevations between 2.5 and 5.5 feet MLLW.

o   Three mounds, with a top elevation of 8.5 feet MLLW shall
        be placed on each island.

The areas defined by tidal elevations represented specific habitats
desired by the wildlife agencies. The areas at -3.0 feet MLLW were
considered primarily fish habitat, the areas between -3.0 and 2.5 feet
MLLW were defined as mudflats and areas above 2.5 feet MLLW were
defined for vegetation and use by endangered species.

A geotechnical investigation and topographic survey of the areas
identified several constraints that were considered during design.
The soils in the new wetland areas were saturated silts and clays,
with poor slope stability. The geotechnical consultant, GEOFON,
determined that slopes should not exceed 1 vertical to 4 horizontal.
The topographic survey by Dulin and Boynton revealed that the existing
wetland channels were narrow and shallow near the border with the new
wetland areas. The existing channels were analyzed in terms of scour
effects of high flow velocities and a limit of 3 feet per second was
established based on studies of similar wetlands in San Francisco Bay
(Pestrong, 1965).

2.2   Wetland Features

Ocean tides provide a natural means of circulating water in wetlands
and maintaining a level of water quality and nutrients that support a
variety of plant and animal life. The hydraulic system was designed
to bring the beneficial effects of the ocean tides into the new
wetlands.

At a particular ground elevation in a wetland, there is an associated
duration of inundation by tides and a salinity in the soil. Wetland
plants grow in areas where they are best suited to resist the
inundation by tides and soil salinity. Mahall and Park (1976)
examined the transition between cordgrass and pickleweed in wetlands
in San Francisco Bay and found tidal inundation to be the most
important factor.

Tide elevations define the various types of habitats that exist in
wetlands, as shown in Figure 4. Tidal sloughs are almost always
submerged and offer habitat for fish and many benthic species.
Mudflats extend from about Mean Sea Level (MSL) to about Mean Lower

High Water (MLHW). These are foraging areas for birds when exposed at
low tides. Cordgrass extends from MLHW to about Mean Higher High
Water (MHHW) and is the main habitat for the endangered Light-Footed
Clapper Rail. Pickleweed grows from about MHHW to about Extreme High
Water and is the main habitat for the Belding's Savannah Sparrow, a
State-listed endangered species. A more complete description of
wetland habitats can be found in Zedler (1982).

In newly created wetlands, the flora can be established by either
natural colonization or by planting. The desired fauna, instead, can
only be introduced by natural colonization. The type of hydraulic
connection between the new wetland and the ocean should consider the
passage of plant seeds and marine animals, such as fish and
crustaceans. This is important were natural plant colonization is
desired and were the wetlands are to function as spawning areas for
fish.

2.3   Hydraulic Design Principles

The parameter that is most influential in determining the hydraulics
of a wetland area is the tidal prism (Myrick and Leopold, 1963). The
tidal prism is defined as the amount of water that flows into the
wetland or out again with the movement of the tide. The amplitude of
the ocean tide, size of channels, and slope of the banks determine the
tidal prism in a wetland.

Wetlands can have tide regimes and sea levels that are different from
those in the open ocean. This is an essential consideration in
wetland design. Friction and flow restriction induce energy losses in
tidal flow that give rise to reduced amplitudes and phase lags in the
tides of inland water bodies. Under some conditions, the mean water
level in the wetland will be different from that in the ocean (Walker,
Bertolotti, Flick and Feldmeth, 1988).

Conservation of mass establishes a direct relationship between the
tide range and the quantity of flow in and out of a wetland subject to
an ocean tide. If a culvert or artificial channel restricts the flow
of water, the inner wetlands will have a tide range smaller than in
the ocean. A reduced tide range, often referred to as "muted", is the
result of head losses that slow the flow of water.

The ground elevations in a wetland should be designed considering
stable soil slopes and the possibility for scour and siltation of the
channels. Myrick and Leopold (1963) applied the concept of
equilibrium profiles developed for inland rivers to describe the
hydraulic geometry of tidal channels. These equations relate the
water surface width, mean depth and mean velocity to the discharge.
In the design of muted wetlands, channels that are initially sized
close to the equilibrium profile can be expected to have less changes
over time that channels that are of other sizes. Adding new wetland
areas can induce scour in channels of existing wetlands.

2.4 Hydraulic Design Implementation

The hydraulic analysis of the new wetlands required the simultaneous
consideration of ocean tides, existing channels, bridges and culverts,
and adjacent marina areas to determine the effect on the tide regime
in the new wetlands. To aid this analysis, a numerical model based on
work by Fisher (1977) was developed to analyze tide elevation and flow
velocities in wetlands subject to an ocean tide. The hydraulic system
is divided into a series of basins and channels. The water level in
the basins and the flow velocity in the channels are related through
the conservation of mass and conservation of momentum equations.
These equations are solved for one-dimensional flow by the method of
finite differences. The model performs a time domain simulation and
calculates tide elevations and flow velocities at specified time step
intervals.

The model, called the Hydrodynamic Circulation Model (HCM), employs
and implicit routine, where the water elevations and velocities are
related to each other and to their present values through a system of
simultaneous non-linear equations. The equations are solved
iteratively using a Newton-Raphson algorithm. One-dimensional
equations are used to express the conservation of mass and momentum in
an open channel.

The hydraulic features of an area are included in the model through a
number of user-selected parameters. These include the length, width
and depth of channels, the friction associated with the type of
channel bed, the slope of the channel banks, the storage curve of the
basins, the tide properties of the ocean, and the type and geometry of
the tide control structure. The friction of the channel beds and
concentrated head losses of restrictions to flow can be derived from
published values. However, calibration of these parameters with field
measurements from a hydraulic area similar to that being modeled
provided more reliable values for analysis of the new wetlands.

The model was used to predict tidal datums in the new wetland areas.
Several alternative configurations of the new wetlands were analyzed.
Culvert sizes, channel routing, and channel sizes were varied to
determine the individual effect of each on the final design and an
optimized hydraulic network was developed.

2.5 Grading Design

The existing ground elevations within the new wetland areas did not
provide the desired habitat zonations. A grading plan was developed
to match the predicted tide elevations with the desired habitat mixes.

With guidance from the POLB, FWS, NMFS, and CDFG, the tide elevations
described in the MOU were converted to the respective tidal datums
rather than referenced to a fixed land elevation. This change was
necessary because the tide ranges and ground elevation specified in
the MOU could not be provided unless very wide channels were cut

through the existing wetlands. Instead, ground elevations referenced
to tidal datums allowed the use of existing channels with minimal
modification.

Two of the four sites were graded to include islands. The other two
were graded to be open bodies of water. The slopes on the islands
were modified to meet the habitat distribution described in the MOU.
As shown in Figure 5, slopes varied between 1 (vertical) on 4
(horizontal) to 1 on 50. Transitions between habitats were done at 1
on 4 which was the maximum allowable slope. Other slopes were varied
to provide the required percentage of each habitat.

Because a change in grade within the new wetland areas modified the
tidal prism, a hydraulic analysis was performed after every grading
change. Modifications to the hydraulic network were done to provide
the desired tide ranges within the new wetland areas. This iterative
procedure was repeated until a final grading design was reached.

2.6   Culvert Design

Railways and roads built on dikes provided a barrier to tidal flow
into the new wetland areas. A total of seven culverts were designed
to connect the wetlands to the ocean. Culverts were selected instead
of bridges because culverts could be built more economically.

The culvert under Bolsa Avenue was designed to be the throttling point
for tidal flow into the Bolsa Cell, Forrestal and Case Road sites.
The culvert, about 15 feet wide by 6 feet deep, was designed to
control the amount of flow in order to avoid scour in the existing
wetland channel. Hence, the culvert was designed to give the maximum
possible tide range in the mitigation sites while maintaining flow
velocities in the existing wetlands below the 3 feet per second limit.
This culvert was fitted with stoplogs at the inlet to allow for
adjustments in water flow. Mussels, barnacles and other marine
organisms will grow at the inlet and inside the culvert, increasing
friction to flow and reducing the flow area. Stoplogs were designed
to be present during the first period of operation and, as marine
organisms grow on the culvert, would be removed over time to maintain
tidal flow through the culvert.

An additional five culverts were provided to carry the tidal flow into
the Forrestal and Case Road sites, as shown in Figure 2. These
culverts were designed large enough to minimize further restrictions
to tidal flow while providing the desired tide range in the new
wetland areas. The culvert for the Seventh Street site was sized to
throttle the flow and give the same tide range as the other wetland
sites. The existing channel into Seventh Street site was too small to
carry the expected flow and was enlarged. By reducing the flow into
Seventh Street, only a small enlargement was necessary.

3.0   PROJECT CONSTRUCTION

3.1   Northern Mitigation Sites

The Forrestal and Case Road sites were separated from the tidal waters of the Seal Beach Wildlife Refuge wetlands by several roadway and railroad embankments. As a result, these sites remained dry during excavation, and were excavated to the final grades by normal earthmoving equipment such as scrapers, bulldozers, and loaders.

3.2   Southern Mitigation Sites

The Seventh Street site was separated from the existing wetlands by only one embankment for a narrow dirt road. The upper portion of the excavation was removed by normal earthmoving equipment. However, because of the close link to tidal water, the lower portion of the excavation contained saturated soils which were too weak to support normal earthmoving equipment. To reach final grade, large backhoes and shovels perched on high ground were used to excavate the remaining soils. This equipment moved across the site in steps, reaching final grade in one location before moving to the next location.

The Perimeter Road site was totally within the existing wetland, with ground elevations above the higher tidal ranges. Saturated soil conditions were found a few feet below the ground surface, requiring the use of backhoes and shovels to reach final grade.

3.3   Permanent Dike in Bolsa Cell

A wetland area with a muted tide range, called Bolsa Cell, is located between the existing wetlands and the two northern sites. To preserve the natural condition of this area, a permanent dike was built across the cell to separate the new water channel from the other areas of the cell. The dike was constructed near the western end of the cell by placing a geotextile fabric on the existing mud bottom. Selected fill was placed over the fabric and was compacted into a dike capable of supporting construction equipment and future maintenance equipment.

3.4   Supply Channel

The supply channels to the northern sites were excavated in the dry using conventional excavating equipment. In one area of the northern supply channel, a communication duct bank crossed at an elevation slightly above the normal channel invert elevation. In this reach, the supply channel was lined with concrete in order to reduce the hydraulic friction and maintain the required flow to and from the Case Road site.

The supply channel for the southern sites was originally planned to be excavated using a hydraulic dredge although this method was not used. The contractor opted to use a temporary embankment along the length of the channel route. Backhoes were used to excavate the channel, starting at the point of connection with existing channels and moving upstream towards the new sites. As the equipment moved along the channel, the temporary embankment was removed.

3.5 Culverts

The concrete culverts which extend the supply channels under the road and railroad embankments were constructed in the dry. The culvert crossing Bolsa Avenue was protected from the tidal waters by a temporary dike, and was constructed in two phases to allow the roadways to remain usable during construction. The culvert into the Seventh Street site was also constructed behind a temporary dike, while the remaining culverts were constructed prior to allowing water into the supply channels and did not require dike protection.

4.0 CONSTRUCTION SCHEDULE AND COST

The construction of the four wetland sites, supply channels, and culverts required approximately 13 months from the time the contractor mobilized to the time when the sites were opened to tidal water.

Construction began with the relocation of utility pipe lines from within the Forrestal and Case Road sites to a location around their perimeter. Once the pipe lines were relocated, excavation of those sites began. As excavation work progressed, overhead power and communication lines crossing Case Road site were also relocated. When excavation of the northern sites began to approach final grade and required smaller equipment, the excavation of the southern sites began.

All excavated material was deposited within the Naval Weapons Station, first to fill low areas between the roadway system and to increase the size of the protective embankment around the practice shooting range. Subsequently, the material was stockpiled behind the Station's harbor in an area previously used for dredge material disposal.

The culvert construction began approximately five months after the start of excavation to allow completion of the water supply system at about the same as the completion of soil excavation. Temporary timber bulkheads were constructed at the outlet end of the culverts connecting the existing wetlands to the northern supply channel and to the Seventh Street site to allow removal of the temporary dikes prior to flooding the sites. Once the site excavation had been completed, the timber bulkheads were removed and tidal water was introduced into the newly created wetland sites.

The total construction cost for the creation of the new wetlands was 5.1 million dollars. Nine contractors submitted bids for the job. Considering a total area of about 116 acres, the project cost was $44,000 per acre. About $3.52 million of the total cost was for site and channel excavation and $1.15 million was for culvert construction. The remaining costs were for mobilization, utility relocations, and permanent dike construction.

5.0 CONCLUSION

Four new wetland sites were created in Southern California in connection with an expansion project in the Port of Long Beach. The size and type of wetlands were determined from a careful assessment of the habitat lost due to the port expansion and the proposed habitat gained from creating new wetlands. A number of design criteria were determined at the onset of the project to help the engineers determine several technically feasible alternatives for flow routing and grading the new sites. A numerical hydrodynamic circulation modes proved very useful for determining the hydraulic response of the wetlands sites with various culvert and channel features.

Construction of the new wetlands consisted primarily of soil removal. Saturated soil conditions near the ground surface required the use of backhoes and shovels to complete the grading. Culverts and channels were built in the dry by temporarily diking the new wetland areas. A channel in the existing wetlands was built by using a temporary embankment to allow access by land-based equipment.

The total project required about 13 months to complete. The construction cost was $5.1 million, or about $44,000 per acre.

6.0 REFERENCES

Fisher, H.B., "A Numerical Model of the Suisun Marsh, California", Proceedings of the 17th Congr. Intl. Assoc. Hydraul. Res., Baden-Baden, Germany, 1977.

Mahall, B.E. and R.B. Park, "The ecotone between Spartina Foliosa Trin. and Salicornia virginica in salt marshes in northern San Francisco Bay. I. Biomass and Production. II. Soil water and salinity. III. Soil aeration and tidal immersion", J. Ecology, 1986, March, 1987.

Myrick, Robert M. and Luna B. Leopold, "Hydraulic Geometry of a Small Tidal Estuary", U.S. Geological Survey, Professional Paper 422 B, U.S. Government Printing Office, Washington, 1963.

Pestrong, R., The Development of Drainage Patterns on Tidal Marshes, Stanford University Publications, Geological Sciences, Vol X, Num 2, 1965.

Walker, J.R., A. Bertolotti, R.E. Flick and C.R. Feldmeth, "Hydraulic Aspects of Wetland Design', Coastal Engineering Conference 1988, pp. 2666 - 2680.

Zedler, Joy B., Salt Marsh Restoration: A Guidebook for Southern California, published by the California Sea Grant College Program, Institute of Marine Resources, University of California, A-032, La Jolla, California, 1984.

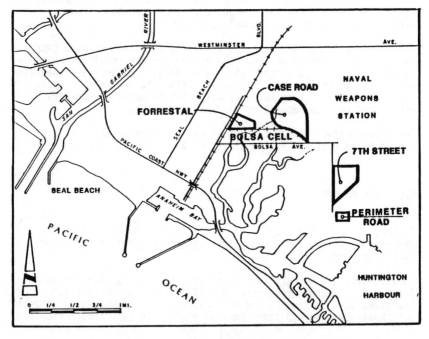

FIGURE 1   PROJECT SITES

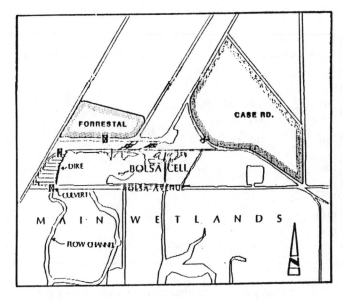

FIGURE 2 FORRESTAL AND CASE ROAD PLAN

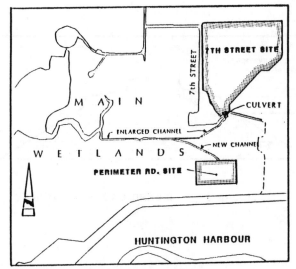

FIGURE 3 7th STREET AND PERIMETER ROAD PLAN

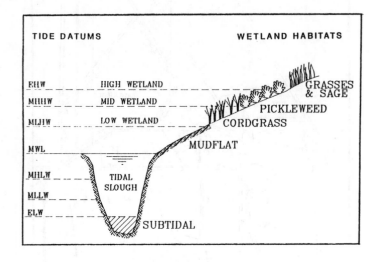

FIGURE 4   TYPICAL WETLAND ZONATION

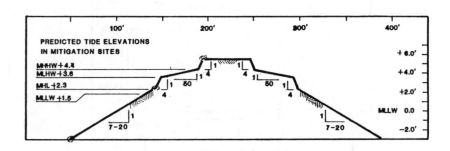

FIGURE 5   ISLAND CROSS SECTION

THE EFFECTS OF SHORELINE EROSION IN GALVESTON BAY, TEXAS

Robert W. Nailon[1]   Edward L. Seidensticker[2]

Abstract

The shoreline of Trinity Bay and East Galveston Bay, Texas, is eroding at an average annual rate of four feet. Land losses to these coastal properties will continue unless low-cost, effective measures are developed and implemented for shoreline erosion control and wetland habitat enhancement. Once established, smooth cordgrass, Spartina alterniflora Lois, provides an effective means of shoreline erosion protection. The Trinity Bay Soil and Water Conservation District and the Texas A&M University Marine Advisory Service initiated a project in 1985 to study the impacts of shoreline erosion and test cost-effective, vegetative erosion control measures in Galveston Bay, Texas. Smooth cordgrass was transplanted in four sites along the affected shoreline and adjacent coastal waters for erosion abatement and re-creation of valuable wetland habitats. To date, approximately 1825 meters of shoreline has been vegetatively treated. Turbidity levels were measured during high and low erosion conditions. Sediment accretion was also measured within the study area. Relative abundance of fish species was documented by site during the study. Transplant survival rates varied from 60-70 percent. New shoot development seemed to depend on the care of the transplant stock. More work needs to be done in selecting disease resistant strains of smooth cordgrass and in plant genetics. Turbidity levels were highest during severe erosion conditions during the study. Average sediment accretion as a result of the newly-

[1]County Marine Extension Agent, Texas A&M University Marine Advisory Service, P.O. Box 699, Anahuac, TX 77514

[2]Resource Conservationist, Soil Conservation Service, P.O. Box 819, Anahuac, TX 77514

194   COASTAL WETLANDS

established stands of cordgrass was 0.14 m in each site.
Results of this study indicate that marine organisms
readily utilize the artificially-created smooth
cordgrass colonies within the study area. A significant
difference in relative abundance of catch was noted
between an unvegetated site and a vegetated site (12.7
percent versus 87.3 percent). Predominant species found
during both sampling periods were white shrimp, Gulf
menhaden, and striped mullet. Technical assistance was
provided by the Soil Conservation Service, Texas
Agricultural Extension Service, and the Texas A&M
University Sea Grant College Program. Funding was
provided through grants from The Moody Foundation, The
Brown Foundation Inc., and the Texas A&M University Sea
Grant College Program.

Introduction

        Local landowners in the Galveston Bay area have
been concerned about erosion for more than 30 years.
Attempts have been made using car bodies, concrete pipe,
rip-rap, and used automobile and tractor tires to halt
the erosion rate. Attempts at transplanting cordgrass
have also been tried, but the failure rates have been
very high. In 1984 the authors began a joint effort to
address the erosion problem and develop a method of
planting smooth cordgrass to increase the rate of
success of transplant survival.

        The erosion not only represents a physical loss of
land, but it also has a detrimental effect on the marine
ecosystem. The resulting sedimentation stymies
beneficial aquatic vegetation through increased water
turbidity. The lack of intertidal vegetation has
resulted in a loss of nursery area for shrimp, Penaeus
sp., redfish, Sciaenops ocellata, speckled trout,
Cynoscion nebulosus, blue crabs,Callinectes sapidus,
Spot, Leiostomus xanthurus, Tidewater Silversides,
Menidia beryllina, and numerous other marine organisms.
These nursery areas play a critical role in providing
sites where juvenile stages of these species can hide,
feed and grow. Continued erosion along Texas bays and
the resulting loss of vegetation can be devastating to
the future of the sport and commercial fishing
industries in the Gulf of Mexico and its surrounding
waters.

        This paper addresses the feasibility of using
smooth cordgrass as an alternative erosion control
method through a pilot project conducted by the
participating agencies. This pilot project also
monitors the short term impact vegetation has on
sedimentation and turbidity within the newly-created

salt marshes in the study area, and describes the
success of three types of wave barrier protection used
to protect young transplants.

Description Of The Study Area

The study area selected for the erosion control
studies is located on the north shoreline of East
Galveston Bay and Lake Stephenson, an adjacent tidal
lake in southern Chambers County (Figure 1). Chambers
County, Texas, is in the extreme southeastern part of
Texas. It is bordered on the south by Trinity Bay,
Galveston Bay, East Galveston Bay and the Gulf of
Mexico. Chambers County covers a total area of 224,000
hectares, of which 152,527 hectares is land and 71,393
is water.

Erosion rates vary with site. Sites #1 and #3
erode at an average annual rate of 1.8 m/yr. Site #2
erodes at an average rate of 2.6 m/yr. Site #4 erodes
at an average rate of 0.5 m/yr (Table 1).An 11.3 km
fetch of water across East Galveston Bay exists at Sites
#1, #2, and #3. Site #4, located on Lake Stephenson,
possesses a 0.8 km fetch of water. The shoreline
geometry is meandering with a silty or a very fine sandy
sediment in the swash zone.

TABLE 1. SITE CONDITIONS

SOIL TYPE	EROSION RATE m/yr	WAVE BARRIER PROTECTIONS
Site #1 Veston Soils (mostly clay)	1.4 - 2.2	Double Row Christmas Trees
Site #2 Veston Soils (mostly clay)	1.9 - 3.1	Double Row Parachutes
Site #3 Veston Soils (mostly clay)	1.6 - 2.0	No Wave Barrier Protection
Site #4 Harris Clay	0.3 - 0.7	Single Row Plastic Snow Fence

Using the U.S. Corps of Engineers, Vegetative
Stabilization Site Evaluation Form, the East Galveston
Bay Site has a cumulative score of 200 and a potential
success rate of 30 percent. Most of the shoreline is
absent of any vegetation. Only in a very few locations

Figure 1
LOCATION OF STUDY AREA

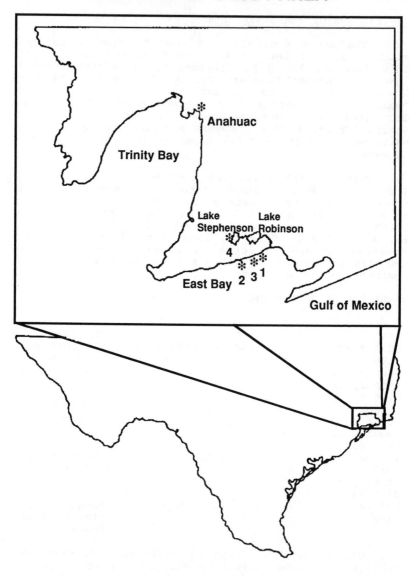

can smooth cordgrass be found. One small area of smooth
cordgrass that was established in 1958 is still
expanding and has remained stable   (USAE, Texas Coast
Shores Regional Inventory Report. 1968).

     The area is predominantly Gulf Coast Saline
Prairie (USDA, Soil Conservation Service, 1981). The
plant community is dominated by Gulf Cordgrass, Spartina
spartina, Lesser concentrations of marshhay cordgrass,
S. patens, are found in the area. The predominant soils
found in the study area are Harris Veston association,
(Crout, 1976). When bare and unprotected by plant
cover, these soils are very susceptible to erosion.
Sparse natural stands of smooth cordgrass are found in
the intertidal zone at only a few sites.   The majority
of the shoreline in East Galveston Bay lacks any
intertidal vegetation. The land-water interface is
characterized by a one-meter-high bare bluff at Sites
#1, #2, and #3.   Site #4 possesses a 0.5 meter high
bluff. The shoreline at Sites #1, #2, and #3 has a
shallow shelf with very little drop off from the shore
seaward. Site #4 has a similar land slope.

     Turbid water conditions prevail at Sites #1,#2, and
#3.   Water turbidity at Site #4 is somewhat less.

     Many species of marine organisms inhabit inshore
waters in the study area, juvenile marine shrimp,
Penaeus sp., Gulf menhaden, Brevoortia patronus, blue
crabs, bay anchovies,Anchoa mitchilli, Atlantic croaker,
Micropogon undulatus, striped mullet, Mugil cephalus,
and Gulf killifish, Fundulus grandis, are the
predominant organisms found at all sites.

Materials And Methods

Transplanting Techniques

     The transplants were acquired from existing native
stands growing in the immediate vicinity on East
Galveston Bay. Cordgrass transplants were dug and
separated. Efforts were made to minimize root
disturbance by keeping wet soil around the plant roots.
Moist transplants were transported in a washtub to the
four study sites. Transplanting methods followed those
outlined by Cutshall, 1985. A 0.66 meter space between
plants and rows was allowed. Transplants were planted at
a water depth of 15.2 centimeters in the intertidal
zone.   Transplants were planted approximately 6.1 meter
seaward of the shoreline. No fertilizers were applied
in this study.

Wave Barrier Protection

Temporary wave barriers were constructed to minimize wave impacts on young transplants. Three different materials were used to construct temporary wave barriers. A double row of used Christmas trees were tied trunk-to-trunk with parachute strap, laid parallel to the shoreline and staked on the seaward side of the transplants at Site #1. A single row of plastic snow fence was used as a temporary wave barrier at Site #4. Fence posts were driven in a line on the seaward side of the transplants, and the plastic snow fence was attached to the posts. The plastic snow fence was obtained from American Excelsior Company and provided a suitable barrier that also can be reused. A double row of used cargo parachutes acquired from Texas Surplus Property Agency were used as a temporary wave barrier at Site #2. A double row of one-meter-wide strips of the parachute material approximately 10 meters long were attached by parachute strap end-to-end to fence posts driven in a line parallel to the shoreline seaward of the young transplants (Figure 2). The parachutes were inexpensive and easy to install. Site #3 with no wave barrier protection for the transplants provided a control to assess the feasibility of the wave barrier structures on transplant survival. Although there may be other wave barrier protection measures, the measures selected by this project were chosen because they are removable. The authors feel that a removable barrier would be aesthetically desirable and would not interfere with tidal flow and marine organism movement patterns.

Sedimentation

Sedimentation was measured at two sites beginning in August 1988. Reference stakes were established at Sites #1 and #2 and cross-sections were taken from the reference stakes seaward at 20-foot intervals. Readings were taken to the nearest .01 foot.

Turbidity

Turbidity was measured at each site weekly. Readings were taken with a standard 20 cm Secchi disk, and recorded to the nearest centimeter of water clarity. Readings were also taken at other locations in Galveston Bay to determine the range of water conditions at other eroding areas.

Fisheries

Fish and shellfish were collected during the study using a 32 m bag seine having a 1.9 cm mesh. The

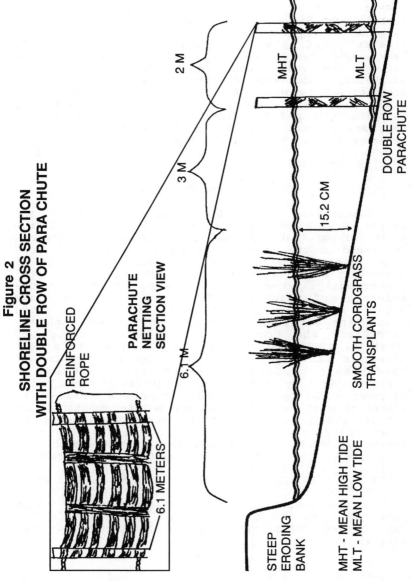

Figure 2
SHORELINE CROSS SECTION
WITH DOUBLE ROW OF PARA CHUTE

seine was pulled approximately 50 meters along the
seaward side of the Spartina stand parallel to the
shoreline at Sites #2 and #3 on August 23-24, 1988 and
October 25-26, 1988. Fish and shellfish species were
also collected by a 1.24 m cast net on August 23-24,
1988 and October 25-26, 1988. Random throws of the cast
net were conducted on each sampling date at Sites #2 and
#3. Relative abundance of fish and shellfish species was
documented and compared by site and sampling date in the
study. Due to the amount of rubble disposed in the
sampling area and the prohibition of the use of gill
nets and trammel nets in coastal waters of Texas, no
other gear types were used in the study.

Results and Discussion

Transplanting Techniques

    Single stem transplant survival at Sites #1,#2, and
#4 was very successful. The survival rates varied from
60-70 percent (Table 2). New shoot development appeared
to depend on the care of the transplant stock. There
also appeared to be some difference in growth of
transplant stock taken from different locations planted
concurrently. More work needs to be done on the
selection of planting materials used in shoreline
erosion control. Further investigations concerning
transplant survival in various soil types might be
warranted. The transplanting methods outlined by
Cutshall, 1985 were adequate for transplanting at all
sites in the study area. The complete failure of
transplants on Site #3 demonstrates that smooth
cordgrass is very difficult to establish at these sites
without some type of wave protection (Table 2).
Transplants at Site #3 were exposed to constant wave
action and lacked sufficient root development to
withstand the wave forces. The U. S. Army Corps of
Engineers Vegetative Stabilization Site Evaluation for
this site gave it a 30 percent potential success rate.

TABLE 2.   TRANSPLANT SURVIVAL COMPARISONS BY SITE AND DATE

DATES TRANSPLANTED	NUMBER OF SHOOTS TRANSPLANTED	SURVIVAL AS OF MARCH 87	SHOOT PERCENT SURVIVAL	NUMBER SHOOTS/ TRANSPLANT
Site #1				
1/13-17/86	850	510	60	2.39
Site #2				
8/18-27/86	620	410	66	2.84
Site #3				
5/20-22/86	250	0	0	0.00

Site #4
7/15-23/86   600   440   70   2.09

Wave Barrier Protection

There appeared to be little difference in the
survival rates of transplants at all sites using any of
the three wave barrier protection methods during the
study. The used Christmas trees worked successfully but
had several drawbacks. The seasonal availability of
Christmas trees is a problem. The utilization of the
trees is the most labor-intensive method   of barrier
protection evaluated in this study. They were also
difficult to collect and transport to the site. The
cost of the snow fence is quite high and may be
expensive for large-scale treatments. By far the used
cargo parachute strips were the most effective wave
barrier material tested in the study. There is little
doubt that the three types   of wave barriers tested in
this study have enhanced the success of smooth cordgrass
transplant establishment.

Sedimentation

The preliminary data show a slight accumulation of
sediment on Site #1 from August 1988 to December 1988
(Figure 3). Studies by (Gosselink and Mitsch 1986)
indicate that one of the primary functions of a salt
marsh is to accumulate sediments from off site and
incorporate them into the marsh ecosystem. It is
anticipated that the artificial salt marsh created
during this study will function in the same way as a
naturally occurring marsh given time. If so,
theoretically, some of the poor water quality conditions
caused by shoreline erosion will gradually improve in
the artificial marsh created during this study.

Turbidity

Results of the turbidity level measurements in this
study show a direct correlation between high erosion
conditions and high levels of suspended sediments in the
water column (Table 3). There was no significant
difference in turbidity levels between Sites #1, #2, and
#4 that were transplanted and Site #3 which has no
vegetation. Since the shoreline of East Galveston Bay is
over 60 km. in length, it is doubtful that the
transplanted vegetation from this study has exhibited
any measurable effect on the overall turbidity of East
Galveston Bay. The entire shoreline of East Galveston
Bay contributes to the high   turbid water conditions
found in all sites during the study.

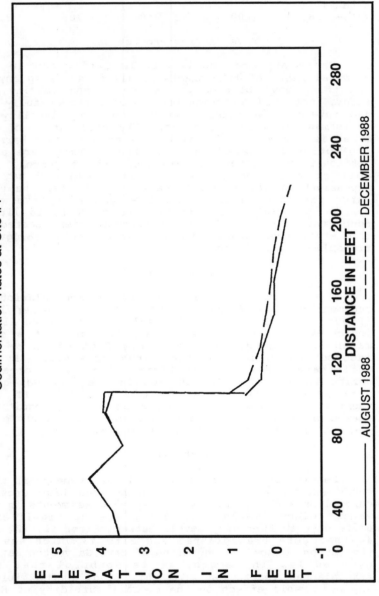

Figure 3. Cross Sectional Profile
Sedimentation Rates at Site #1

Table 3. Turbidity Levels by Site from 9-88 through 3-89.
Secchi Disk Readings (cm.)   Site
Date   Wave Conditions   #1   #2   #3   #4

Date	Wave Conditions	#1	#2	#3	#4
09/09/88	slightly choppy	19	20	20	31
10/14/88	calm	43	44	44	29
11/11/88	very choppy	19	18	18	13
12/13/88	choppy	26	26	26	13
01/30/89	calm	30	30	30	43
02/17/89	slightly choppy	21	21	21	27
03/17/89	very choppy	13	13	13	9

Fisheries

Results of this study indicate that marine
organisms readily utilize smooth cordgrass colonies
within the study area. A total of 87.3 percent of the
marine finfish and shellfish in this study were
collected from Site #2, a transplanted smooth
cordgrass colony. A significant difference in relative
abundance of catch was noted in Site #2 versus Site #3
(87.3 percent and 12.7 percent, respectively). The
predominant species found at Site #2 during both
sampling periods were white shrimp (33.6 percent), Gulf
menhaden (31.6 percent), and striped mullet (15.2
percent) of the total catch (Table 4).

Table 4. Relative Abundance (%) of Marine Organisms
Collected by Seine and Cast Net During the August 23-24
and October 25-26, 1988 Sampling Periods--Site #2

Species	Number	Percent Abundance	Rank
White Shrimp	195	33.6	1
Gulf Menhaden	183	31.6	2
Striped Mullet	88	15.2	3
Gulf Killifish	34	5.8	4
Atlantic Croaker	20	3.4	5
Blue Crab	18	3.1	6
Tidewater Silverside	18	3.1	7
Bay Anchovy	14	2.4	8
Spot	9	1.6	9
N = 580		100.0	

The remaining 12.7 percent of the marine finfish
and shellfish in this study were collected from Site #3,
a control with no vegetation present. Gulf menhaden
(53.6 percent) and striped mullet (31.0 percent)
comprised the majority of the catch at Site #3 (Table
5). Although the sample size is fairly low, these data
likely indicate a representative sample of both
habitats.

Table 5.   Relative Abundance (%) of Marine Organisms
Collected by Seine and Cast Net During August 23-24 and
October 25-26, 1988 Sampling Periods--Site #3

Species	Number	Percent Abundance	Rank
Gulf Menhaden	45	53.6	1
Striped Mullet	26	31.0	2
Blue Crab	6	7.1	3
Gulf Killifish	3	3.5	4
Tidewater Silverside	2	2.4	5
White Shrimp	2	2.4	5
N = 84		100.0	

          Species   composition   remained   remarkably
similar   between Sites #2 and #3 during the study.
Several bay anchovies were collected only from Site #2
and one Gulf kingfish was collected only from Site #3.

Conclusions

     The utilization of wave barrier protection methods
outlined in this paper to protect smooth cordgrass
transplants until colonies can establish themselves
appear to play a major role in transplant survival and
growth.   The cargo parachute strips were the most
effective wave barrier protection tested in this study.
     Overall estuarine enhancement, a by-product of
this vegetative erosion control project, may take years
to develop.   It is well documented that fish and
shellfish resources utilize smooth cordgrass colonies in
the Galveston Bay salt marsh for food, cover, and
nursery areas.   (McHugh 1966; Gosselink et al.   1973;
Zimmerman et al. 1984; Zimmerman and Minello 1984; Texas
Parks and Wildlife Department. 1989.).

     The role of the estuary in supporting fisheries
productivity depends on the endemic species and other
factors, including developmental stage, time of year,
geographic location, and physical and chemical
characteristics of the estuarine waters. All estuarine
habitats are utilized to some extent, and each may be
critical to some life stage of a marine organism.
Shallow open-water areas, like those found in Galveston
Bay, support large numbers of juvenile marine organisms.
Small bayous and channels also provide access   to
intertidal marshes and refuge from stranding during
periods of low water.

     High turbidity levels in East Galveston Bay
drastically affect the primary productivity of the
estuary (Texas Parks and Wildlife Department 1964).   The
high turbidity levels reduce water quality by shading

out beneficial phytoplankton and zooplankton organisms in the water column and therefore reduce the total food web productivity in the estuary.

Sedimentation in East Galveston Bay caused by shoreline erosion has historically had a negative effect on water quality. Shoreline erosion has been a problem since the 1930's (Paine and Morton 1986). The preliminary results of this study show some benefits in reducing shoreline erosion. The short-term results of the cross-sectional data indicate a slight increase in sediment accumulation within the transplanted grass colony. It is not known whether this trend will continue or if the data reflect a seasonal variation. Long-term sedimentation studies are needed to determine potential impacts and benefits in sediment accumulations in the intertidal zone. High sedimentation rates are known to cause mortality in oyster reefs in East Galveston Bay (Texas Parks and Wildlife Department 1964). Sediment collects in the gills of the oyster, drastically reducing pumping efficiency and oxygen uptake capability.

Between wetlands surveys of 1956 and 1979 of Galveston Bay, the estuary lost approximately 16 percent of its marshes and an estimated 95 percent of its submergent vegetation (Galveston Bay Seminar Executive Summary. 1988). The importance of this estuary for commercial and recreation fisheries, combined with accelerations in habitat loss due to shoreline erosion and other factors in Galveston Bay, will require further attention in the future. The disruption of the marsh ecosystem in the study area, through erosion created by man's intervention or natural processes, drastically affects the overall marsh production as is brought forth in this paper. Additional research is needed on how estuarine habitats function for marine organisms.

The long-term effects of re-creating marsh habitat lost through erosion are obvious. The landowner ceases to lose valuable bayfront property to the forces of erosion. Concurrently, commercial fishermen benefit greatly as a result of marsh revegetation through increased harvests of commercial seafood products, and recreational fishermen benefit by experiencing more successful fishing trips.

References

Crout, Jack D. 1976. U.S.D.A. Soil Conservation Service. "Soil Survey of Chambers County, Texas. 53 pp.

Cutshall, Jack. 1985. U.S.D.A. Soil Conservation
    Service. "Vegetative Establishment of Smooth
    Cordgrass (Spartina alterniflora) For Shoreline
    Erosion Control. Technical Note. 6 pp.
Gosselink, James G. and W.J. Mitsch. 1986. Wetlands.
    Van Nostrand Reinhold Company. New York.
Gosselink, James G., Eugene P. Odom, and R.M. Pope.
    1973. "The Value of the Tidal Marsh". Institute
    of Ecology Work Paper Number 3. University of
    Georgia. Athens, Ga. 31 pp.
McHugh, J.L. 1966. "Management of Estuarine Fishes.
    In: A Symposium on Estuarine Fisheries. American
    Fisheries Society Special Publications Number 3.
    pp. 133-154.
Paine, J.G. and R.A. Morton. 1986. Historical
    Shoreline Changes in Trinity, Galveston, West and
    East Bays, Texas Gulf Coast. Bureau of Economic
    Geology. University of Texas at Austin.
Sheridan, P.F., R.D. Slack, S.M. Ray, L.D. McKinney,
    E.F. Klima. 1988. Biological Components of
    Galveston Bay. Executive Summary--NOAA Estuary of
    the Month.
Texas Parks and Wildlife Department. 1964. "The Texas
    Oyster Fishery. Bulletin 40. 6pp.
Texas Parks and Wildlife Department. 1989. "A Report
    to the Governor--The Texas Shrimp Fishery". Coastal
    Fisheries Branch.
U.S. Army Engineers. 1968. Texas Coast Shores Regional
    Inventory Report. Galveston District. Galveston,
    Tx. 26 pp.
U.S. Army Engineers. 1983. "Shore Stability With Salt
    Marsh Vegetation". P.L. Knutson, and W.W.
    Woodhouse, Jr. Special Report Number 9. Coastal
    Engineering Research Center. Ft. Belvoir, Va.95 pp.
U.S.D.A., Soil Conservation Service. 1981. "Land
    Resource Regions and Major Land Resource Areas of
    the United States". Agriculture Handbook 296.
    156 pp.
Zimmerman, R.J. and T.J. Minello. 1984. Densities of
    Penaeus aztecus,Penaeus setiferus, and other natant
    macrofauna in a Texas salt marsh. Estuaries 7(4A):
    421-433.
Zimmerman, R.J., T.J. Minello, and G. Zamora, Jr. 1984.
    Selection of vegetated habitat by brown shrimp,
    Penaeus aztecus,in a Galveston Bay salt marsh.
    Fish. Bull. 82: 325-336.

The "Geologic Review Process" an Evolution of a
Successful Interagency Program

Brian J. Harder [1], James D. Rives III [2],
Lynn H. Wellman [3]

Introduction

The coastal wetlands of Louisiana represent 40% of all coastal wetlands in the United States including Alaska. Each year Louisiana's marshes produce 30% of the nation's fisheries catch and $20 billion in crude and natural gas production is extracted from wells drilled in the wetlands. This represented 16% of the U. S. total production of petroleum and 29% of the total production of natural gas.

To gain access to these mineral resources canals are dredged through the fragile wetlands to establish drill-sites for recovering oil and gas reserves. In 1982 over 200 new well-access canals were dredged, which directly altered over 775 acres of coastal wetlands. The dredging of canals to access oil and gas drilling sites has not only caused substantial direct wetland loss, but coupled with long-term secondary impacts, it may have been responsible for up to half of wetland loss that has occurred in coastal Louisiana. It is estimated that if present land-loss trends continue one million acres of Louisiana's coastal wetlands will be lost by the year 2040, reducing commercial fish and wildlife harvests by 70%. This will have a significant national impact.

To prevent and reduce the loss of valuable wetland habitat caused by oil and gas activities, the Coastal Management Division of the Louisiana Department of Natural Resources (CMD) instituted the Geologic Review Process in 1982. Through this process the losses of wetlands due to oil and gas activities have been reduced when it is feasible by requiring directional drilling of wells from existing waterways and slips or the use of other less environmentally sensitive sites. The process also allowed all the state and federal environmental agencies to comment on each proposed drilling site at the same meeting (Johnston and Rives, 1985.) The final result of this process is that the applicant extracts mineral resources at minimal environmental cost.

The Geologic Review process, through time, has reduced the total disturbed area impacted by oil and gas access operations from 767 acres per year in 1982 to just under 77 acres per year in 1989 as shown in **Figure 1**. Average canal lengths have decreased 79%, and the total area

---

[1] Research Engineer, Louisiana Geological Survey, P.O. Box G, Baton Rouge, LA 70893

[2] Assistant Director, Coastal Management Division, Louisiana Department of Natural Resources, P.O. Box 44487, Baton Rouge, LA 70804

[3] Manager, Permit Section, Coastal Management Division, Louisiana Department of Natural Resources, P.O. Box 44487, Baton Rouge, LA 70804

disturbed per location has been reduced 45% in the same time period. This process has provided the applicants and the agencies with a forum to settle differences in a constructive manner while continuing to allow access to the needed resources and preventing needless destruction of wetland habitat. The success of the project lies in the fact that oil & gas recovery and wetland preservation are proceeding simultaneously with beneficial results for all concerned.

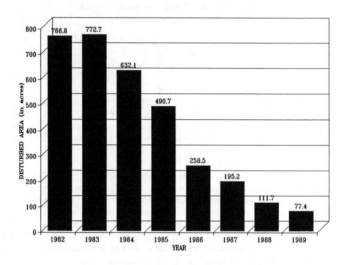

Figure 1: Acres of wetlands permitted for oil & gas canals by CMD.   LGS/LSU

History

        Since the middle of the 1930's, oil and gas companies have used board roads and canals to transport drilling rigs to wetland sites in Louisiana. Board roads are usually x through wetlands by excavating and filling to create an elevated surface which is covered by boards for vehicular traffic. The "footprint" of a board road is a levee-like elevated area with an adjacent borrow ditch. At the end of the road is a large square or rectangular ring levee to accommodate a drilling rig. Canal construction usually involves the dredging of wetlands to create a waterway and the deposition of the excavated material from the new waterway onto the adjacent wetlands. The purpose of the canal is to float a barge-mounted drilling rig to a drilling site. The "footprint" of a canal is a long rectangular body of water with adjacent shrub covered levees. At the end of such a canal is a wider waterbody, the "slip", which is larger in size to accommodate a drilling rig, accessory barges, and supply vessels. Both roads and canals cause impacts to wetlands by the direct conversion of wetland to water and shrub/scrub habitats as well as by indirect impacts through alteration of surface hydrologies. However, all things being equal, a board road causes fewer direct and indirect impacts than a canal, and restoration of a road is generally more successful.

        The Louisiana Coastal Resources Program (LCRP), which was established by the State and Local Coastal Resources Management Act of 1978 (SLCRMA), was federally approved in 1980 and began to process Coastal Use Permits on October 1, 1980. The state's coastal management program tracks the regulations of the federal Coastal Zone Management Act by incorporating procedures for citing energy facilities (including oil and gas wells) into the permitting process. This was originally done by the Coastal Management Division through the Coastal Use Guidelines that were established as regulations pursuant to SLCRMA (FEIS, 1980). These guidelines provided that the best practical techniques should be used for minimizing the impacts of oil and gas activities. The guideline most relevant to this concept is Guideline 10.2, which states:

"To the maximum extent practicable, the number of mineral exploration and production sites in wetland areas requiring floatation access shall be held to the minimum number, consistent with good recovery and conservation practices and the need for energy development, by directional drilling, multiple use of existing access canals and other practical techniques."

However, although the guideline language was clear, it soon became apparent to the CMD staff that the resources allocated to the CMD were not sufficient to insure compliance with the guideline, and that to do so would require the employment of a special level of expertise. Therefore, the Louisiana Department of Natural Resources, Coastal Management Division (CMD) created the permit review process now called Geologic Review. This process came about because of CMD's need for an independent evaluation of the geologic and engineering data presented by applicants to justify proposed well canals and board roads. In 1982, CMD obtained the assistance of the Louisiana Geological Survey (LGS) to provide technical advice and assistance on all aspects of petroleum operations in the wetland areas. This technical advisory role over time became the Geologic Review Procedure. The Geologic Review Procedure is a thorough review by an LGS petroleum engineer and petroleum geologist of an applicant's geologic, engineering and economic information. This data is presented at a meeting that includes the CMD analyst, the state commenting agencies, i.e. Department of Environmental Quality (DEQ) and Department of Wildlife and Fisheries (LDWF), federal commenting agencies, i.e. U.S. Fish and Wildlife Service (USFWS), National Marine Fisheries Service (NMFS), and the Environmental Protection Agency (EPA), the LGS consultants and the permit applicants personnel. After examining the data, the LGS consultants make a technical recommendation to the CMD on the feasibility of directionally drilling the well from an existing or less damaging site, of moving the surface location to a geologically equivalent site that is less damaging, or of changing the method or route of access to minimize wetland impacts. This forum allows applicants to meet the commenting agencies to discuss and explain their situation and technical requirements while being informed of the agencies' concerns. Typically, at the conclusion of a Geologic review meeting, a single recommendation on access and well placement is accepted by all participants which reduces losses to wetland habitat and permit delays (Johnston et al, 1989).

The CMD's original criteria for deciding whether or not a proposed drilling access route would be subject to the Geologic Review process were based upon the New Orleans District Corps of Engineers' (NOCOE) criteria for determining whether a proposed use was a major or minor activity. Under those criteria, an oil and gas canal of 500' (150m.) or less and a board road of 1500' (450m.) or less were in the minor category. Later, the NOCOE used these same criteria as the maximum allowable dimensions of the NOD-22 (oil and gas canal) and NOD-13 (board road) general permits.

Interagency cooperation between state and federal regulators was enhanced by the decision in 1984 of the NOCOE to join the Geologic Review process to evaluate permits under Section 404 of the Clean Water Act. This action brought the federal commenting agencies including the USFWS, NMFS, and EPA into the program. This action expanded the areas covered by the Geologic Review process outside the boundaries of the Louisiana Coastal Zone. The entry by the

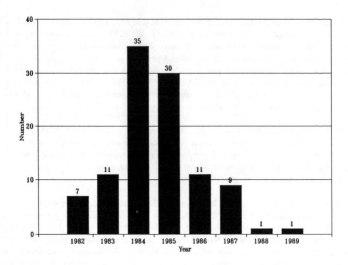

Figure 2: Frequency of requests for proposed 500' (150 m.) canals to CMD.   LGS/LSU

NOCOE into the program has allowed permit applicants to discuss proposed projects with both permitting agencies in a single meeting. In most instances, the agencies have agreed upon a single recommendation at each Geologic Review meeting.

The inclusion of the NOCOE has had a significant impact on the guidelines for having a Geologic Review Meeting. In a case of litigation involving the NOCOE, the policy of allowing a 500' (150 m.) canal without holding a Geologic Review Meeting was challenged, along with many other issues, in federal court by local and national environmental groups. As a result of this challenge in June 1987, both the NOCOE and CMD made a major change in determining when Geologic Review meetings were needed. The new policy stated that all proposed well canal locations and all board road locations that directly impacted wetland areas would require a Geologic Review Meeting.

This policy change had a significant and positive impact from the environmental view on proposed canal operations. **Figure 2** clearly shows that after this change, operators stopped requesting 500' (150 m.) canals. For instance, oil and gas operators only requested one 500' (150 m.) canal each in 1988 and 1989 and after Geologic Review, the permitting agencies required that both wells be drilled without dredging. This data infers that oil and gas companies were requesting a 500' (150 m.) canal whether or not they needed that length for access. When the NOD-22 General Permit was reauthorized in 1988, a new provision was added that required Geologic Review meetings for all wells that directly impact wetland areas. Also in 1987, CMD introduced its first general permits for oil and gas wetland access. The water access general permit (GP-2) allows the dredging of a slip 375' x 120' (112.5 m. x 36 m.) parallel to an existing oil and gas canal. The land access general permit (GP-3) allows the construction of a 300' x 300' (90 m. x 90 m.) ring leveed site where one side of the new ring levee is an existing road, levee or other ring levee. Authorizations issued pursuant to these general permits result in significant reductions in wetland

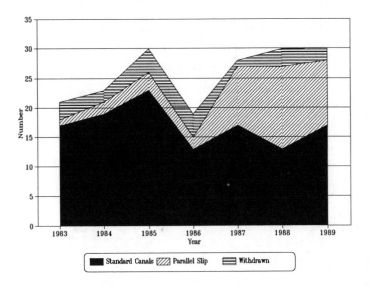

Figure 3: Breakdown by type CMD oil & gas access permits 1983-1989   LGS/LSU

impacts because the use of the Geologic Review process ensures that the least damaging access method and the smallest well-site location will be used. The use of general permits has created incentive for applicants to use the least damaging alternative for well access because general permits are approved rapidly compared to standard permit applications. The normal permitting time for a standard permit is 60 to 180 days, while a general permit can be issued in less than 7 days providing all requirements are met. The increased use of general permits is reflected by the increase in the number of parallel slips permitted by CMD (see **Figure 3**) since 1987. In 1988, 26% of all canal permits in the coastal zone directly impacting marsh areas were CMD general permits.

LGS, through research carried out in conjunction with the Geologic Review program, has recommended numerous changes in how oil and gas operations are carried out in wetland areas. These recommendations have led to significant cumulative reductions in wetland impacts. **Figure 4** shows the comparison of an 1982 well location that had a ring levee dimension of 500' x 500' (150 m. x 150 m.), impacting 5.74 acres, compared with the largest ring levee currently permitted, a 400' x 350' (120 m. x 105 m.), that affects only 3.21 acres and still allows the same access to oil and gas operations. Not all ring levees permitted are this large with the typical permitted ring levee being much smaller at 300' x 300' (90 m. x 90 m.) or 2.07 acres. Standardization of ring levee and slip sizes has led to savings in impacts on wetland habitat. **Figure 5** demonstrates the fact that the average size of wetland areas disturbed by dredging canals to oil and gas well locations has decreased significantly since this program was instituted.

The experience of the Geologic Review procedure has helped the permitting agencies to develop requirements for practices that further reduce impacts. The LGS technical consultants are

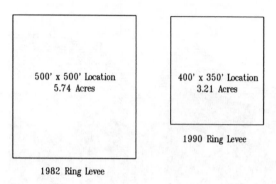

1982 Ring Levee

Figure 4: Comparison of ring levee sizes.   LGS/LSU

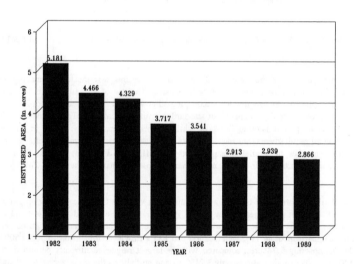

Figure 5: Average acres of wetlands permitted for oil & gas canals by CMD.   LGS/LSU

engaged in ongoing research to reduce the impacts of oil and gas operations in wetland habitats. This research includes alternate access methods, i.e. hovercraft and helicopter access, and improved

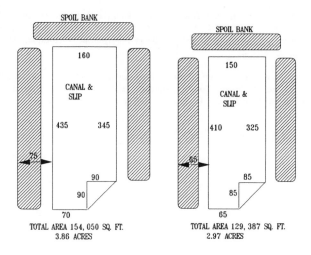

Figure 6: Comparison of proposed changes in slip configuration.   LGS/LSU

directional drilling methods, i.e. horizontal drilling. **Figure 6** represents a proposed change in slip configuration and canal width that is presently being reviewed by the commenting and permitting agencies before implementation.   This minor change will save .90 acres per well location and may save 25-30 acres of wetland habitat per year in South Louisiana.

Conclusions

The Geologic Review has been a very successful interagency program. Through continued cooperation and research, the Geologic Review Procedure is gradually reducing the impacts on a valuable wetland ecosystem.

REFERENCES

Johnston J.E.,III and J.D. Rives,III 1985. GEOLOGIC   REVIEW: A SUCCESSFUL
    PROCEDURE FOR THE MINIMIZATION OF THE IMPACT OF OIL AND
    GAS EXPLORATION AND PRODUCTION ON LOUISIANA WETLANDS.
    Proceedings of the Fourth Symposium on Coastal and Ocean Management,
    Baltimore, Maryland. pp 2521-2529

Johnston J.E.,III, J.D. Rives,III and D.M. Soileau. 1989. GEOLOGIC REVIEW: BETTER
    REGULATION THROUGH INTERAGENCY COOPERATION. Proceedings
    of the Sixth Symposium on Coastal and Ocean Management, Charleston, South
    Carolina. pp. 4264-4277.

Louisiana Coastal Resources Program. Final Environmental Impact Statement (FEIS). 1980.
    NOAA Office of Coastal Zone Management and LA. Dept. of Natural Resources,
    Washington, D.C. and Baton Rouge, LA. 190 pp. and Appendices.

RESULTS OF GEOLOGIC PROCESSES STUDIES OF BARRIER ISLAND EROSION AND WETLANDS LOSS IN COASTAL LOUISIANA

S. Jeffress Williams[1], Shea Penland[2], and
Asbury H. Sallenger, Jr.[3]

Abstract

The U.S. Geological Survey (USGS), as part of its Coastal Geology Research Program, is conducting geologic research throughout the delta plain of south central Louisiana. The objective is to provide basic information necessary to improve our understanding of the geological processes responsible for coastal erosion and wetlands deterioration. The Louisiana Geological Survey (LGS), Louisiana State University (LSU), U.S. Fish and Wildlife Service (FWS), and the Argonne National Laboratory (ANL) are cooperating partners in two studies which the USGS has undertaken.

The first study examines the function of coastal barrier islands in the protection of bays, estuaries, and wetlands from ocean waves, storm surge flooding, and saltwater intrusion. The USGS and LGS are mapping and interpreting the physical changes that have occurred over the past several thousand years on the inner continental shelf, along the barrier coastline, and in the wetlands landward of the barriers, paying particular attention to geomorphic evolution during the past century. A series of atlases and reports presenting the results are available. Additionally, a data base of 15,000 line-km of offshore geophysical profiles, 400 continuous vibracore samples, 565 grab samples, and digital hydrographic and shoreline data from maps and surveys of the deltaic plain over the past 136 years (1853-1989) have been incorporated into a computer Geographic Information System (GIS).

The second study is a comparative investigation of critical physical processes in the wetlands of representative sediment-deficient and sediment-rich basins in Louisiana. The Terrebonne basin, sediment-deficient with badly deteriorated wetlands, and the Atchafalaya basin, sediment-rich with an emergent and recently vegetated delta and healthy wetlands, typify the two

---

[1] U.S. Geological Survey, 914 National Center, Reston, VA 22092
[2] Louisiana Geological Survey, Box G, University Station, Baton Rouge, LA 70893
[3] U.S. Geological Survey, 600 Fourth St. South, St. Petersburg, FL 33701

extremes and were selected as study sites.   Joint field studies by USGS, LSU, and LGS were recently completed on both regional and local levels in Terrebonne basin, with the focus on:

   o Storm effects,

   o Freshwater and saltwater dispersal,

   o Fine-grained sediment dispersal,

   o Marsh deterioration,

   o Soils development, and

   o Subsidence and sea-level rise.

Similar field investigations in the Atchafalaya basin are underway in 1991. Included in this effort is a compilation of baseline data into a GIS network for use by coastal resource and management agencies.

Field investigations are also underway to evaluate the potential effectiveness of small-scale freshwater diversions from the Mississippi River as a means to mitigate wetlands deterioration. This study is being conducted at White's Ditch, just downriver from New Orleans.   Measurements of dispersion and retention of freshwater and sediments in the wetlands of Plaquemines Parish were initiated in February 1990 by Coastal Environments, Incorporated.

Introduction

Coastal erosion and wetlands deterioration are serious and widespread problems affecting all regions of the Nation.   All 30 of the coastal States are experiencing erosion, and the long term social, environmental, and economic consequences are major concerns to the coastal zone population as well as to resource managers.   As shown on the plate (Figure 1) depicting coastal erosion and accretion in the U.S. Geological Survey's *National Atlas of the United States of America* (1988), Louisiana has the distinction of experiencing the highest coastal erosion rates in the United States (Morgan and Larimore, 1957; Penland and Boyd, 1981; Morgan and Morgan, 1983; McBride and others, 1989).   Coastal erosion rates average - 4.2 m/yr with a standard deviation of 3.3; the rates of shoreline change range between + 3.4 m/yr and - 15.3 m/yr. The majority of erosion is concentrated on the barrier islands that front the Mississippi River delta plain.

Louisiana's average coastal erosion rate of 4.2 m/yr represents long-term conditions and is not representative of individual storm events.   Coastal erosion is not a uniform process; especially rapid erosion rates occur with the passage of major cold fronts, tropical storms, and hurricanes (Penland and

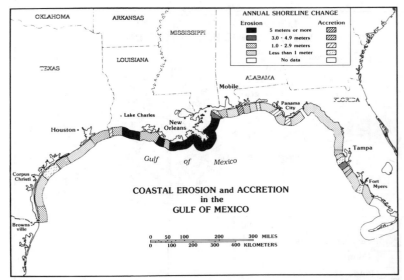

Figure 1. - *Map showing the extent and magnitude of coastal erosion around the five-state region of the Gulf of Mexico. (Based on U.S. Geological Survey Atlas, 1988.)*

Ritchie, 1979; Boyd and Penland, 1981; Ritchie and Penland, 1988; Dingler and Reiss, 1990). Field measurements have documented 20 - 30 m of coastal erosion during a single storm event lasting 3 - 4 days. In addition to erosion at the shoreline, the total area of Louisiana's barrier islands is decreasing rapidly. According to Penland and Boyd (1982), the coastal barriers lost nearly 50 percent of their surface area between 1880 and 1980.

The natural processes of wetlands degradation and wetlands destruction and alteration by public agencies and private landowners have resulted in the loss of more than 50 percent of the wetlands that existed in the contiguous United States at the start of European settlement over 200 years ago. These wetlands losses are continuing, and nowhere is the problem greater than in the Mississippi River delta plain of Louisiana, an area which accounts for an estimated 25 percent of the vegetated wetlands and 40 percent of the tidal wetlands in the 48 conterminous States. Louisiana is undergoing the greatest amount of wetlands loss and deterioration of any State in the Nation; an estimated 80 percent of the Nation's tidal wetlands loss has occurred in Louisiana, and by current estimates, approximately 100 km^2 are lost each year (Figure 2). These losses are the result of a combination of physical erosion by waves and currents as well as conversion of vegetated marsh to open water estuaries due to disintegration of the marshlands, sea-level rise, and regional subsidence (Williams and Sallenger, 1990).

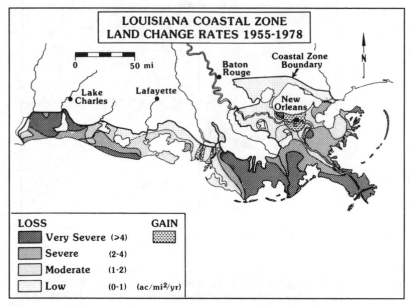

Figure 2. - *Map of the distribution and extent of coastal wetlands loss in Louisiana. (Based on van Beek and Meyer-Arendt, 1982.)*

Louisiana Delta Plain

The Mississippi River has been the dominant force controlling delta plain development and evolution along the north central coastline of the Gulf of Mexico. Interpretations of the geologic record from the coast and continental shelf of Louisiana show that over the past 6,000 to 8,000 years, large shifts in the course of the Mississippi River have occurred at roughly 1,000-year intervals (Figure 3). Such changes in the river's channel have been responsible for repeated cycles in the evolutionary development of the delta plain. Each cycle consists of coastal progradation during the delta building process followed by rapid sediment compaction, subsidence, and widespread coastal erosion and wetlands deterioration after the channel shift has occurred (Penland and others, 1990).

As part of these cyclic changes in the delta, sandy barrier islands form at the seaward margins of the delta plain. These barriers, in many cases, provide a buffer to the wetlands and estuaries from ocean waves and currents. With continued subsidence and a lack of adequate coastal sediment, however, the barriers undergo rapid erosion, at rates up to 20 m/yr, and eventually are broken into smaller, less protective segments exhibiting very low profiles and separated by wide tidal inlets. Ultimately, the coastal barriers are unable to maintain their subaerial geomorphology and become submerged sand bodies.

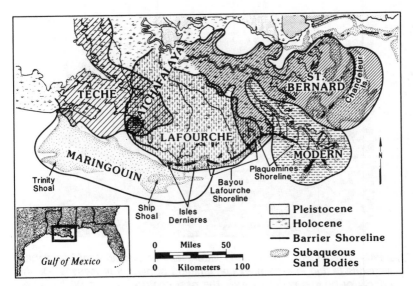

Figure 3. - *The Louisiana delta plain is the product of a series of deltaic lobes which resulted from shifts in position of the Mississippi River over the past 7,000 years. (From Frazier, 1967)*

A discussion of how the delta plain coast has rapidly changed and evolved since the mid-1800's is contained in McBride and others (1989). Examples of ancestral barrier coastlines can be seen on offshore seismic-reflection profiles and in core samples as buried sand bodies on the Louisiana continental shelf.

Effects of Human Activities

In addition to the natural geologic processes that cause coastal erosion and wetlands loss, human activities during the past century and, especially, in the past 50 years have had dramatic effects (Figure 4). For example, dam building on the Mississippi River and its tributaries since the 1930's has reduced the volume of sediment being transported by the river and, therefore, available to the wetlands (Meade and Parker, 1985). The massive levees that channel the flow of the Mississippi River for more than 1500 km are intended to enhance navigation and reduce flood potential, but they also have significant detrimental effects on the downstream delta plain and adjacent wetlands. Widespread and seasonal flooding that once provided sediments to build and maintain the wetlands no longer occurs. The sediment that is carried by the Mississippi River is now discharged far out on the continental shelf, rather than being widely distributed throughout the delta plain as it was before the levees and other engineering control structures were built.

Figure 4. - *Coastal land loss and the widespread deterioration of Louisiana's wetlands is due to a complex combination of natural processes and the effects of human activities over the past century. (From Williams and Sallenger, 1990)*

An additional human activity that contributes to wetlands loss in Louisiana is an extensive system of dredged canals and waterways. These serve as pipeline paths, access routes for hydrocarbon exploration and production, and waterways for boat traffic (Figure 4). Not only does dredging and maintaining these canals impact the wetlands, but many of them that open to the Gulf of Mexico enable saltwater to intrude brackish and freshwater wetlands, accelerating their deterioration. Other causes that are suspected to be important, but not well documented as yet, involve subsidence that is associated with the extraction of hard minerals and fluids in the shallow subsurface. For example, sulfur mining over salt domes has resulted in localized subsidence of several meters in just the past few decades (Hunt, 1990). Forced drainage, where marsh areas are diked and pumps are used to draw down the ground water, is a widespread agricultural and developmental practice which seems to contribute greatly to wetlands loss by increasing soil compaction and subsidence.

Information Needs - Research Results

The physical processes that cause coastal erosion and wetlands deterioration are extremely complex, highly varied, and still not particularly well defined or understood. The rates and magnitudes of future land loss, therefore, are not predictable with any high degree of confidence. Also, much debate still exists in the technical and scientific community about which of the natural and human-induced causes are most responsible for coastal erosion and wetlands loss.

Various measures and recommendations have been proposed to mitigate natural and man-made causes. However, considerable controversy also exists over some of the mitigation measures such as marsh management, river diversions, barrier island nourishment, and wetlands restoration. Much of the debate has to do with uncertainties in predicting the long-term success of these measures, all of which require varying expenditures of time and money to design, construct, and maintain.

Since 1986, the U.S. Geological Survey, working cooperatively with the LGS, coastal researchers at LSU, ANL, and the FWS, has been conducting research on erosion and wetlands loss in Louisiana. As shown in Figure 5, the region included in the study extends from the Chandeleur Islands west to the Atchafalaya River, a distance of approximately 300 km. The objective of these efforts is to improve our knowledge and scientific understanding of the geologic processes responsible for the erosion and land loss. Such information is critical for predicting future conditions and for developing a strategy to conserve and in some cases restore coastal Louisiana. To meet these information needs, field investigations are focused on:

o   Mapping and interpreting the recent geologic framework and evolutionary history of the delta plain, onshore, along the coast, and throughout the inner continental shelf. Obtained as part of this effort are a data base of 15,000 line-km of high resolution seismic profiles, 400 vibracore samples, 565 sediment grab samples, precision bathymetric data from 1880 to 1989, digital shoreline positions from 1853 to 1989, and tide-gauge records for measuring historic changes in relative sea level. Analyses of the shoreline data show that the barriers are eroding very rapidly by a process of in-place narrowing and inlet enlargement. Extrapolation of the historic record suggests that the Isles Dernieres barriers will convert to a submerged shoal by the early 21st century;

o   Measuring a wide array of meteorological and oceanographic parameters which influence sediment transport at the coast and in the nearshore zone. A long term monitoring station on the Isles Dernieres barriers provided high quality information, vital to deciphering the sediment budget for the Louisiana coast, during both fair weather and storm conditions (Sallenger and Williams, 1989a);

o   Comparing the geologic character and processes of the sediment-deficient Terrebonne basin in the Louisiana delta plain with the sediment-rich Atchafalaya basin. These comparative investigations are focusing on sediment compaction, sea-level rise, and land subsidence (Penland and Ramsey, 1990; Bailey and Roberts, 1990); effects of meteorological events such as hurricanes; dispersal of fine-grained sediments; movement of fresh and saline water; processes of physical erosion; and conditions required for soils to develop in wetlands. Field investigations on all these topics are complete for the Terrebonne basin, and analyses and evaluations are underway. Experiments to measure sediment flux in salt and brackish marshes in Terrebonne basin were carrier out at the same sites during winter cold front conditions and summer fair weather

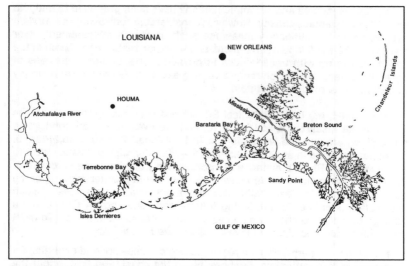

Figure 5. - Location map of the southern Louisiana delta plain encompassing areas included in the USGS coastal erosion and wetlands studies.

conditions. To decipher rates and variability of regional subsidence in the basin, five strike and five slip sections are being developed by a combination of deep borings and cores and instrumented subsidence monitoring stations. Similar field investigations are underway in the Atchafalaya basin.

o Developing a comprehensive coastal data base and using these data in a network of computer-based geographic information systems (GIS) available in Federal, State, and local agencies and private companies. All available baseline data are being incorporated. A listing of data sets judged by the Federal and State agencies and the research community to be important in addressing Louisiana's land loss problems is included in Table 1. Incorporating many of these data sets into a Louisiana GIS network is underway.

o Field investigations by scientists at Coastal Environments, Incorporated, were carried out over a one-year cycle on the Mississippi River at White's Ditch in Plaquemines Parish. The purpose was to assess and evaluate the potential effectiveness of small scale freshwater diversions on mitigating wetlands deterioration. Systematic measurements and observations were made on the dispersion and distribution of river water and suspended sediment; the results will aid in the design and operation of diversions under construction or planned (van Beek and others, 1990).

A listing of interim reports is contained in Sallenger and Williams (1989b).

Table 1. Ranking of most important data sets in coastal Louisiana.

1. 7.5 minute USGS quadrangle maps for the coastal zone
2. Spatial index of available photos and remote sensing data
3. U.S. Fish and Wildlife Service habitat change maps
4. Hydrography and hydrology
5. Geology and engineering framework
6. Mineral extraction information: textual and spatial
7. Land use maps
8. Biological surveys
9. Land elevation data, less than 5-foot contour intervals
10. Detailed soils maps
11. Point source discharge sites and records
12. Land loss maps
13. Weather records
14. Canal and pipeline data
15. Federal and State regulatory information specifically related to the physical environment
16. Shoreline history and geomorphology
17. Land cover information along with thematic mapper data
18. Land ownership records
19. Ecologically sensitive areas
20. Soil Conservation Service national resource
21. Census data from the TIGER files
22. Potable water sources
23. Recreational uses of the coastal zone
24. Economic zones and industrial sites

Summary

Coastal erosion and the deterioration and loss of valuable wetlands are widespread problems of national concern. The delta plain region of Louisiana, however, leads all other regions in land loss due to a combination of natural processes and a long history of human intervention and development. To provide the high quality scientific information necessary to understand and quantify the processes responsible, the USGS, working jointly with LGS, LSU, FWS and ANL, is undertaking studies of the barrier islands and inner shelf region and the coastal wetlands west of the Mississippi River delta. The 5-year study of the barriers is complete, and a variety of reports and atlas products are available. Field investigations in the wetlands of the sediment-starved Terrebonne and Barataria basins are complete, and results are being compared with similar field studies in the sediment-rich Atchafalaya basin. As interim and final results are obtained, they are published and made available to the coastal resource community.

224COASTAL WETLANDS

REFERENCES

Bailey, A.M. and Roberts, H.H., 1990, Geochemical indicators of subsidence in sediment, Terrebonne coastal plain, Louisiana (abs.): Transactions of the Gulf Coast Association of Geological Societies, v. 40, p. 1.

Boyd, R. and Penland. S., 1988, A geomorphic model for Mississippi delta evolution: Transactions of the Gulf Coast Association of Geological Societies, v. 38, p. 443-452.

Dingler, J.R. and Reiss, T.E., 1990, Cold-front driven storm erosion and overwash in the central part of the Isles Dernieres, a Louisiana barrier arc: Marine Geology, v. 91, p. 195-206.

Frazier, D.E., 1967, Recent deposits of the Mississippi River, their development and chronology: Transactions of the Gulf Coast Association of Geological Societies, v. 17, p. 287-311.

Hunt, J.L., 1990, Impact assessment of offshore sulfur mining subsidence on oil and gas infrastructures in M.C. Hunt, S. Doenges, and G.S. Stubbs, eds., Proceedings - 2nd Symposium on Studies Related to Continental Margins - A Summary of Year-Three and Year-Four Activities, p. 54-60.

McBride, R.A., Penland, S., Jaffe, B., Williams, S.J., Sallenger, A.H., and Westphal, K.A., 1989, Erosion and deterioration of the Isles Dernieres barrier island arc, Louisiana, U.S.A.: 1853 to 1988: Transactions of the Gulf Coast Association of Geological Societies, v. 39, p. 431-444.

Meade, R.H. and Parker, R.S., 1985, Sediments in rivers of the United States, National Water Supply Summary: U.S. Geological Survey Water-Supply Paper 2275.

Morgan, J.P. and Larimore, P.B., 1957, Changes in the Louisiana shoreline: Transactions of the Gulf Coast Association of Geological Societies, v. 7, p. 303-310.

Morgan, J.P. and Morgan, D.J., 1983, Accelerating retreat rates along Louisiana's coast: Louisiana Sea Grant College Program, Louisiana State University, 41 p.

Penland, S. and Ritchie, W., 1979, Short-term morphological changes along the Caminada-Moreau coast, Louisiana: Transactions of the Gulf Coast Association of Geological Societies, v. 29, p. 342-346.

Penland S. and Boyd, R. 1981, Shoreline changes on the Louisiana barrier coast: Oceans, v. 81, p. 209-219.

Penland, S. and Boyd, R., 1982, Assessment of geological and human factors responsible for Louisiana coastal barrier erosion, In Boesch, D.F., ed., Proceedings of the conference on coastal erosion and wetland modification in Louisiana: causes, consequences and options: Joint publication FWS/OBS82159, Baton Rouge, Louisiana Universities Marine Consortium/U.S. Fish and Wildlife Service, p. 20-59.

Penland, S. and Ramsey, K.E., 1990, Relative sea-level rise in Louisiana and the Gulf of Mexico: Journal of Coastal Research, v. 6(2), p. 323-342.

Penland, S., Roberts, H.H., Williams, S.J., Sallenger, A.H., Cahoon, D.R., Davis, D.W., and Groat, C.G., 1990, Coastal land loss in Louisiana: Transactions of the Gulf Coast Association of Geological Societies, v. 40, p. 685-699.

Ritchie, W. and Penland, S., 1988, Rapid dune changes associated with overwash processes on the deltaic coast of south Louisiana: Marine Geology, v. 81, p. 97-122.

Sallenger, A.H., Jr., and Williams, S.J., 1989a, Processes of barrier island erosion (abs.): Transactions of the Gulf Coast Association of Geological Societies, v. 39, p. 541.

Sallenger, A.H., Jr., and Williams, S.J., 1989b, U.S. Geological Survey studies of Louisiana barrier island erosion and wetland loss: an interim report of status and results: U.S. Geological Survey Open-File Report 89-372, 17 p.

U.S. Geological Survey, 1988, Map of coastal erosion and accretion, In National Atlas of the U.S.A.: Department of the Interior, U.S. Geological Survey, Reston, VA.

van Beek, J.L. and Meyer-Arendt, K.J., 1982, Louisiana's eroding coastline: recommendations for protection: Louisiana Department of Natural Resources, Baton Rouge, 49 p.

van Beek, J.L., Roberts, D.W., and Fournet, S., 1990, Abatement of wetland loss through diversions of Mississippi River water using siphons (abs.): Transactions of the Gulf Coast Association of Geological Societies, v. 40, p. 837.

Williams, S. J. and Sallenger, A.H., Jr., 1990, Loss of coastal wetlands in Louisiana - a cooperative research program to fill in the information gaps: U.S. Geological Survey Yearbook for Fiscal Year 1989, p. 48-52.

Landscape Simulation of Coastal Wetlands

Thomas W. Doyle[1]

Abstract

The U.S. Fish and Wildlife Service (Service) is steward of our nation's wildlife refuges, many of which include coastal wetlands. These coastal preserves are threatened by a host of human-induced and natural disturbances capable of disrupting the function and balance of these systems and the wildlife they support. Landscape simulation models are an innovative tool for assessing physical and biological changes in wetland systems. The National Wetlands Research Center (NWRC) has played an integral role in developing the Coastal Ecological Landscape Spatial Simulation (CELSS) model of major wetland basins in coastal Louisiana. This dynamic process-based model predicts the ecological succession of basin vegetation as altered by changes in regional hydrology and geomorphology. Technological developments are also being pursued to improve the portability and utility of CELSS. Applications include assessments of the impact of global climate change, coastal pollution, water control projects, and marsh management. The model is also being used to aid in the review of wetland permits and wildlife management plans.

Introduction

The U.S. Fish and Wildlife Service serves the public interest in conserving, protecting, and enhancing our nation's fish, wildlife, and their critical habitat. Over 90 million acres of lands and water are managed for wildlife in the National Wildlife Refuge System. A significant portion of the more than 450 refuges include coastal wetlands. Today, coastal environments are being threatened by a host of natural and anthropogenic

---

[1] U.S. Fish and Wildlife Service, National Wetlands Research Center, 1010 Gause Boulevard, Slidell, LA 70458.

disturbances. Human impacts from oil and gas activities, flood control projects, and urban development affect the natural function and balance of fragile ecosystems. Natural processes, such as subsidence, tropical storms, and impending climate change, may further amplify the detrimental effects of human intervention. Recognizing the gravity of wetland problems, the Service created the National Wetlands Research Center to spearhead research and development that address resource issues related to the protection, restoration, and management of wetlands. The Center has embarked on an aggressive program to develop and use simulation models to investigate the causes of wetland loss and to assess management alternatives for mitigating loss. These models predict how existing vegetation will respond to environmental change and assess how valued wildlife habitat may be best managed under changing environmental conditions. Our program is not limited to but includes modeling projects to examine the impacts of global climate change, coastal pollution, water control projects, and marsh management practice.

Global climate change, a prime threat to our biospheric reserves, best illustrates the value of landscape modeling. Global climate change is expected to severely modify present environmental conditions such that immigration, emigration, and extirpation of plant species may alter the quality and distribution of wildlife habitat in coastal environments. Of special concern to the USFWS are potential impacts to Federal refuges along our nation's coastline that provide food and shelter for millions of migratory birds. Coastal wetlands are at risk because of the unique complex of environmental factors and their projected change expected with global climate change. Located on the interface of land and water, wetlands will be influenced by sea-level rise, altered salinity conditions, increased tropical storms, and regional shifts in temperature and precipitation patterns (Titus 1988). The Southeastern United States is of major concern because of its relative susceptibility to these environmental consequences and its large proportion of our nation's wetland habitat, wildlife, and other economically important coastal resources. Gulf coast wetlands are especially vulnerable because of high subsidence rates and predicted sea-level rise that will increase flooding and saltwater intrusion into large freshwater expanses. Among the more vulnerable states, Louisiana contains nearly 30% of the coastal marshes in the conterminous United States (Turner and Gosselink 1975). For these obvious reasons, our modeling applications to date have focused on major hydrologic basins in coastal Louisiana. This paper reviews the development and utility of landscape models to address climate change and other environmental effects on coastal wetlands.

Ecological Simulation Models

Computer simulation models are an important adjunct to field and laboratory studies by providing a comprehensive means of integrating our knowledge and understanding of natural and physical systems into a holistic framework. They also offer the ability to interpolate and extrapolate between and beyond empirical observations that are constrained by space, time, and cost. Their greatest utility, however, is in addressing problems that are spatially and temporally extensive and dynamic. Because of the complexity and magnitude of some environmental problems, models offer our best available technology with which to assess the problem and solutions. For example, simulation models are particularly useful in evaluating environmental impacts involving either a variable or long-time span that otherwise cannot be tested reliably with experimental methods. The same holds true for spatial models that can account for processes and phenomena of varying scales that interact to control ecosystem structure and function within a given landscape.

Over the past two decades the computer models designed to simulate the dynamics of ecological succession have proliferated. At present, there are several hundred succession models for forest ecosystems alone (Shugart 1984). Succession models differ with respect to the biological unit that is simulated, be it the organism, population, community, or ecosystem. They also differ in mathematical design whereby the rate and direction of vegetation change is projected either as a function of probabilities or processes. Probability-based approaches rely on extensive data sets or stochastic functions to determine the statistical probability that one vegetation type will proceed another. In contrast, process-based models describe ecosystem change based on biological relationships, biotic and abiotic, that are independent of the resultant state condition and field data. Model applications, regardless of type, are generally limited to localized site and system-specific use and interpretation. To achieve landscape simulation requires that intra- and inter-community dynamics across an enlarged land area be linked with the physical environment at appropriate spatial and temporal scales.

Ecological Scale and Hierarchy

The need to consider ecological scale and hierarchy is paramount to addressing the local, regional, and global responses of wetlands to environmental change. Predicting the ecological impact of any disturbance is complicated by the myriad of interactions and processes coupling the biotic and abiotic components of natural systems. This problem is further compounded in that these mechanisms of change are also operating interdependently at different spatial and temporal scales. Every level of system organization, whether organism, population, community, ecosystem, or biome, is subject to a given suite of factors and processes that are unique to corresponding spatial and temporal conditions, and in turn influence ecological response at higher or lower levels of biotic organization. It is this complex mix of controlling factors that are operating at different space and time scales that confounds efforts to identify unifying concepts and methods for classifying and analyzing natural systems. Elucidating the impact of climate change or other environmental disturbances on wetland systems requires explicit consideration of the many spatial and temporal scales at which system responses occur and incorporation of the links and feedbacks between these scales as can be expressed with the development of landscape simulation models.

Spatially expansive and long-term scalar phenomena and the linkages involved at each system level cannot be readily addressed by applying experimental methods alone and thus require a hierarchial modeling approach. Experimental studies are limited by design to addressing questions at a single space or time scale. Models, on the other hand, offer the capability to integrate across all spatial and temporal scales. For example, large-scale disturbances such as hurricanes and fire are likely to increase with global warming trends (Emanuel 1987; Rind et al. 1989). However, the influence of these macro-scale events on the spatial and temporal heterogeneity in ecosystem response will be significant yet somewhat independent of localized species response to ambient climate. Therefore, a hierarchial approach to developing and linking ecological models is required to simulate time- and space-varying abiotic predictions and biotic responses of wetland species, communities, ecosystems, and landscapes. Accomplishing this link and capturing processes important at different spatial and temporal scales demand a landscape approach that combines ecological models with spatial models.

Linking Ecological and GIS Spatial Models

Geographic information systems (GIS) have become a versatile tool for capturing, analyzing, and displaying landscape features and attributes in a digital mapping format. Their utility has been expanded where GIS data bases are linked with dynamic simulation models to predict changes in the spatial and temporal heterogeneity of landscapes. The introduction of powerful desktop and main-frame computers has opened the way for equally powerful GIS, remote sensing, and modeling applications for

investigating broad-scale ecological problems. Dynamic
ecosystem models alone are restricted to a one-dimensional
view of single site applications. The combination of
dynamic simulation models within the spatial template of a
GIS data structure allows a two- or three-dimensional
approach to problem solving. Linking dynamic ecological
models with GIS spatial models provides a powerful tool for
predicting the succession of ecosystems as influenced by
landscape pattern and prevailing environmental and
anthropogenic influences.

        Typically, GIS data structures represent space as
polygons of defined origin, size, and shape, achieved in
either a raster- or vector-based format. Raster designs
employ a fixed grid scheme, while vector formats usually
describe complex polygons of irregular shape. Raster-based
models are easiest to develop because of their facsimile to
matrix and array structures common in computer programs.
Raster-based landscape simulation models estimate land unit
changes as a function of discrete events in time and space.
Few vector-based simulators have been developed because of
the difficulty of handling vector data structures and
establishing criteria for altering polygon (i.e., biological or land unit) size and shape. The National Wetlands
Research Center in cooperation with Louisiana State
University has been involved in developing a landscape
simulator that links a dynamic process-based ecosystem
model with a GIS data structure called the Coastal
Ecological Landscape Spatial Simulation (CELSS) model.

CELSS Model Description

        The CELSS model is a landscape simulator designed
for coastal wetlands (Costanza et. al. 1990; Sklar et. al.
1991). The model has been used to simulate the landscape
dynamics of two major estuarine complexes in coastal
Louisiana. CELSS has been applied to evaluate the impacts
of large water-control projects on wetland loss and change.
It employs a raster-based GIS spatial configuration linked
with a process-based ecosystem model of wetland dynamics.
The model projects system changes in both the physical and
biological characteristics of the basin landscape. This
dynamic process-based spatial simulation model predicts the
ecological succession of basin vegetation as altered by
changes in regional hydrology and geomorphology.

        Model Specifications and Initialization. Study
basins were parceled into a matrix of 1-km^2 cells. Eight
primary variables were monitored for each cell unit
including: water level, standing biomass, habitat type,
elevation, salinity, detritus, suspended sediments, and
nutrient levels. The Service's GIS data base of wetland

1983, and 1988 was used for model application and
validation. The basic biological unit of change was
habitat types that corresponded to the major vegetation
classes from forested swamps, emergent marsh (fresh,
intermediate, salt), upland, and open water. Climate data
included temperature, precipitation, river discharge, wind
direction and speed, tide, and salinity. Water budgets
were calculated for the basins based on runoff, climate,
and evapotranspiration estimates. A historical profile of
levee and canal construction within the basins was created
from oil and gas permitting applications to determine year,
location, and size of projects. Separate maps were
generated for each year of simulation from 1956 to present
to capture the changing landscape on a historical basis.
Individual cells were initialized at the start of each
simulation with respect to habitat type, salinity, and
elevation based on the 1956 data. Weekly time steps were
used to integrate changes in the physical and biological
components.

Hydrologic Functions. Water volume, flow, and distribution are projected from riverine, tidal, urban, and
atmospheric input on a cell-by-cell basis by using a "mass
balance" approach of water exchange across neighboring cell
boundaries. Levees impede overland flow as a function of
location and height, while canals facilitate flow as a
function of size and gravitational and wind forces. The
volume of water exchanged between neighboring cells is
calculated as the sum difference of forces specifying the
net directional flow of water and the resistance to overland flow by vegetation type, levee, and canal density.
Salinity is modeled largely as a function of water volume
and proximity from the Gulf of Mexico and tidal range.

Geodynamic Processes. Sedimentation, erosion,
resuspension, nutrient availability, subsidence, and
accretion influence the degree of predicted wetland growth,
degradation, and change. Suspended sediment concentrations
were calculated as a proportional estimate of the hydrologic flux between adjacent cells. Both organic and
inorganic sediment concentrations were monitored in
separate categories of deposited and resuspended material.
Each habitat type was assigned specific resuspension and
sedimentation rate coefficients. CELSS also calculates a
detrital component that is used in remineralization and
accretion processes that affect plant productivity and site
elevation. Net sedimentation and resuspension estimates
were used to adjust cell elevation with each iteration.
Levee structures and cell elevation were subject to
regional subsidence estimates.

Plant Productivity and Habitat Succession. Plant
productivity was tracked for each cell as influenced by

nutrient levels, remineralization, temperature, water level, and salinity. Optimal production curves were established for each habitat type. Both primary production and biomass accumulation were tracked by site location. Persistent water level and salinity conditions influenced the likelihood that any given cell and vegetation type might be succeeded by another vegetation type or revert to open water.

Model Validation and Application. Model results were validated by goodness of fit measures of matching habitat predictions from simulated output with that of actual records from latter year GIS data sets. Findings indicated that model fit was substantially high (80%-90%) and statistically valid. Model predictions of salinity, suspended sediment, water volume, nutrients, and primary production were verified on average to correspond with available field observations and historical trends.

Model applications included an evaluation of the impact of proposed U.S. Army Corps of Engineers levee, weir, and controlled freshwater diversion structures on basin water and vegetation dynamics. Simulations of these Federal water control projects allowed site-specific placement and assessment of their impact on wetland loss, gain, or change within the basin. Comparative analyses of model runs with and without the proposed structural facilities provided a basis for evaluating the potential effectiveness or drawbacks of project location and objectives.

Technology Development. The goals of the NWRC modeling program also include the development of new technologies that allow more comprehensive and realistic landscape models. The CELSS model already presents a computationally intense computer program that, in its original application, expended extensive processing time on supercomputer hardware. Current applications and codes have been modified to achieve supercomputing power from a microcomputer platform by using parallel processing technology, which increases the portability and accessibility of this model. Recent efforts have focused on assembling an expert system to allow non-modelers and end-users to capitalize on the utility of this research and management tool.

Research and Management Implications

Spatially explicit landscape simulation models like the CELSS model offer many more research and management applications than have been explored to date. Because landscape models offer the comprehensive ability to summarize spatial and temporal events into a single framework, they can be used to investigate the effect of past and modern land use and specific location, extent, and magnitude of disturbance events on ecosystem development and change. As an example, the effect of landscape pattern, fragmentation, and connectivity on habitat quality and distribution can be investigated to establish criteria concerning wildlife needs such as carrying capacity and management of a given area or refuge. Also, many attempts have been made to define and quantify cumulative impacts of combined activities and events that disturb wetland function and balance. Many landowners of coastal marsh are increasingly applying localized marsh management techniques (i.e., constructing levee and weir structures) to control land loss and to preserve land value. Federal wetland regulators will undoubtedly benefit from the availability of this technology for reviewing wetland permits and practices. The combined effect of numerous localized and independent marsh management projects may eventually become a serious regional problem requiring models to assess cumulative impacts and abatement measures. The impact of sediment dredging and spoil disposal in coastal environments can likewise be assessed by using integrated spatial modeling that identifies areas where spoil removal and deposits might be most beneficial. Models can also be used to determine the role of macroscale disturbance events (e.g., fire and hurricanes) in that the timing, intensity, and frequency of such events may influence species composition and distribution. The effect of oil spills whether inshore or offshore can be mimicked to assess the gravity and long-term impacts of spill size and location under given environmental conditions. Water quality issues related to urban and agricultural runoff on wetland response can also offer another application. The indirect effects of wildlife by selective- or over-grazing (e.g., such as nutria eatouts) might be important depending on the resiliency of preferred plant species and their role in ecosystem succession. These are but a few of the generic applications that are expected with the advent of tested and reliable landscape simulators.

Summary

The utility of landscape models for research and management illustrates their effectiveness and value. Spatial landscape models are a tool for linking the spatial and temporal variation of environmental change and vegetation dynamics. Coastal wetlands are under siege from both human-induced and natural disturbances. Landscape models can equip researchers, managers, regulators, and landowners with the means to understand our environmental problems and provide realistic means to preserve these valued resources--our nation's wetlands and wildlife.

234   COASTAL WETLANDS

References

Costanza, R., F.H. Sklar, and M.L. White. 1990. Modeling coastal landscape dynamics. Bioscience 40:91-107.

Emanuel, K.A. 1987. The dependence of hurricane intensity on climate. Nature 326:483-485.

Rind, D., R. Goldberg, and R. Reudy. 1989. Climate change and disturbance. Climate Change 14:5-37.

Shugart, H.H. 1984. Theory of forest dynamics: an investigation of the ecological implications of several computer models of forest succession. Springer, New York.

Sklar, F.H., M.L. White, and R. Costanza. 1991. The coastal ecological landscape spatial simulation (CELSS) model: user's guide and results for the Atchafalaya-Terrebonne study area. National Wetlands Research Center Open File Report 91.02. In Press.

Titus, J.G. 1988. Greenhouse effect, sea level rise and coastal wetlands. U.S. Environmental Protection Agency, EPA-230-05-86-013, Washington, DC.

Turner, R.E., and J.G. Gosselink. 1975. A note on standing crops of Spartina alternifolia in Texas and Florida. Contrib. Mar. Sci. 19: 113-118.

Managing Louisiana Marshes--An Experimental Approach

E. C. Pendleton, A. L. Foote,
and G. R. Guntenspergen[1]

Abstract

Although marsh management is practiced on a large scale in coastal Louisiana, the effects of various practices on wetland productivity and maintenance and on fish and wildlife are not well understood and have been widely disputed. The typical management plan is designed to maintain or improve deteriorated brackish to freshwater marshes through semi-impoundment and manipulation of water levels by using weirs and flapgated structures. Few studies to date have systematically assessed how processes critical to marsh maintenance are influenced by these practices; thus, little quantitative information is available on how effective these measures actually are. This paper will present an approach being used by the U.S. Fish and Wildlife Service to address the effectiveness of marsh management on Louisiana's coastal wetlands.

Introduction

Louisiana is experiencing the largest loss of coastal wetlands in the nation. Forty percent of the coastal wetlands in the contiguous 48 states are in Louisiana; these are fragmenting and converting to open water at a rate estimated as high as 150 km^2 per year (Salinas et al. 1986). Losses are greatest in brackish and fresh marsh zones and are the result of a complex matrix of natural and human causes that includes regional subsidence and sea level rise (Dozier et al. 1983; Barnett 1984); alteration of natural hydrology by an extensive network of canals (Turner 1987; Turner and Rao 1990); disruption of natural patterns of sediment delivery by Mississippi River and canal levees (Allen and Hardy 1980; Templet and Meyer-

[1]U.S. Fish and Wildlife Service, National Wetlands Research Center, 1010 Gause Boulevard, Slidell, LA 70458.

Arendt 1988); increased water salinities (Salinas et al.
1986); waterlogging (Mendelssohn and McKee 1988); and
herbivory (Fuller et al. 1985).

Restoration, creation, and maintenance of coastal
wetlands to curtail wetland loss have been identified
through the Coastal Wetlands Conservation and Restoration
Plan (Governor's Wetlands Task Force 1990) as some of the
most important environmental issues in Louisiana today.
This joint State and Federal plan is recognition that
continued deterioration and decline of these wetlands are
problems of national significance as well. Wetland losses
threaten nationally important waterfowl, fishery, and
furbearer resources.

Marsh management has been proposed as the primary
means of reversing wetland loss in coastal Louisiana,
particularly in areas that are not affected by large-scale
diversions of Mississippi River water planned by the U.S.
Army Corps of Engineers. At least 165 applications for
permits to manage over 200,000 ha of wetlands have been
submitted to the State of Louisiana since 1980 (Cahoon et
al. 1990). Management proposed or underway involves
passive or active control of water levels in impounded or
semi-impounded marshes. Other practices include
introduction of fresh or sediment-laden waters from
upstream areas, marsh burning to encourage the growth of
plants beneficial to furbearers and waterfowl, and
construction of sediment trapping and containment devices.

Although many now believe the only way to save
Louisiana's deteriorating wetlands is through intensive
marsh management programs, the effects of various
management practices on wetland productivity and marsh
maintenance, and on fish and wildlife, are not well
understood and have been widely disputed (Duffy and Clark
1989). A controversy exists within the ranks of marsh
managers and scientists on the extent to which marsh
management practices are effective at retarding or
overcoming wetland loss.

Few investigations have systematically assessed how
processes critical to marsh maintenance are influenced by
management practices. A notable exception is contained
within the most comprehensive study to date of the status
of marsh management in Louisiana, conducted by
investigators at Louisiana State University (Cahoon and
Groat 1990). Their findings, based on analysis of
sequential aerial photographs, indicate that some marsh
management projects are working while others are not.
Field monitoring of two plans, one in the Mississippi River
Deltaic Plain and one in the Louisiana Chenier Plain,
indicated different and complex responses to similar

management approaches. According to their findings, management success (i.e., the maintenance and increase of vegetated emergent marsh area) may depend on location, local hydrology, sediment type, type of management, degree of control of water levels, meterological conditions, and other factors.

Wetland loss is so acute in Louisiana that management actions have necessarily preceded comprehensive research. Managers are required by State and Federal permits only to measure periodically water levels and salinities within and outside managed areas, and to visually assess changes in plant communities. Thus, little quantitative information is available on the effectiveness of these measures. If a policy of large-scale marsh management in Louisiana is to be followed, it is imperative that any marsh management plans be implemented on a firm scientific basis with a complete understanding of the ways in which management actions alter natural processes and wetland vegetation.

Description of Research

The U.S. Fish and Wildlife Service's National Wetlands Research Center has embarked on a multiyear, multidisciplinary research program to study the effectiveness of active marsh management in the brackish marshes of the Mississippi River Deltaic Plain. The broad objective of the research is to characterize critical ecological processes, structure, and functions of marshes and to compare these aspects in marshes under water level management to those in adjacent unmanaged marshes. Although marshes in Louisiana are managed for multiple purposes, our emphasis will be on determining the extent to which revegetation of deteriorated areas and retarding of habitat conversion to open water occurs as a result of hydrologic manipulation.

Site selection and description. A steering committee of academic researchers and Federal, State, and private managers assisted us in identifying areas of brackish (meso- to oligohaline) marsh in the Mississippi River Deltaic Plain that showed evidence of widespread marsh conversion to open water but no prior management. These sites were screened for availability of suitable control areas similar in vegetative cover, salinity, hydrologic patterns, sediments, subsidence, and recent habitat change history; lack of near-term plans for disruptive or compromising development activities; and logistical considerations.

After intensive screening and site visits, four pairs of sites (Figure 1) were selected. Two replicate

pairs of study sites are located in each of two hydrologic
basins, one relatively sediment-enriched because of its
proximity to the Atchafalaya River (the Terrebonne Basin;
Figure 2), and one relatively sediment-poor (the Barataria
Basin; Figure 3). One of each pair of study sites will be
completely enclosed with levees with a water control
structure (Figure 4) sited and installed to maximize the
effectiveness of water removal during drawdown periods.
The other site in each pair serves as an unmanipulated
control.

The sites are brackish marshes dominated by
Spartina _patens_ and _Scirpus_ _olneyi_. They are microtidal,
with tidal ranges averaging less than 0.6 m, and are highly
influenced by dominant southeasterly winds in the summer
and northwesterly winds in the winter. We have calculated
land loss rates on these sites averaging 0.46 ha/yr between

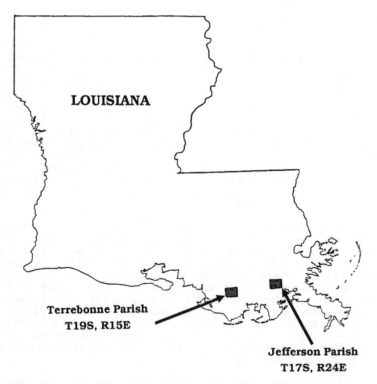

Figure 1.   Marsh management study sites in Louisiana.

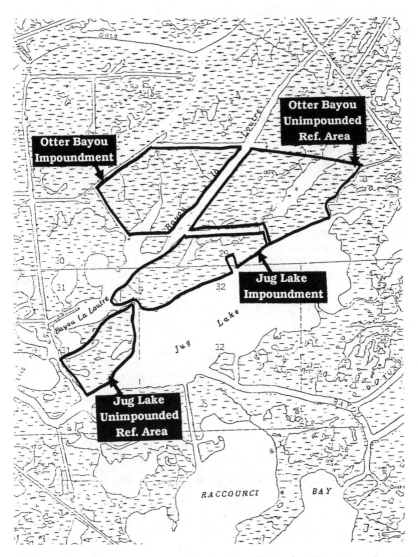

Figure 2. Marsh management study sites in the Terrebonne Basin, Louisiana.

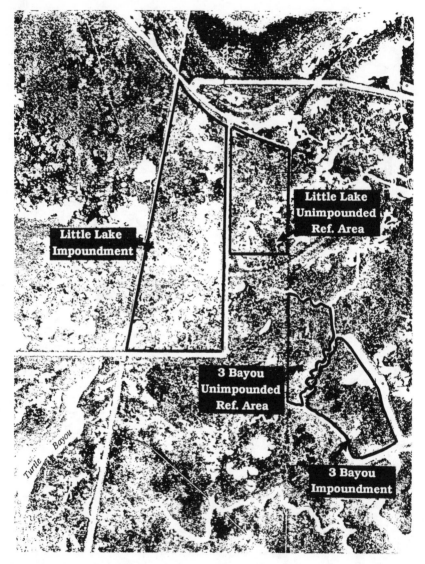

Figure 3.   Marsh management study sites in the Barataria
Basin, Louisiana.

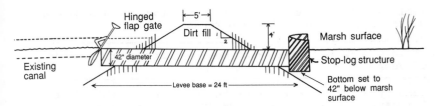

Figure 4.   Design of water-control structure used for controlled drawdowns of marsh management units.

1956 and 1978, or about 1% loss per year. Management units cover from 49 to 178 ha; because they are smaller than most public or private units, they may be properly considered to be mesocosms of current management activity.

Management protocol. Increasingly, the management method of choice for Louisiana is active water level control. As opposed to passive management involving the placement of fixed-crest weirs in canals and levees to retard the intrusion of higher salinity waters, active management allows water levels to be manipulated through the seasonal adjustment of the water control structure. In 1992 and for the following 2 years we will follow a protocol of water-level manipulation that incorporates features common to marsh management plans currently receiving regulatory approval through the Louisiana Department of Natural Resources Coastal Management Division and the U.S. Army Corps of Engineers (Table 1).

Our water-control schedule reflects the desire of marsh managers and regulators to accommodate multiple uses and multiple goals. Water will be drawn down by opening the water-control structures in the spring to allow drainage during the passage of frontal events and at other times of local low-water conditions. Drawdowns are maintained during peak plant growth periods in spring and summer. In the winter, water-control structures will be closed and water levels maintained at marsh level to promote waterfowl use. Control structures will be reopened in late winter to maximize access to fishery organisms using the wetlands as nurseries. Our protocol of annual drawdowns represents "Phase I" marsh management; in "Phase II," drawdowns are limited to 1 year in 3.

Experimental approach. Both sites are being monitored and characterized in 1991 for 1 year prior to

Table 1. Water level manipulations for wetland impoundments in Louisiana deltaic marshes.

Time of year	Water Management Goal	Structure position
January - February	Maximum water exchange Estuarine organism exchange. Little saltwater intrusion.	Weir crest 12" below marsh level. Flapgates open.
March - June	Draw down for emergent plant germination, growth and reestablishment.	Weir crest 12" below marsh level, flapgates down, water flowing out only.
July - September	Continue plant growth, minimize turbidity.	Weir crest remains 12" below marsh level but flap-gate open unless salinity exceeds 10 ppt.
October-December	Retain sufficient water in ponds for waterfowl use.	Weir crest set 6" below marsh level. Flap-gates open if salinity below 5 ppt.

beginning management, a feature unique in research on marsh
management to date. Baseline data on environmental variables affecting plant growth, marsh loss, and vertical
marsh accretion are taken along five 15-m transects at each
site. We measure factors related to hydrology (including
water levels and precipitation), plant growth (including
surface and sediment pore water salinity, sulfide concentrations, sediment reduction-oxidation potentials, nutrient
concentrations, and sediment temperature) and sediment
accretion (including bulk density, organic matter content,
and surface elevations) to establish temporal and spatial
variability as a basis for ascertaining and comparing
changes wrought during later management. Measurements are
made at various times along transects or intensively at
discrete sampling stations with continuously recording
probes. These measurements will continue for the duration
of the study, through a management period beginning in
spring 1992 and currently proposed to end in fall 1994.
Further studies may be developed to document longer term
effects.

    During the premanagement and management periods, we
will focus on processes vital to maintaining or restoring

emergent vegetation: short- and long-term sedimentation processes and mechanisms, aboveground and belowground primary production and decomposition rates that are related to sediment formation, and plant community change that may be a consequence of changing hydrologic regimes. Habitat used by wintering waterfowl and wading birds will also be assessed by semi-monthly aerial monitoring. As with environmental variables, we are comparing these processes and functions both before and during management to those in unmanaged areas.

Comparisons of managed and unmanaged areas form the framework of our study; however, several additional smaller scale experimental manipulations will be made on the sites to address various aspects of marsh management and wetland loss. We will examine impacts of herbivory by nutria on plant production using a series of exclosures. In conjunction with this experiment, small plots of marsh will be burned to investigate the relation among burning, herbivory, and organic matter accumulation. We also hope to manipulate sediment additions to other small plots and track plant production responses.

Analysis of variance will be used to determine whether changes in any of the dependent variables are associated with treatment effects. The experimental design to be used appears in Table 2.

In addition to field activities, impacts of our experimental marsh management will be assessed through larger-scale syntheses provided by aerial photography and simulation modeling. Aerial overflights will be made annually, and wetland changes will be related to short-term monitoring of ecosystem processes and assessed by photo-interpretation and analysis with an ARC-INFO geographic information system (GIS). GIS data bases and field data collected during the course of the study will form the basis for construction of site-specific simulation models (Costanza et al. 1990) to predict impacts of future management actions and identify modifications or improve-ments of existing management practices.

Summary

Wetland loss in Louisiana is an acute problem with State and national consequences. Marsh management by impoundment and hydrologic manipulation is viewed as a possible solution; management is proceeding, however, with a serious lack of comprehensive, long-term scientific data. Management may be much improved by a better understanding of the complex interplay of physical and biological effects of profound hydrologic alterations. A comprehensive re-search program undertaken by the National Wetlands Research

Table 2.   Statistical model used to compare managed and unmanaged sites.

MODEL

$$Y_{ijklm}=u + S_i + E_j + T_k + (S*E)_{ij} + (S*T)_{im} + (E*T)_{jm} +$$

$$(SET)_{ijk} + P_{l(ij)} + X_{m(ijl)} + Q_{n(ijlm)} + e_{ijklmn}$$

where Y = the dependent variable, u = the grand mean, S = regions, P = replicate site within region, E = treatment, X = transect within site, Q = quadrat within transect, T = time and e = error. The appropriate degrees of freedom associated with this model are:

SOURCE OF VARIATION		DF
Geographic region	(S)	S-1
Marsh management treatment	(E)	E-1
S*E		(S-1)(E-1)
T		(T-1)
S*T		(S-1)(T-1)
E*T		(E-1)(T-1)
S*E*T		(S-1)(E-1)(T-1)
P(SE)		(P-1)(SE)
X(SEP)		(X-1)(SEP)
Q(SEPX)		(Q-1)(SEPX)
Total		SETPXQ-1

Center, U.S. Fish and Wildlife Service, is one step toward achieving that understanding.

Acknowledgments.   Don Cahoon, Darryl Clark, Alan Ensminger, Greg Linscombe, Irv Mendelssohn, Ron Paille, John Reddoch, David Reece, Ric Ruebsamen, Robert Twilley, James Winston, and John Woodard serve on our steering committee and provided many valuable suggestions and review comments.   In-house reviews of this project were made by Gerald Grau, Janet Keough, Thomas Michot, Hilary Neckles, Robert Stewart, and Mary Watzin.   Bruce Pugesek provided a valuable statistical review.   Ron Paille provided special assistance in visiting and evaluating field sites.

References

Allen, K.O., and J.W. Hardy. 1980.   Influence of navigational dredging on fish and wildlife: a literature review.   U.S. Fish Wildl. Serv. Biol. Serv. Program FWS/OBS-80/07. 81 pp.

Barnett, T.P. 1984. The estimation of "global" sea level change: a problem of uniqueness. J. Geophys. Res. 89(C5):7980-7988.

Cahoon, D.R., and C.G. Groat, eds. 1990. A study of marsh management practice in coastal Louisiana, Vol. III, Ecological evaluation. Final report submitted to Minerals Management Service, New Orleans, LA. Contract No. 14-12-0001-30410. OCS Study/MMS 90-0075. 423 pp.

Cahoon, D., R. Hartman, and F. Zeringue. 1990. Chapter 7, Marsh management plan profile. Pages 205-249 in D.R. Cahoon and C.G. Groat, eds. A study of marsh management practice in coastal Louisiana, Vol. II, Technical description. Final report submitted to Minerals Management Service, New Orleans, LA. Contract No. 14-12-0001-30410. OCS Study/MMS 90-0075.

Costanza, R., F.H. Sklar, and M.L. White. 1990. Modeling coastal landscape dynamics. Bioscience 40:91-107.

Dozier, M.D., J.G. Gosselink, C.E. Sasser, and J.M. Hill. 1983. Wetland change in southwestern Barataria Basin, Louisiana, 1945-1980. LSU-C9L-83-11. Coastal Ecology Laboratory, Center for Wetland Resources, Louisiana State University, Baton Rouge, LA. 102 pp.

Duffy, W.G., and D. Clark, eds. 1989. Marsh management in coastal Louisiana: effects and issues--proceedings of a symposium. U.S. Fish and Wildlife Service and Louisiana Department of Natural Resources. U.S. Fish Wildl. Serv. Biol. Rep. 89(22). 378 pp.

Fuller, D.A., W.B. Johnson, C.E. Sasser, and J.G. Gosselink. 1985. The effects of herbivory on vegetation on islands in the Atachafalaya Bay, Louisiana. Wetlands 4:108-114.

Governor's Wetlands Task Force. 1990. Coastal Wetlands Conservation and Restoration Plan (Fiscal Year 1990-91). Resubmitted to the House and Senate Committees on Natural Resources. April 1990. Office of the Governor, State of Louisiana, Baton Rouge. 149 pp.

Mendelssohn, I.A., and K.L. McKee. 1988. *Spartina alterniflora* die-back in Louisiana: time-course investigation of net primary production. Ecology 49:147-149.

Salinas, L.M., R.D. DeLaune, and W.H. Patrick, Jr. 1986. Changes occurring along a rapidly submerging coastal area: Louisiana, USA. J. Coastal Res. 2(3):269-284.

Templet, P.H., and K.J. Meyer-Arendt. 1988. Louisiana
wetland loss: a regional water management approach to the
problem. Environ. Manage. 12(2):1167-1179.

Turner, R.E. 1987. Relationship between canal and levee
density and coastal land loss in Louisiana. U.S. Fish
Wildl. Serv. Biol. Rep. 85(14). 60 pp.

Turner, R.E., and Y.S. Rao. 1990. Relationships between
wetland fragmentation and recent hydrologic changes in a
deltaic coast. Estuaries 13:272-281.

Overview of the Federal Coastal Wetlands
Mapping Effort

Sari J. Kiraly *

Abstract

   The overall Federal effort to map the Nation's coastal
wetlands was examined by the National Ocean Pollution
Policy Board's interagency Habitat Loss and Modification
Working Group. This technical committee was formed by the
National Ocean Pollution Policy Board pursuant to
priorities set forth in the Federal Plan for Ocean
Pollution Research, Development, and Monitoring: Fiscal
Years 1988-1992, and was charged to address habitat-related
recommendations presented in that Plan. In addressing the
Plan's recommendation that the frequency of mapping coastal
habitats should be increased at the Federal level in order
to adequately document rate of loss, the Working Group
surveyed a series of Federally-funded coastal wetland
mapping programs during a two-day workshop. A report
consisting of nineteen manuscripts describing these
programs plus an overview assessment and set of
recommendations is currently in press. This paper
discusses the Working Group's project and summarizes the
programs and recommendations contained in that report.

Introduction

   Fundamental to studying and managing the Nation's
coastal wetland habitats is the development of a
comprehensive data base which documents the extent of
remaining wetlands and also indicates rate of wetland loss.
Using various technologies, the mapping of wetlands
provides an important basis for developing such an
inventory and assessing trends in this loss. This paper
describes a project undertaken by the National Ocean
Pollution Policy Board's Habitat Loss and Modification

----
* NOAA, Office of the Chief Scientist, National Ocean
Pollution Program Office, Room 625, Universal Building,
1825 Connecticut Avenue, NW., Washington, DC 20235

Working Group to examine what the Federal agencies are
doing to map the Nation's coastal wetland habitats and to
provide an assessment of that effort plus a set of
recommendations for improving it.

Background

    In the spring of 1989, the National Ocean Pollution
Policy Board (the Board) established the Habitat Loss and
Modification Working Group, an interagency technical
committee, pursuant to priorities set forth in the Federal
Plan for Ocean Pollution Research, Development, and
Monitoring: Fiscal Years 1988-1992 (Federal Plan). The
Working Group is jointly chaired by representatives from
the National Oceanic and Atmospheric Administration's
(NOAA) National Marine Fisheries Service and the Department
of the Interior's U.S. Fish and Wildlife Service (USFWS).
The activities of the Working Group are coordinated through
NOAA's National Ocean Pollution Program Office, which also
directed the preparation of the Federal Plan.

    In the Federal Plan, six broad goals of the National
Marine Pollution Program are identified. One of these
goals is to understand the effects of losing or modifying
marine habitats as a result of human activities. To assist
the Federal agencies in attaining this goal, information
gaps are identified and a set of recommendations presented.
The Board charged the Working Group to address the habitat-
related recommendations set forth in the Federal Plan by
performing implementable projects that would result in
products useful to the Federal agencies in planning and
conducting their habitat programs to meet the goal.

    Recommendations which address habitat loss and
modification in the Federal Plan include the need to
frequently update the Nation's coastal wetland maps in
order to accurately assess trends in wetland habitat loss
in a timely fashion. Recognizing the importance of a
fundamental coastal wetland acreage data base in studying
and managing coastal habitats, the Working Group proceeded
to examine the Federal effort in this area.

    In accomplishing the project, a workshop was conducted
that included persons representing Federally funded coastal
wetlands mapping programs. The workshop took place in
December 1989 at the USFWS National Wetlands Research
Center in Slidell, Louisiana. The papers presented at the
workshop are contained in a report currently in press at
the USFWS to be published as Biological Report 90 (18)

(Kiraly et al., eds., in press). The report contains nineteen manuscripts describing a series of Federally funded coastal wetland mapping efforts, plus the Working Group's conclusions and recommendations as to how the overall Federal effort in coastal wetlands mapping could be improved so that the status and trends of the Nation's coastal wetlands are comprehensively documented in a timely fashion.

Summary of Major National Programs

The Working Group concluded that of the Federal coastal wetland mapping efforts surveyed, two are currently designed to map coastal wetlands on a comprehensive, Nation-wide basis. These are the USFWS National Wetlands Inventory and NOAA's habitat mapping under its Coastal Ocean Program.

USFWS National Wetlands Inventory (NWI)

The most extensive on-going National wetlands mapping program is the USFWS NWI. This program provides the most comprehensive inventory of the Nation's inland and coastal wetlands, and serves as the basis for many other Federal and state efforts. The NWI began in 1975 to generate detailed wetland maps (based on Cowardin et al., 1979), and reports on wetland status and trends. Using conventional aerial photography, the NWI has produced over 30,000 wetland maps, including over 5,300 detailed 1:24,000 scale maps covering 100 percent of the coastal wetlands in the lower 48 states. One percent of the coastal wetland maps have been digitized for inclusion as a national mapping data base category in the National Digital Database under the supervision of the U.S. Geological Survey.

Based on a national sampling grid of stratified random samples, the NWI has developed a status and trends analysis to document losses and gains in the Nation's wetlands. As legislatively mandated by the Emergency Wetlands Resources Act of 1986, a national status and trends report for the mid-1960's to mid-1970's has recently been updated, with future updates to be prepared every 10 years.

NOAA Coastal Ocean Program

NOAA is developing a comprehensive, nationally standardized information system for land cover and habitat change in the coastal region of the United States as part of its Coastal Ocean Program. Under this program satellite

imagery, aerial photography, and surface geographic data
will be interpreted, classified, analyzed and integrated
within a Geographic Information System (GIS) to delineate
coastal wetland habitats and adjacent uplands, and monitor
changes on a cycle of one to five years. Maps will be
spatially registered digital images, allowing land cover
change to be detected in a pixel by pixel comparison of
different time periods. Also, maps for a given period will
be synoptic, based on satellite images or aerial
photographs collected over short (days-weeks) time
intervals. This type and frequency of information will
allow linkages between wetlands and the distribution,
abundance, and health of living marine resources to be
determined.

Operational protocols for delimiting emergent and
submergent coastal wetlands are being developed through a
series of interagency workshops and meetings. Current
efforts include a change analysis (1984-1988/1989) for
emergent coastal wetlands and adjacent uplands of
Chesapeake Bay using satellite imagery, and mapping
submerged aquatic vegetation (SAV) in North Carolina using
aerial photography at a scale of 1:12,000 and 1:24,000.
The intent of the program is to eventually map all coastal
regions of the United States.

Summary of Other Major Federal Programs

Environmental Protection Agency (EPA)

EPA wetland mapping activities rely, to a large
extent, on the mapping convention developed by the USFWS
NWI program, and in most cases directly use NWI maps and
NWI mapping capabilities. Wetland mapping has been
supported by the EPA through the Clean Water Act Section
404 and Superfund programs. There are two basic types of
wetland mapping activities under these programs: 1)
comprehensive planning under the Section 404 program,
referred to as "advance identification" (ADID), which is
intended to direct development away from the most valuable
wetlands, and 2) specific studies of certain identified
Superfund sites, which focus on wetland boundary changes
over time, with historical data often providing the goal
for restoration of the site to its original condition.

The EPA is initiating an Environmental Monitoring and
Assessment Program (EMAP) to characterize the condition of
the Nation's ecological resources on regional and national

scales and over long periods of time. The wetland resource component of EMAP will develop a program to assess the status and trends of wetland condition and extent. The proposed EMAP sampling design calls for selection of 30 representative 40 km^2 sites within each of 11 near-coastal geographic regions. Each year, 25 percent of these sites will be visited; sampling will be taken from plots within each site to determine habitat condition.

NOAA Coastal Wetlands Inventory

In 1989, NOAA's National Ocean Service and National Marine Fisheries Service completed a comprehensive Coastal Wetlands Inventory of estuarine drainage areas of the United States. Using a 45-acre grid sampling technique to quantify existing NWI wetlands maps which were based on aerial photographs from 1971 to 1985, data was entered into a GIS data base which displayed and calculated acreage summaries by NWI map, county, state and/or estuary. The data base, which contains 5,290 NWI maps and presents data on 507 counties and 92 estuaries, has proved useful in providing a summary of wetland distribution and abundance across large geographic areas.

U.S. Geological Survey (USGS)

Although the USGS does not conduct a wetlands mapping program per se, its National Mapping Division provides standard topographic maps at specified scales, as well as a diversity of cartographic, geographic, and remotely sensed data, products and services, including wetlands information. Many Federal and state programs rely on its primary map series as a basis for site-specific wetland and other environmental studies and on the technical assistance it provides to Federal agencies in establishing their GIS capabilities for the development of wetlands data bases. The Division has prepared 1:24,000 scale topographic maps covering most the Nation and is considering developing a new series of land use and land cover maps at the 1:100,000 scale. In addition, cooperative efforts with the U.S. Department of Agriculture Soil Conservation Service and the NWI should result in image base maps and state-of-the-art GIS's to aid in the study of wetlands.

The USGS Water Resources Division collects and disseminates, in written and digital formats, ground-water and surface-water hydrologic information pertaining to tidal and non-tidal wetlands in order to complement the two-dimensional information provided by wetland maps. In

addition, the Geologic Division collects, interprets, and
disseminates basic geologic information on inland and
coastal wetlands. Much of this field data is displayed on
thematic maps and includes information on the three-
dimensional structure of wetlands.

Summary of Major Regional and Federal/State Cooperative Mapping Programs

Chesapeake Bay Program

Under the Chesapeake Bay Program, which is a joint
effort among a number of Federal agencies and the states
bordering the Bay, SAV in the Bay has been surveyed by the
Virginia Institute of Marine Sciences (VIMS). Using
standard aerial photographic techniques at a scale of
1:24,000, VIMS has mapped SAV on a bay-wide basis five
times between 1978 and 1987. In addition, data from
photointerpretation of the imagery have been entered and
stored on a VIMS GIS. The result of these efforts has been
a temporal delineation of SAV's, providing the basis for
long-term trends analysis on the distribution and abundance
of this resource in Chesapeake Bay.

Florida

Habitat mapping and trend analysis are key components
of the Florida Department of Natural Resources, Marine
Research Institute's digital ecosystems data base, which
was begun through NOAA's Coastal Zone Management Program.
Based on a combination of conventional aerial photography
and satellite images, an efficient cost-effective mapping
program has been developed. State-of-the-art techniques
are used for image analysis, resulting in highly accurate
maps. In addition, a data base for trend analyses is being
created by incorporating historical and other contemporary
data with that collected under the program. All
information will be incorporated in to a GIS for use in
implementing an ecosystem approach to coastal resource
management.

Michigan

A detailed land cover/land use inventory that includes
a set of wetland maps has been prepared by the Michigan
Resource Inventory Program. The inventory, which uses
conventional infrared aerial photography for its mapping
effort, provided the data base for digital products which

have been incorporated into the Michigan Resource Information System. The data collection and digital processing methodology are being used by the International Joint Commission to map the remainder of the U.S. Great Lakes shoreline.

North Carolina

Under the Albemarle-Pamlico Estuarine Study (funded by EPA and the State of North Carolina) North Carolina State University's Computer Graphics Center is conducing a land use inventory of Albemarle and Pamlico Sounds and their tributary basins. This inventory, which includes over two-thirds of North Carolina's coastal wetlands, will be prepared from remotely sensed satellite data and include the SAV data generated by NOAA under its Coastal Ocean Program. The goal is to develop a digital land use and land cover inventory of the entire Albemarle-Pamlico drainage area which can be maintained and updated as needed as part of the state's GIS.

U.S. Fish and Wildlife Service (USFWS)

The USFWS National Wetlands Research Center has an on-going program in habitat mapping of wetlands, seagrasses, and uplands whereby projects are developed in cooperation with other Federal and state agencies. The Center uses NWI procedures for photointerpretation, quality control, and quality assurance, producing maps at several scales which depict both wetland and upland classification so that habitat change analyses can address what type of uplands replaced wetland areas. Information gathered under the program has been used to develop digital data bases for various coastal areas of the country. These can be entered into the Center's GIS to implement natural resources inventories, habitat trend analyses, and cartographic modeling projects.

U.S. Geological Survey (USGS)

The USGS has an active National Coastal Geology Program which includes a variety of research field investigations related to wetlands. Currently focusing on the severe wetland loss problems in Louisiana, the USGS, in cooperation with the State of Louisiana and the USFWS, is conducting field investigations on wetland loss to identify natural and man-made causes. Additionally, the USGS is establishing a GIS network of providers and users of wetlands data in Louisiana. This system is likely to

be expanded to include all of the Gulf of Mexico region.
For FY 1991, wetland studies are planned for Louisiana,
Florida, the Great Lakes, and San Francisco Bay, all of
these to be done in cooperation with state agencies.

Conclusions and Recommendations

Based on the information provided at the workshop, the
Working Group arrived at the following conclusions and
recommendations.

The Federal Effort

**A standardized, centralized national digital mapping data
base of coastal wetlands is not available and needs to be
developed.**

A geo-referenced, computerized data base which
provides information on wetland habitat change in a variety
of forms (e.g., statistical and mapped) would serve as an
important tool for decision-makers in administering coastal
programs. Although various Federal agencies conduct
programs to document coastal wetland acreage, some on a
nation-wide scale and others regional, methodology,
frequency, and degree of resolution may vary, primarily
based on purpose, technology availability, and intended
use of the products. A standardized, centralized data base
would allow data collected by different programs under
varying legislative mandates to be incorporated into
individual GIS's to suit user needs, and the data readily
updated to reflect current information.

**Because of the value of the USFWS National Wetlands
Inventory (NWI), it is important that it continues and that
an effort be made to digitize the available coastal wetland
information.**

The USFWS NWI is the most comprehensive nation-wide
mapping program and is a valuable resource which serves
not only as a useful data base, both within the Federal and
private sectors, but also serves as the basis of many other
Federal and state mapping programs. Although the status
and trends analysis component, which is based on a
stratified random sample, perhaps may not be suitable for
assessing trends at the local level, it is useful for
assessing trends on a national scale.

Because it is critical that changes in coastal wetland acreage be monitored in a timely fashion so that appropriate management strategies can be implemented or existing strategies modified, the national mapping effort needs to accelerated.

While documenting the location and acreage of the Nation's coastal wetlands, it is essential that the implications of change on coastal ecosystems, including living marine resources, be evaluated, particularly in areas of rapid habitat loss. Such timely documentation should be done both at the national level, to assess the overall status and trends of the Nation's coastal wetlands, and at the regional or local level so that more detailed assessments can be made. NOAA's habitat mapping under the Coastal Ocean Program is designed to address some of these needs.

More focused research is needed to support the development of cost-effective, state-of-the-art mapping technologies using detailed digital satellite images an aerial photographs.

It is anticipated that development of newer technologies can make it possible to map coastal areas more frequently and accurately. This would provide up-to-date information to Federal, state, and local wetlands mangers for decisions related to wetland habitat issues.

It is particularly important that high-resolution, geo-referenced digital data bases for critical habitat types, based on synoptopic images, be developed.

The development of such data bases would allow comparison of chronological digital data to assess both national and local trends in wetland coverage. Fundamental to such as effort is the development of a standardized set of protocols for extracting digital information on wetlands coverage from satellite images and aerial photographs. It is recommended that these needs be addressed in order to improve the overall Federal coastal wetland mapping effort.

NOAA's Coastal Ocean Program is developing standardized protocols to produce geo-referenced digital data bases and digital maps from satellite images. This effort, which complements rather than duplicates the USFWS NWI, should continue.

Because of the relative importance of submerged aquatic
vegetation (SAV) to coastal ecosystems, a national
initiative is needed to map SAV along the Nation's coasts.

Although SAV mapping is currently being conducted by
the State of Florida, the Chesapeake Bay Program, and
NOAA's Coastal Ocean Program for the North Carolina coast,
SAV is not being mapped in a consistent manner on a
national scale. Standard protocols for mapping SAV, such
as those being developed as part of NOAA's Coastal Ocean
Program, are needed and should be instituted in future
mapping programs. Additionally, it is recommended that a
Federal initiative be developed to standardize SAV mapping
and provide a national SAV data base.

Fundamental cartographic information, such as that
developed by the USGS and NOAA, needs to be updated.

Updating fundamental cartographic information is
particularly critical in areas where shorelines have eroded
or accreted substantially over time. Coordination and
maintenance of the national data bases that provide
standardized and uniform quality photographic coverage of
the 48 conterminous States on a five-year acquisition cycle
should continue. Meanwhile, existing cartographic,
geologic, and hydrographic information should be digitized
and collated into a coastal wetland GIS.

Interagency Federal and State Coordination

Coordination among the Federal agencies should continue
and efforts made to identify additional opportunities for
cooperative programs.

Considerable coordination exists among the Federal
agencies. Examples include the cost-sharing agreements
between the USFWS NWI and many Federal and state agencies,
interagency coordination for the preparation of NOAA's
Coastal Wetlands Inventory, and habitat mapping under
NOAA's Coastal Ocean Program. It is important that such
efforts be coordinated so that duplication of effort,
wasteful resource allocation, and incompatibility of data
can be minimized.

Coordination among the Federal, state, regional and local
levels needs to be improved.

A number of Federal mapping projects are already coordinating efforts with state agencies (e.g., USFWS National Wetlands Research Center coastal mapping projects, USGS coastal erosion and wetland loss projects, EPA's Albemarle-Pamlico Estuary Program and NOAA's Coastal Ocean Program). However, Federal cooperation with state mapping projects should be increased and national protocols developed so that state mapping efforts can be integrated with, and complement, other regional and national projects. Ultimately, coordination and cooperation at this level will allow the development of a comprehensive data base of coastal wetland habitats. A possible mechanism for improved coordination of programs at the state level is the Coastal Zone Management Program administered by NOAA. Also, the coastal mapping and change analysis protocols being developed by NOAA's Coastal Ocean Program could provide the vehicle by which standardization of methodology and data generation could occur.

<u>Existing mechanisms for coordinating agency programs could be used for developing a consensus on state-of-the-art wetlands mapping technology. In addition, these coordination mechanisms should be used to facilitate the identification of additional mapping efforts which are needed, and to promote even closer coordination and interaction among the coastal mapping programs.</u>

Existing mechanisms include the interagency National Ocean Pollution Policy Board, which addresses Federal agency coordination for marine (including coastal) pollution research and monitoring; the Office of Management and Budget Circular No. A-16 which, as revised, provides for the establishment of an interagency committee to promote the coordinated development, use, sharing, and dissemination of surveying, mapping and related spatial data; and the President's Domestic Policy Council Task Force on Wetlands, which is charged with developing a national policy for attaining no-net-loss of wetlands. In addition, multi-agency initiatives for developing coastal management strategies, built upon existing programs, legislative mandates, and management expertise, should provide a framework for identifying and implementing coordinated mapping efforts at the national and regional levels.

References

Cowardin, L.M., V. Carter, F.C. Golet, and E.T. LaRoe.
1979. Classification of Wetlands and Deepwater Habitats
of the United States. U.S. Fish and Wildlife Service.
FWS/OBS-79/31. 130 pp.

Kiraly, S.J., F.A. Cross, and J.D. Buffington, Editors. in
press. Federal Coastal Wetland Mapping Programs: A Report
by the National Ocean Pollution Policy Board's Habitat
Loss and Modification Working Group.   U.S. Fish and
Wildlife Service Biological Report 90(18).

NOAA's COASTWATCH: CHANGE ANALYSIS PROGRAM

James P. Thomas, PhD.[1]
Randolph L. Ferguson, PhD.[2]
Jerome E. Dobson, PhD.[3]
Ford A. Cross, PhD.[2]

Abstract

Timely documentation of the location, abundance and change in coastal wetlands is critical to their conservation and to effective management of marine fisheries. The rapid changes occurring in these valuable wetlands require monitoring on a 1 to 5 year cycle. Therefore, NOAA within its Coastal Ocean Program, has initiated a cooperative interagency and state/federal effort to map coastal wetlands and adjacent upland cover and change in the coastal region of the U.S. on a 1 to 5 year repeating basis.

During the first two years, 1990-91, the program is concentrating on protocol development, and prototype studies in Chesapeake Bay, coastal North Carolina, and elsewhere. Through a series of workshops and working group meetings, a documented standard protocol for classifying and mapping habitat location, abundance, and change in the coastal region of the U.S. is being developed. The Chesapeake Bay prototype is using Landsat Thematic Mapper imagery and collateral data to

[1] NOAA, National Marine Fisheries Service, 1335 East-West Hgwy., Silver Spring, MD 20910.
[2] NOAA, National Marine Fisheries Service, Beaufort Laboratory, Beaufort, NC 28516.
[3] Oak Ridge National Laboratory, P.O. Box 2008, MS 6237, 4500 N, Oak Ridge, TN 37831.

map emergent coastal wetlands and adjacent upland cover and
change.   The coastal North Carolina study is using aerial
photography to map and determine change in submerged
aquatic vegetation.   In outyears coastal wetlands and adjacent
upland cover and change maps will be generated for various
coastal regions of the U.S. Extant land use and habitat mapping
data bases in other Federal and state agencies will be used to
minimize data acquisition and processing costs, supplement
ground truth, and assist in verification and interpretation.

Introduction

        One of the principal impacts on estuarine and coastal living
marine resources and their habitats is development in the
coastal zone.   A United Nations Environment Program (IMO
1990) report on The State of the Marine Environment states,
"The coastal strip, encompassing the shallow-water and
intertidal area along with the immediately adjacent land, is
clearly the most vulnerable as well as the most abused marine
zone.   Its sensitivity is directly tied to the diversity and
intensity of activities which take place there, and the threat to
its future is related to the increasing concentration of the world
population in this area.   The consequences of coastal
development are thus of the highest concern.   They arise not
only from the variety of contaminating inputs associated with
great concentrations of people, commerce and industry, but also
from the associated physical changes in natural habitats,
especially salt marshes, sea-grass beds, coral reefs and
mangrove forests."

        The Wetlands Policy Forum (Kean et al., 1988) says, "The
United States...needs much better information on the condition of
its wetlands resources, [and] the rate at which they are being
altered....we need to make information more widely available to
those involved in wetlands protection and
management....accurate maps depicting where wetlands
exist...[are needed]."   Further the report states,"...current survey
efforts are too infrequent....Particularly in regions where
wetlands are being lost rapidly, where they are under
substantial threat, or where they are of unusual value, more
frequent assessments...preferably every one or two years...are
essential to an effective protection and management program...."

In response to these issues NOAA in 1990, began a program to develop a comprehensive, nationally standardized, information system for land/habitat cover and change in the coastal region of the U.S. (Dobson and Bright, 1991; Thomas and Ferguson, 1990; Ferguson and Wood, 1990) The program will be standardized based on a series of regional, protocol workshops being held around the country. The coastal region covered by this program includes those land and water components of the various watersheds within the U.S., its possessions and territories, that most directly influence estuarine and coastal marine habitats utilized by living marine resources (LMR). The land/habitat cover includes those classes of vegetation and physical cover of ecological significance to LMR and/or their habitats. The major classes include submerged aquatic vegetation (SAV), emergent coastal wetlands and adjacent uplands. By change we mean all differences in land/habitat cover of 1 acre or greater that occur between times, T_1 and T_2. Our planned time interval for repeated looks at the coastal region of the U.S. is every 1 to 5 years. Regions with little change or interest will be monitored every 5 years; areas of intense development, every 2 or 3 years; and areas disturbed by extreme events (e.g., oil spills, hurricanes), annually.

NOAA is responding to other issues as well. The Wetlands Policy Forum (Kean et al., 1988) notes the need to "develop methods enabling trend information to reflect losses of certain functions, particularly in regions subject to significant stress." A component of the program is being designed so that not just areal coverage is determined, but also functional health, whereby we could see a decline in the functioning of a coastal habitat prior to its loss in area. Admittedly, this is a topic requiring research, but there are indications that it is doable using remote sensing (Bartlett, 1987; Crouse, 1987; Gross et al., 1987).

The purpose of the program is to build a digital data base that when integrated with other data within a GIS ultimately will enable us to link development in the coastal region to the ecological and economic productivity of the coastal zone/coastal ocean, and particularly in the case of the NOAA/National Marine Fisheries Service, the abundance, distribution and health of LMR. Our rationale is that changes in land use and cover affect critical habitat required by LMR for spawning, feeding and survival.

We believe that as these critical habitats are affected, so is the potential recruitment of larval and juvenile stages to estuarine and coastal fisheries, including those in the Exclusive Economic Zone. Further, we believe that it is necessary to monitor upland cover, as well as that of the critical habitats, because knowledge of the upland cover provides a significant determinant to the water quality affecting the critical habitats of LMR. As relationships between uplands, fishery habitat and LMR are better understood, improved long-term planning (and perhaps regulation) can be accomplished to sustain the productivity of the coastal zone/coastal ocean system.

Quicker, more frequent updates of land/habitat cover and change (i.e., location, type and magnitude) for the coastal region of the U.S. and the ability to see such data geographically arrayed in a geographical information system (GIS) in relation to other data will allow earlier warning and earlier, more focussed management actions regarding loss or change in coastal wetlands and adjacent uplands, and potential impacts on coastal zone/coastal ocean productivity-- ecological, including fisheries, and economic. The ability to see specific mapped changes will enhance the research effort toward understanding the relationships between coastal wetlands (including SAV), adjacent uplands and LMR. These quicker, more frequent updates, it is hoped, someday will allow much better projections (i.e., very early warnings), based on predictive models, of the effects of future changes in coastal wetlands and adjacent uplands on LMR and coastal zone/coastal ocean productivity.

Approach

The NOAA CoastWatch Change Analysis Program is a cooperative effort with other federal and state agencies. Extant land cover data bases in other Federal and state agencies will be used where possible to avoid duplication, supplement ground truth, and assist in verification and interpretation. In many cases it will be necessary to obtain the very detailed data and other information needed by this program from state and local programs.

Our approach is to use satellite imagery, aerial photography and surface level data that will be interpreted, classified, analyzed and integrated within a GIS. The data will be collected as synoptically as possible to facilitate national and regional change analysis. Satellite imagery will be the primary data source for looking at emergent coastal wetlands and adjacent uplands, because it is less expensive than aerial photography for large areas, is already in a digital format and is highly manipulable for processing and analysis. The sensor of choice, combining both spectral and spatial resolution, is the Landsat Thematic Mapper, although other satellite sensors may be used in special circumstances (e.g., to penetrate cloud cover or to provide higher spatial resolution). Aerial photography will be the primary method used to determine the distribution and abundance of SAV, because of the stricter requirements (e.g., higher spatial resolution, low sun angle, low tide, no white caps, little or no turbidity) for the acquisition of SAV data. These data will be digitized for inclusion in a GIS.

Program Activities

During the first year of the program we engaged in three major activities -- protocol development, a Chesapeake Bay Prototype, and a North Carolina SAV study. These are continuing this year.

A standard protocol, applicable to both aerial photography and satellite imagery, is being developed for 1) monitoring land/habitat cover and change in the coastal region of the U.S. on a 1 to 5 year cycle, and 2) producing data and information useful to managers and researchers alike. The protocol is being developed through a series of workshops and working group meetings with input from personnel in Federal and state agencies producing and/or using habitat classified data in the coastal zone and individuals with related expertise in remote sensing, geographical information systems, or wetlands in each coastal region of the U.S. The protocol emanating from these meetings must be valid for all coastal regions and be able to accommodate various types of remote sensing and ground based data. It will provide the operational procedures to be followed by those accomplishing or assisting the habitat classification and change analysis within or cooperatively with the program.

The first of these workshops was held at the University of South Carolina, May 29-31, 1990. It concentrated on remote sensing and classification of emergent coastal wetlands and adjacent uplands. Approximately 40 individuals participated, representing a number of Federal and State agencies and academic institutions located in the southeast (North Carolina to Texas). The workshop was chaired jointly by Jerome E. Dobson, ORNL, and Kenneth D. Haddad, Department of Natural Resources, State of Florida. A draft protocol based on earlier work in Chesapeake Bay was provided to the participants prior to the workshop. This document served as the focus of discussion during the workshop and afterward was revised to reflect the consensus reached by workshop participants. The revised document was sent out to all participants for their review and comment. Subsequently their comments were incorporated into the evolving draft protocol.

The second workshop focussed on SAV. It was held July 23-25, 1990, in Tampa, Florida, and about 40 people from the coastal region around the U.S. participated. Robert J. Orth of the Virginia Institute of Marine Science and Randolph L. Ferguson of the NOAA/National Marine Fisheries Service chaired the meeting. A series of questions were mailed out to the participants prior to the workshop in order to guide the discussions. Based on the consensus reached by the workshop participants, a draft protocol was produced and mailed out to the participants for their review and comment.

The third workshop, held January 8-10, 1991, at the University of Rhode Island, again chaired by Dobson and Haddad, concentrated on emergent coastal wetlands and adjacent uplands. As before the draft protocol was mailed out to the participants prior to the workshop. About 40 individuals participated, this time from Federal and state agencies and academic institutions in the northeast (Maine to Virginia). The fourth and fifth workshops will concentrate on emergent coastal wetlands and adjacent uplands of the west coast (~April, 1991) and the Great Lakes (~July, 1991).

Our plan is to merge the separate protocols for SAV and emergent coastal wetlands and adjacent uplands so that one single product is produced. Users will continue to be part of the protocol development process. The protocol will receive a final review and hopefully become operational by the end of 1991.

A second activity, the Chesapeake Bay Prototype, was begun in 1990 (Dobson and Bright, 1991). Its purpose was to establish a base of knowledge and experience for the CoastWatch Change Analysis Program and to provide wetlands data to the user communities including the NOAA Coastal Ocean Program and Chesapeake Bay Program. Land cover classifications (i.e., emergent coastal wetlands and adjacent uplands) for the entire Chesapeake Bay area (below fall line) for two time periods (1984 and 1988/89) were completed based on Landsat TM imagery. Additionally, a change analysis between those two times was accomplished. During 1991, these results will undergo a rigorous field validation, correction and statistical evaluation. In previous fiscal years a classification of wetlands and uplands was completed for four Landsat MSS scenes for one time period (1978) in the Chesapeake Bay (Dobson and Bright, 1989), and change detection was completed for three time periods (1974, 1978, 1982) in the vicinity of Metomkin Inlet, VA. This earlier experience is being utilized in the present effort.

The third programmatic activity in 1990, was the North Carolina SAV study (Ferguson and Wood, 1990). Aerial photography for Currituck and Albemarle Sounds, North Carolina was accomplished. The area from Cape Lookout to Ocracoke Inlet was photointerpreted for SAV based on aerial photography from 1988, and the region from Drum Inlet to Ocracoke Inlet (also based on 1988 photography) was digitized and compiled for publishing as a NOAA chart. The area from Cape Lookout to Drum Inlet based on 1985 photography was previously published as a NOAA chart (Ferguson, et al., 1988). During 1991, the SAV mapping and change analysis project will continue in North Carolina, and hopefully begin expansion into other regions of the U.S. as well. We also anticipate publishing our first change analysis map for SAV of southern Core Sound, North Carolina, based on 1985 and 1988 photography.

Several new activities are beginning. They include a number of regional projects (e.g., Galveston Bay, Texas; coastal Louisiana, Tampa Bay, Florida; coastal S.C.; coastal N.C.; and R.I.) designed to assist in resolving some of the protocol issues (Dobson and Bright, 1991). It is anticipated that west coast areas will be included next year. Also the program is beginning to look at the feasibility of accomplishing assessments to determine biomass, productivity and functional health of emergent coastal wetlands using remote sensing. The initial effort will be a literature survey and report describing previous research, status of technology and knowledge, and directions for future research. Someday we hope large areas of wetlands can be surveyed and assessed by satellite or aircraft much more rapidly and easily for biomass, productivity and functional health than is done today.

Beyond 1991, our plan is to begin monitoring land/habitat cover and change (including SAV, emergent coastal wetlands and adjacent uplands) over broad areas of the coast. Our products will include digital data in standard formats and hard copy maps. The digital data will be distributed on a cost-recoverable basis by the NOAA, National Oceanographic Data Center, Washington, D.C. We are still exploring the distributional issue of the hard copy maps. Ultimately we plan to integrate these data with other data in a GIS for the purpose of developing predictive models to help us project the affects of coastal development on coastal zone/coastal ocean productivity.

References

Bartlett, D.S. 1987. Remote sensing of tidal wetlands. Pages 145-156 in V. Klemas, J.P. Thomas, and J.B. Zeitzeff, eds. Remote sensing of estuaries, proceedings of a workshop. U.S. Dept. Commerce, NOAA and U.S. Gov't Printing Office, Washington, D.C.

Crouse, V. 1987. Gauging the health of a salt marsh through remote sensing. Univ. Delaware Sea Grant Reporter. 6(1): 1-5.

Dobson. J.E. and E.A. Bright. 1991. CoastWatch--Detecting change in coastal wetlands. Geo Info Systems. Jan./Feb. 1991.

Dobson, J.E. and E.A. Bright. 1989. Cover photograph of Chesapeake Bay. Photogrammetric Engineering & Remote Sensing 55(6).

Ferguson, R.L., J.A. Rivera and L.L. Wood. 1988. Seagrasses in Southern Core Sound, North Carolina. NOAA-Fisheries Submerged Aquatic Vegetation Study, Southern Core Sound, North Carolina. NOAA-Fisheries, Beaufort Laboratory, Beaufort, N.C. 28516. 3'x4' chart with text and illustrations.

Ferguson, R.L. and L.L. Wood. 1990. Mapping submerged aquatic vegetation in North Carolina with conventional aerial photography. Pages 113-121 in S.J. Kiraly, F.A. Cross, and J.D. Buffington, tech. coords. Federal coastal wetland mapping programs. U.S. Fish and Wildl. Serv., Biol. Rep. 90(18).

Gross, M.F., M.A. Hardisky, V. Klemas and P.L. Wolf. 1987. Quantification of biomass of the marsh grass (Spartina alterniflora, Loisel) using Landsat Thematic Mapper imagery. Photogrammetric Engineering & Remote Sensing 53(11): 1577-1583.

IMO/FAO/Unesco/WHO/IAEA/UN/UNEP. 1990. Joint Group of Experts on the Scientific Aspects of Marine Pollution (GESAMP): The state of the marine environment. Rep. Stud. GESAMP No. 39. 111 pp.

Kean, T.H., C. Campbell, B. Gardner and W.K. Reilly. 1988. Protecting America's Wetlands: An Action Agenda. The Final Report of the National Wetlands Policy Forum. The Conservation Foundation, Washington, D.C. 69pp.

Thomas, J.P. and R.L. Ferguson. 1990. NOAA's Habitat mapping under the Coastal Ocean Program. Pages 44-54 in S.J. Kiraly, F.A. Cross, and J.D. Buffington, tech. coords. Federal coastal wetland mapping programs. U.S. Fish and Wildl. Serv., Biol. Rep. 90(18).

United Nations Environment Program. 1988. The State of the Marine Environment. GESAMP Working Group No. 26.

NATIONAL MAPPING DIVISION PROGRAMS, PRODUCTS, AND SERVICES THAT CAN SUPPORT WETLANDS MAPPING

Franklin S. Baxter[1]

Abstract

The U.S. Geological Survey programs can play an important role in support of the President's policy of a "no net loss of wetlands." This paper will discuss the programs, products, and information services of the Survey's National Mapping Division, the tools available to determine where wetlands exist, and the capability for periodic measurements of wetlands to help in assessing compliance with the concept of no net loss of wetlands.

Introduction

The President's Wetlands Initiative of "No Net Wetlands Loss" will result in a refocusing of priorities for the collection, processing, and publishing of cartographic data. The National Mapping Division has been collecting wetlands information as part of its National Mapping Program for a number of years. The refocusing will ensure that data collection will directly support the President's Initiative.

Programs, Products, and Services

The National Mapping Division provides a diversity of cartographic, geographic, and remotely sensed data, products, and services in support of Federal, State, and public interests through the National Mapping Program. The

---

[1] Chief, Office of External Coordination, National Mapping Division, U.S. Geological Survey, 590 National Center, Reston, Virginia 22092

products and services include cartographic and geographic information about the Earth's natural and cultural features, basic and special maps in several scales, digital cartographic data, and remotely sensed data. The Division prepares standard topographic maps at specified scales and revises existing maps to provide current and accurate cartographic data.

The cartographic data needs of Federal and State programs are identified and prioritized under the Office of Management and Budget Circular A-16 process. Circular A-16, revised in 1990, names the Department of the Interior (delegated to the Geological Survey) as responsible for the National Mapping Program and for exercising governmentwide leadership "in assuring the coordinated planning and execution of these functions and related surveying, mapping, and related spatial data activities of Federal agencies, including activities funded in whole or in part by such agencies" so that "the surveying, mapping, digital cartographic, and related spatial data, earth science, and public land information needs of Federal, State, and local government agencies and the general public can be met in the most effective, expeditious, and efficient manner possible with the available resources." Circular A-16 also assigns the Department the responsibility of assuring that "aerial photography, and other remotely sensed imagery, survey data, topographic mapping, geologic mapping, digital cartographic and other related spatial data, produced by Federal agencies can be conveniently accessible for use in meeting the cartographic, earth science, and public land information needs of other Federal agencies, State and local governmental authorities, and the general public."

The primary map series comprises the largest scale information available on a nationwide basis. This series includes the 7.5-minute topographic quadrangle maps of the conterminous United States, Hawaii, and Puerto Rico and, in Alaska, 15-minute topographic quadrangle map coverage. Many Federal and State programs rely on this map series as a base for site-specific environmental studies or as the primary series for recording information relative to program needs. A major goal of the National Mapping Division, to achieve initial once-over national coverage in this map series, has been achieved. Maps covering approximately 98 percent of the United States have been published, and advance manuscript copies are available for an additional 2 percent. Orthophotoquads at 1:24,000 scale also are available for about two-thirds of the United States.

The currentness, accuracy, and usefulness of the primary map series will be maintained through an expanded map revision program. The National Mapping Division has begun a comprehensive plan for the identification and scheduling of map revisions that are most beneficial to the Federal and State agencies and the general public. The most efficient methods for revising the primary scale maps are being tested, and procedures for incorporating user requirements identified through the Circular A-16 process are being devised. Projects that reflect the most urgent needs of the user are being designed for short-term production.

The use of graphic maps is being supplanted rapidly by the use of base digital cartographic data because such data have wider utility and are cost efficient to maintain and apply. The National Mapping Division is collecting digital cartographic data to meet the needs of a wide variety of users, and is producing data in both digital line graph and digital elevation model formats. Digital data revision methods facilitate the recording of changes to the natural and cultural environment. Also, digital cartographic data are essential in analyzing the impact of environmental problems in a geographic information system (GIS) context. The National Mapping Division has devoted considerable effort to developing and promoting the standards and specifications necessary to ensure accessibility and usability of base cartographic data throughout the Federal Government. The responsibilities assigned by the revised Circular A-16 include coordinating the collection of spatial data to minimize duplication of effort where practical and economical to do so; facilitating the exchange of information and the transfer of data; and establishing and implementing standards for quality, content, and transferability. Working through the Federal Geographic Data Committee and the Interior Digital Cartography Coordinating Committee these activities will ensure that digital cartographic data contribute to the national geographic data infrastructure and may be used with data from other Federal agencies, such as wetlands data coordinated by the Fish and Wildlife Service and soils data coordinated by the Soil Conservation Service.

Other map products that have been useful in Federal
and State programs are the intermediate-scale maps at
1:50,000 and 1:100,000 in quadrangle and county formats.
The National Mapping Division plans to complete the
1:100,000-scale topographic map series in FY 1995. At
present (1991), there are about 1,205 maps available as
topographic editions, about 450 maps as planimetric
editions, about 950 available as Bureau of Land Management
editions (surface and subsurface mineral overlays), and
another 23 available as advance manuscript copies. The
1:50,000-scale quadrangle maps are produced to meet a
Defense Mapping Agency requirement but are made available
to the general public. The county formatted maps are
produced as needed on a cooperative basis with individual
States.

The intermediate-scale maps also are being digitized
to support the planning needs of Federal and State
agencies. Currently, all hydrographic and transportation
data at the 1:100,000 scale are available. Other
categories of data, such as hypsography (contours), public
land survey system information, boundaries, and digital
elevation models, are produced in response to Federal and
State agency requirements.

Many Federal and State agencies use a combination of
intermediate-scale products and data for planning purposes
and larger scale data for more detailed analyses, while
some agencies are satisfied with the level of information
at intermediate scale for land management and environmental
studies. For example, time and cost benefits can be
realized by locating study areas from a regional
perspective and obtaining source material (photographic
coverage, base maps, or appropriate digital data) for only
those areas needed at a larger scale.

The National Mapping Division is conducting pilot
projects to investigate the benefits of producing a large-
scale orthophotoimage product and of revising the land use
and land cover map series at a larger scale. In both
instances, the Division is responding to specific
requirements expressed by Federal and State agencies
through the Circular A-16 process.

The National Mapping Division is assessing the value
of land use and land cover map revisions at 1:100,000

scale. Now that completion of the topographic editions at
this scale is within sight, the use of these maps as a base
for land use and land cover mapping based on an enhanced
classification system seems quite promising. A number of
Federal and State agencies have expressed interest in
designing the classification system for use in a GIS
environment. Currently, the land use and land cover
mapping program results in maps and associated information
(political boundaries, hydrologic units, and census county
subdivisions) at 1:250,000 scale. The lower 49 States are
covered at this scale, and about 85 percent of these maps
have been digitized using the Geographic Information
Retrieval and Analysis System. The maps and data are
becoming out of date and use of the Geographic Information
Retrieval and Analysis System is not widespread. In
addition to the benefits to State users, a larger scale
mapping and digital data program will benefit wetland
analyses and other studies such as global change research.

     The National Aerial Photography Program (NAPP)
provides standardized and uniform quality photographic
coverage of the 48 conterminous States on a planned 5-year
acquisition cycle. Color-infrared (CIR) or black-and-white
(B/W) photographs, at a scale of 1:40,000, are centered on
quarter sections of each standard 7.5-minute U.S.
Geological Survey quadrangle. Because of the availability
of high quality B/W film and the economics of using this
type of film, NAPP acquisition is moving toward B/W
photography. The following States or portions thereof were
covered under NAPP contracts awarded in 1990: Connecticut
(CIR), Rhode Island (CIR), Massachusetts (CIR), New Jersey
(CIR), western Washington (B/W), Iowa (B/W), Missouri
(B/W), central Texas (B/W), eastern Wyoming (B/W), Montana
(B/W), Nevada (B/W), and eastern North Dakota (B/W). Those
Federal agencies or States that participate in the NAPP
program receive a discount on all NAPP products. NAPP
products are available from the Geological Survey's EROS
Data Center in Sioux Falls, South Dakota, and the U.S.
Department of Agriculture's Aerial Photography Field Office
in Salt Lake City, Utah.

     Image maps, primarily orthophotoquads, are prepared in
response to specific requirements of Federal and State
agencies. Orthophotoquads are scale-rectified image bases
that meet National Map Accuracy Standards, are produced
from NAPP photographs and are prepared at 1:24,000,

1:63,360, or 1:12,000 scales. These image bases can be produced in about one-third the time required for topographic maps; however, they contain no contours and only a limited number of feature names. During FY 1990 the National Mapping Division prepared 1,010 orthophotoquads at 1:24,000 scale, 96 at 1:12,000 scale and, in Alaska, 342 at 1:63,360 scale. Presently, about 41,000 orthophotoquads are available, covering more than two-thirds of the conterminous United States, all of Hawaii, and a portion of Alaska. All orthophotoquads are produced under cooperative (cost and work-share) agreements with other agencies; much of the work is produced directly from requests from the Bureau of Land Management and the Soil Conservation Service. On May 15, 1990, a Forum on Orthophotography was held to expand the understanding and use of orthophoto products among the user community, as well as to provide an opportunity for discussion of future needs concerning orthophoto use and coordination. This Forum was jointly sponsored by the Soil Conservation Service, the National Governors' Association, the National Association of Counties, and the National Mapping Division of the U.S. Geological Survey. Participants included representatives from Federal, State and local government, as well as from the private sector.

Of particular importance and application to wetlands mapping, the Soil Conservation Service proposed, at the Forum, that a national cooperative orthophotoquad program be established for the conterminous United States using aerial photography acquired from the NAPP. The proposal recommends 1:12,000-scale digital orthophotoquads for approximately 75 percent of the United States and 1:24,000scale for the remaining land areas that may not need coverage at the larger scale. Both digital and graphic products could be produced at each scale. To assist in the evaluation of this proposal, an Orthophoto Product User Survey was subsequently formulated by the National Mapping Division of the U.S. Geological Survey, and distributed in conjunction with the Office of Management and Budget Circular A-16 annual (1990) solicitation of agency requirements for mapping and digital cartographic data. Results from the user survey are currently being analyzed.

The cornerstone of the National Mapping Division's information delivery network is the Office of Information and Data Services. This office manages the Earth Science Information Center (ESIC) network composed of 13 ESIC offices, 1 Federal and 61 State ESIC affiliates. Developed through the merging of National Cartographic Information Centers (NCIC) and the Public Inquiries Offices (PIO), ESIC's responded to approximately 567,000 inquiries last year. ESIC offices maintain data records in such publications as the Cartographic Catalog, the Map and Chart Information System, and the Aerial Photography Summary Record System.

The EROS Data Center produces high-quality map products from satellite data for a variety of Federal and international organizations. The Data Center archives more than 800,000 Landsat scenes and, in 1990 will have more than 150,000 Thematic Mapper scenes and Advanced Very High Resolution Radiometer data for the entire country. The Data Center has established agreements with the commercial companies EOSAT and SPOT Image Corporation to serve as a single point of contact to purchase Landsat and SPOT data for Federal agencies. In the mid-1990's, the Data Center will process, archive, and distribute remotely sensed land data acquired by selected sensors flown on the proposed National Aeronautic and Space Administration's Earth Observing System.

Applications to Wetlands Studies

Following are several examples of National Mapping Division products and their use by other Federal agencies in wetlands studies.

Primary Map Series (1:24,000-scale maps in the conterminous United States and Hawaii and 1:63,360-scale maps in Alaska)

U.S. Fish and Wildlife Service:

Conterminous United States - to serve as a base map of the National Wetlands Inventory conducted by the U.S. Fish and Wildlife Service.

Coastal Louisiana - to revise maps to better reflect the loss of coastal wetlands to support studies on impact of habitat loss.

National Park Service:

Cape Cod National Seashore – to update and correct topographic maps to better portray coastal wetland areas.

Wrangell-St. Elias National Park and Preserve – to update maps to portray current topographic situation, including wetlands development due to rapid glacial retreat.

Environmental Protection Agency:

Horry County, South Carolina – to revise maps to assist in the study of the creation, maintenance, and impact of environmental problems on Carolina Bays.

Intermediate-Scale Maps (1:50,000-scale and 1:100,000-scale maps in quadrangle and county formats, and 1:250,000-scale quadrangle maps)

National Park Service:

Kenai Fjords National Park – to update maps to record the creation of wetlands, among other conditions, due to glacial recession.

U.S. Fish and Wildlife Service:

Coastal Areas of the Southern U.S. – to update maps to reflect the rapid loss of coastal wetlands in Texas, Louisiana, Mississippi, and Alabama.

Digital Data (digital line graphs and digital elevation models at several scales)

National Park Service:

Cape Cod National Seashore – to support use of a GIS in the monitoring of land use changes, including wetlands.

Cumberland Island National Seashore – to study a variety of activities along the coast, including emergency preparedness and habitat studies, with several other agencies.

U.S. Fish and Wildlife Service:

Coastal Areas of the Southern and Eastern U.S. – to be
used as ancillary data bases for wetlands habitat
data.
Lake Okeechobee – to support wetlands studies being
conducted by Region 8 Florida Cooperative Fish and
Wildlife Research Unit.

Other areas where digital data are to be used to
support wetlands studies:

    Okefenokee National Wildlife Refuge, Georgia-
      Florida
    The Everglades, Florida
    Galveston Bay, Texas
    San Joaquin Valley, California
    Malheur National Wildlife Refuge, Oregon
    Charles M. Russell National Wildlife Refuge,
      Montana
    Hawaiian Islands
    Mobile Bay, Alabama

Environmental Protection Agency:

Areas where hydrographic data are needed to support
related habitat and wetlands studies.

    Albermarle/Pamlico Environmental Study, North
      Carolina
    Merrimack River, Massachusetts-New Hampshire
    Massachusetts Bays, Massachusetts
    Slidell, Louisiana
    Edisto River, and Horry County, South Carolina
    Chesapeake Bay Study, Virginia-Maryland-
      Pennsylvania
    Narragansett Bay, Rhode Island

Land Use and Land Cover – The National Mapping Division is
considering the creation of a new series of land use and
land cover maps at 1:100,000 scale. The wetlands
classifications would be developed in coordination with the
U.S. Fish and Wildlife Service. Areas where land use and
land cover data are needed include:

National Park Service:

    Big Thicket National Preserve, Texas
    Saratoga National Historic Park, New York

Environmental Protection Agency:

    Chesapeake Bay Study, Virginia-Maryland-
      Pennsylvania
    Albemarle/Pamlico Study Area, North Carolina
    Pearl River Basin, Louisiana
    Savannah River Basin, Georgia
    Georgetown and Beaufort, South Carolina
    Delaware Bay, Delaware-New Jersey

Image Maps - The National Mapping Division is investigating the utility of new maps that could enhance the study of wetlands. This includes the development of a new series of image-based maps at 1:12,000 scale called orthophoto quarter-quadrangles. Preliminary analysis of the aforementioned Orthophoto Product User Survey indicates that a nationwide requirement for orthophoto products is evolving with an increased need for 1:12,000-scale orthophoto products in conventional and digital form. With over 60 percent of the survey respondents identifying a preference for a higher resolution, larger scale orthophoto product, such as 1:12,000 scale, it is evident that this newer series will soon be in demand. The use of these maps could result in the more precise recording of the existence and extent of wetlands. Because the orthophotoquad production process is much shorter than the production of standard topographic map revisions and the positional accuracy is comparable to the revised map, the orthophotoquad could become an essential tool in the study of wetlands.

The Soil Conservation Service, Department of Agriculture, has been working with the National Mapping Division to design an image base map at 1:12,000 scale in support of the national soils inventory. These image maps are used by Soil Conservation Service field personnel and are produced under a joint funding arrangement and use NAPP photographs as source materials. With interest increasing in cooperative orthophoto production, the U.S. Geological Survey has initiated several efforts to develop standards for both conventional and digital 1:12,000-scale orthophoto

products. It is believed that with the expanding use of
computers and geographic information systems, the latter
(digital) product will become the U.S. Geological Survey's
most requested image-based product.

Coordination Efforts in Wetlands Research and Technical
Assistance

     Several research studies, program initiatives, and
coordination ventures that relate to wetlands are being
pursued in cooperation with Federal and State agencies.
Examples include Mystic, Connecticut; Elizabeth River,
Virginia; James River, Virginia; and the prairie pothole
regions in the Midwest.

     As the use and acceptance of GIS technologies become
more widespread at all levels of government, the reliance
on computer-based environmental studies in the Geological
Survey will increase accordingly.   There are several
studies being conducted by the National Mapping, Water
Resources, and Geologic Divisions that investigate the
quality of water in a wetlands environment, the creation
and maintenance of wetlands, and the impact of manmade
activities on wetlands. GIS projects that are underway
include management of hazardous waste sites, some of which
have direct impact on nearby wetlands; the movement of
toxins through ground and surface water; and continuing
research on the environmental health of the Chesapeake Bay
drainage area.   The wetlands ecosystem provides
investigators with a natural laboratory in which complex
environmental processes can be investigated. GIS modeling
allows scientists to further expand the horizon of
scientific inquiry by permitting effective visualization
and the intersection of the many complex data sets
involved, while simultaneously providing capabilities for
quantified investigation of spatial and temporal patterns
in the data.

     The National Mapping Division and the National
Wetlands Inventory staffs have agreed to conduct mapping
activities that will result in mutually beneficial data
production and use. The three principal objectives of
these cooperative ventures are sharing of personnel,
technology, and data.

With regard to personnel, the National Mapping Division proposes to assign a cartographer for a maximum of 2 years to the NWI, St. Petersburg to provide assistance in the production processes for the generation of national wetlands maps. A remote sensing specialist also will be available on an as-needed basis to identify the most effective use of NAPP photographs, the procedures for handling large amounts of new source materials, and the conventions required to classify wetlands from NAPP source materials.

With regard to technology, the National Mapping Division proposes to identify state-of-the-art software and hardware systems to assist in the standardization of scanning procedures, to assist National Wetlands Inventory personnel in developing techniques to convert NWI graphic products to digital products, and to assist in quality assurance procedures so that wetlands data can be incorporated into the National Geographic Data System. This technical assistance will reduce duplicative efforts, ensure data collection to national standards, maximize program efficiencies, and encourage technology transfer.

With regard to data, a cooperative effort is necessary to transfer wetlands data from the National Wetlands Inventory directly to the National Geographic Data System. These data will be important to the efficient conduct of the national map revision program and the land use and land cover mapping effort. Procedures could be developed to assist in the revision and updating of wetlands data currently recorded on the 1:24,000-scale topographic map series and to update the land use and land cover maps that require current wetlands classification and mapping. This effort also will assure that the most current wetlands data are available to the general public through the National Geographic Data System.

Summary

The National Mapping Division produces and disseminates a variety of cartographic, image, and digital maps and data that are useful to Federal and State agencies involved in wetlands research. The most often used is the primary map series, in both graphic and digital form. For project planning, the intermediate-scale maps and data provide a regional perspective. Some Federal agencies are now increasing their support of even larger scale maps and data, primarily in an image format. The National Mapping Division is investigating the utility of quarter-quad orthophotographic products to respond to this growing need. Data dissemination networks are in place and accessible to Federal and State agencies nationwide.

Currently, the National Mapping Division is providing support to the National Park Service, the U.S. Fish and Wildlife Service, the Environmental Protection Agency, the Soil Conservation Service, the U.S. Army Corps of Engineers, the National Oceanic and Atmospheric Administration, and the U.S. Forest Service in a number of studies related to wetlands research. One of the most effective research tools in the study of wetlands is GIS, a technology in which the National Mapping Division has considerable expertise. The use of maps and data in GIS's is increasing and is expected to continue in the future. One of the major initiatives the National Mapping Division is pursuing is technical assistance to any Federal or State agency that seeks cooperative development of wetland research projects.

The need for a coordinated approach to support the President's Wetland Initiative is being addressed by the National Mapping Division and the National Wetlands Inventory through the establishment of formal cooperative agreements. These agreements will involve the sharing of expertise, the development of a wetlands component in the National Geographic Data System, and the exchange of wetland thematic and base cartographic data between the agencies.

Monitoring Seagrass Distribution and Abundance Patterns

Robert J. Orth[1]

Randolph L. Ferguson[2]

Kenneth D. Haddad[3]

Abstract

Seagrasses are an important natural resource present in many shallow coastal habitats of the world including the continental margins of the United States. These habitats support wildlife, high primary and secondary production, and serve as a nursery for many different species. This importance, coupled with their decline in recent decades in many areas of the United States has led to increasing interest in monitoring their distribution and abundance.

A national seagrass mapping and trends analysis program, using standard aerial photographic and cartographic methodologies, is being developed as part of NOAA's Coastal Ocean Program. This program must deal with a number of technical issues which are critical to effective mapping and trends analysis. These issues can be grouped in the following steps: 1. acquisition of aerial photography, 2. photointerpretation, and development of a hierarchical classification system, 3. surface level surveys, 4. compilation of seagrass habitat information, choice of base maps, and construction of cartographic products, and 5. analysis of historical trends. We discuss some of these issues and their implication for monitoring seagrass distribution and change.

----

[1]Virginia Institute of Marine Science, School of Marine Science, College of William and Mary, Gloucester Pt., VA 23062

[2]NOAA, National Marine Fisheries Service, Beaufort Laboratory, Southeast Fisheries Center, Beaufort, NC 28516

[3]Florida Marine Research Institute, Department of Natural Resources, 100 Eighth Ave., SE, St. Petersburg, FL 33701

Introduction

        Seagrasses are an important natural resource present in
many shallow coastal habitats of the world, including the United
States. There are approximately 50-55 species worldwide (Den
Hartog, 1970), eleven of which occur in the continental U.S.
Seagrasses are restricted to the photic zone, and penetrate from
greater than fifty meters in the clear waters of the Caribbean to
two meters or less in turbid estuarine areas. These are
Thalassia testudinum, Zostera marina, Z. japonica, Syringodium
filiforme, Halodule wrightii, Halophila engelmanni, H. decipiens,
Ruppia maritima, Phyllospadix serrulatus, P. scouleri, P.
torreyi.

        Seagrass systems contribute to coastal zone stability and
productivity (Phillips, 1984; Thayer, et al., 1984; Zieman, 1982;
Zieman and Zieman, 1989). The biological and physical resource
value of seagrasses are similar to those of emergent wetlands.
Seagrasses baffle waves and currents and stabilize sediments.
Seagrasses support coastal food chains through high levels of
primary and secondary production, and serve as nursery habitat
for numerous species. Many species associated with seagrasses
are of commercial significance (e.g. bay scallops and juvenile
blue crabs, finfish, and shrimp), recreationally significance
(e.g. ducks and geese), or are considered endangered (e.g.
manatees and certain sea turtles).

        Seagrasses are presently known to occur in all coastal
states of the continental United States except Delaware (where
they were historically present), South Carolina, and Georgia.
A review of available information on seagrass abundance for the
middle and south Atlantic and the Gulf regions, indicated that
seagrasses may be equally or more abundant than salt marshes in
some states (Table 1) (Orth, et al., in press a). However,
quantitative assessments of the present distribution and
abundance of seagrasses, as well as assessments on historical
changes, are rare or have never been conducted in almost all
states.

        The National Oceanographic and Atmospheric Administration
(NOAA) has recognized the losses of wetlands, both emergent and
submergent, and the rapidly occurring changes in coastal and
watershed development that will place additional stresses on
these fragile ecosystems. Through its Coastal Ocean Program
(COP), it has initiated a cooperative interagency and
state/federal effort to map coastal wetlands and adjacent upland
coverage on a two to five year time scale (Thomas and Ferguson,
in press; Ferguson, in press)). Through a series of workshops

Table 1. Salt marsh and seagrass coverage (hectares) by state*
(from Orth, et al., in press, as modified from Orth and
van Montfrans, 1990).

	Salt Marsh	Seagrass
New York	10,810[1]	78,100[11]
New Jersey	83,989[2]	12,624[1,10]
Delaware	26,183[3]	0
Virginia-Maryland	145,813[3,4]	17,353[12]
North Carolina	64,291[1]	80,972[13]
South Carolina	149,580[5]	0
Georgia	151,538[1]	0
Florida-Atlantic Coast	38,826[1]	2,800[14]
Florida-Gulf Coast**	137,455[6,7**]	913,700[14]
Alabama	11,855[8]	12,300[14]
Mississippi	24,919[9]	2,000[14]
Louisiana	720,648[9]	4,100[14]
Texas	174,899[6]	68,500[14]

* Wetland areas identified as containing salt tolerant
vegetation (categorized as 'salt marsh' or 'non-fresh' in data
reports or published papers) were used and listed in the totals
above. Estimates for seagrass or submerged vegetation coverage
in most states may be underestimated because of the lack of
adequate mapping surveys.

** Includes 34,540 hectares of mangroves listed in Perry, 1984.

1. Field, et al., 1988; 2. Tiner, 1985a; 3. Tiner, 1985b
4. Silberhorn, VIMS, pers. comm.; 5. Tiner, 1977; 6. Reyer et
al., 1988; 7. Perry, 1984; 8. Roach et al., 1987;
9. Pendleton, USF&WS, Slidell, LA, pers. comm.; 10. Macomber and
Allen, 1979; 11. Dennison, et al., in press; 12. Orth et al.,
1989 (Lower and Middle Chesapeake Bay + Chincoteague Bay);
13. Ferguson et al., 1989; 14. Iverson and Bittaker, 1986

involving participants from state, federal, and university
laboratories, efforts are being made to develop a framework to
assess the status of wetlands resources in a standardized mapping
format. NOAA convened a seagrass mapping workshop in Tampa,
Florida, in July, 1990, for the purpose of developing a standard
mapping protocol for seagrasses (Ferguson, et al., in prep).
Some of the methodological information shared in that workshop
and information available in the literature is summarized here.

Background

The need for documenting seagrass distribution and
abundance is becoming increasingly more critical because of
recent rapid, dramatic, and large scale losses of seagrass beds
in many areas of the world, including the U. S. (Orth and Moore,
1983a; Cambridge and McComb, 1984). These losses are directly
related to the degradation of water quality through both point
and non-point sources of pollution resulting from changing
demographics of the coastal zone.

A recent workshop, sponsored in part by NOAA-COP, was
held in November, 1990, at the headquarters of the South Florida
Water Management District, in West Palm Beach (Kenworthy and
Haunert, in prep.). The topic of the workshop was to discuss and
make recommendations on the capability of water quality criteria,
standards, and monitoring programs to protect seagrass from
deteriorating water transparency. The workshop concluded that
prevailing and continuing seagrass decline point out the
inadequacy of present water quality criteria, standards, and
management practices as they are now implemented.

Human population increases are resulting in greater
effects, both direct (dredging and filling, scouring from boat
propellers, and erosion from boat wakes) and indirect (non-point
nutrient and sediment input), that are eliminating or seriously
affecting seagrass habitat. Protection of existing seagrass
areas, and restoration of areas that have experienced seagrass
losses, will require a full understanding of those factors
contributing to seagrass declines, as well as an accurate
delineation of current and historical abundances of seagrasses.

Seagrass monitoring programs are almost non-existent
because of inherent difficulties and high cost of obtaining
synoptic coverage of seagrass in any given region (Orth and
Moore, 1983b; Walker, 1988). Most state and federal programs,
such as the U. S. Fish and Wildlife Service National Wetland
Inventory, have been aimed at assessing emergent wetlands, while
ignoring the presence, role, and value of submergent vegetation

(Reyer, et al., 1988). Some seagrass systems have been successfully mapped or monitored, using aerial photography to inventory seagrasses, and monitor trends (Haddad and Hoffman, 1985; Ferguson, et al., 1989a, b; Orth, et al., 1989; Orth and Nowak, 1990; Ferguson, in press; Ferguson, et al., in press). Aerial photographic techniques are well established, have been successfully used to map different wetland types, and, most importantly, provide a highly resolved and synoptic view of seagrass coverage. Aerial photography is the present choice of mapping seagrasses.

Although satellite imaging is currently available and can provide much greater spatial synopticity and spectral resolution, large scale aerial photography, at present, is the best accepted methodology for mapping seagrasses. Advantages of aerial photography are flexibility of timing of image acquisition, and of scale, and high resolution to identify smaller or patchy seagrass beds. Remote sensing of seagrasses from both satellite and aerial photography require extensive local knowledge based on surface level data. Surface level surveys are essential to interpret seagrass signatures from the imagery and for verification of photointerpretation. Surface level surveys alone, however, are too costly and impractical for synotic mapping of large coastal areas. Large, open-water areas without visible cultural features for horizontal control, at present, are problematic to map with large scale aerial photography. These areas may be best mapped in combination with small scale photography or satellite imagery to spatially orient the large scale photography. Thus, the primary focus of this paper is on aerial photography for mapping seagrass habitat.

Mapping Methodology

Seagrass mapping involves a number of key elements, each one involving stages or steps necessary for the successful completion of that element. The critical elements are 1. acquisition of aerial photography, 2. photointerpretation, and development of a hierarchical classification system, 3. surface level surveys, 4. compilation of seagrass habitat information, choice of base maps, and construction of cartographic products, and 5. analysis of historical trends.

Acquisition of Aerial Photography

The acquisition of aerial photography is the most crucial component of the mapping program. The constraints for visualization of SAV habitat and adjacent unvegetated bottom in photographic images (seasonal window, sun angle, turbidity, sea

state, stage of tide, inclusion of land features, area of
coverage, photographic scale, haze, clouds, film type, etc.) are
more restrictive than most other aerial photographic missions.
Imagery without clearly delineated seagrass boundaries cannot be
adequately photointerpreted and would lead to significant errors
in estimates of seagrass abundance.

Adequate pre-planning is a necessary first step for a
successful aerial photographic mission. This will require
knowledge of the study area and should include seagrass
phenology, water depth limits of seagrass habitat, location of
known and potential seagrass habitat, and seasonality of
turbidity, weather, and atmospheric conditions. Planned
flightlines must include all areas known to have seagrass habitat
as well as those areas that could potentially support seagrass.
Flightlines should be prioritized based on seagrass habitat
distribution, difficulty of access to airspace (e.g. proximity to
airport and military airstation and practice ranges), and
probability of meeting photographic constraints, and should
include sufficient land features in each photograph to provide
accurate local horizontal control. Flightlines should be flown,
if possible, during the period of peak standing crop for the
dominant species in the area. In subtropical and tropical areas
seasonal consideration may not be so critical.

General guidelines for mission execution are given in
Table 2. They address tidal stage, plant growth, turbidity, sun
elevation, wind, water turbidity, and atmospheric transparency,
sensor operation and plotting. Sixty percent endlap (overlap of
sequential exposures along a flightline) and twenty percent
sidelap (overlap of exposures in parallel flightlines) is the
minimum requirement to permit stereo photointerpretation and
assure complete coverage. These guidelines are being designed to
insure acquisition of photography under optimal conditions for
detection of seagrass. Quality control and assurance of the
exposure, development, and printing of film are critical
considerations but beyond the scope of this manuscript.

A variety of film types and filter combinations have been
used in past seagrass mapping efforts (Orth and Moore, 1983b).
Under ideal atmospheric and hydologic conditions virtually any
high resolution color, black and white, or infrared film may be
effective to visualize seagrass habitat in waters less than three
meters deep. The additional spectral information in color film
is recommended for initial mapping programs in new or unfamilar
areas and may permit discrimination of species in tropical areas.
Color film may be only marginally better than black and white
film for visualization of seagrass in regions with turbid water

Table 2. Guidelines followed during acquisition of aerial
photographs (from Orth, et al., in press).

1. Tidal Stage - Photography will be acquired at low tide, +/-0-
1.5 ft., as predicted by the National Ocean Survey Tables.

2. Plant Growth - Photography will be acquired when growth
stages ensures maximum delineation of SAV, and when
phenologic stage overlap is greatest.

3. Sun Angle - Photography will be acquired when surface
reflection from sun glint does not cover more than 30 percent
of frame. Sun angle should generally be between 20 and 40
degrees to minimize water surface glitter. At least 60
percent line overlap and 20 percent side lap will be used to
minimize image degradation due to sun glint.

4. Turbidity - Photography will be acquired when clarity of
water ensures complete delineation of grass beds.

5. Wind - Photography will be acquired during periods of no or
low wind. Off-shore winds are preferred over on-shore winds
when these conditions cannot be avoided.

6. Atmospherics - Photography will be acquired during periods of
no or low haze and/or clouds below aircraft. There should be
no more than scattered or thin broken clouds, or thin
overcast above aircraft, to ensure maximum SAV to bottom
contrast.

7. Sensor Operation - Photography will be acquired in the
vertical mode with less than 5 degrees tilt.
Scale/altitude/film/focal length combination will permit
resolution and identification of one meter squared area of
SAV (surface).

8. Plotting each flight line will include sufficient
identifiable land area to assure accurate plotting of grass
beds.

or increased atmospheric haze. Infrared film loses penetration at depths greater than five meters, requires special pre-exposure storage and special processing. Negative/positive films can be both used according to preference. Cost differentials between color or black and white films, as well as user preference for specific products, may dictate film type used.

Alternative technologies to aerial photography are in the research and development stage, each with potential advantages and disadvantages. These include new satellite sensors, airborne spectral scanners, and video recorders. Acceptance of new technologies should be based on relative costs, demonstrated ability to detect and map seagrass, and suitability for a national monitoring program.

Photointerpretation

Photointerpretation of seagrass habitat uses all available information including knowledge of aquatic grass signature on film, surface level surveys, and low level aerial reconnaissance surveys. The complexity and regional variability of the benthic environment demands that principal investigators have a comprehensive knowledge of the local study area including local seagrasses. Surface level surveys are essential for training photointerpreters and veryifying photointerpretation of seagrass habitat from aerial photography. Repeated mapping of the same area by the same personnel can improve reliability and cost effectiveness.

The most important prerequisite to photointerpretation is the development of a classification system which is hierarchical and versatile in levels of use. The classification categories must be definitive and not promote subjective interpretation, particularly if trends analyses are to be conducted. Such a classification system has not yet been finalized.

Photointerpretation by itself will not be able to accurately partition seagrass habitat according to all information needs. Information obtained from aerial photography with similar scale, film type, filters, etc., will not necessarily be the same from region to region. Seagrass habitat information required by researchers and managers can be divided into three levels, 1. presence or absence delineation, 2. number of species, bed form, and index of cover, and 3. species composition, biomass, productivity, functional status, and health.

Seagrass habitat is delineated by drawing a polygon consistent with the edge of a distinct area of seagrass signature in the photography. This polygon is traced on stable base film (either the base map or an overlay of the film) with the aid of optical devices (magnifying glass or stereo photointerpretation equipment). The designation of a given area of seagrass as habitat will be a function of resolution (or minimum detection unit), minimum mapping unit, signature of seagrass in the immediate area, and relative proximity to other seagrass beds (Fig. 1). Under optimal conditions a photographic scale of 1:24000, a minimum detection unit of 1 meter diameter features can be resolved. The minimum mapping unit, however, is the smallest area to be mapped as seagrass habitat. At a scale of 1:24000, the minimum mapping unit is 0.04 hectares (or about 0.1 acre). At this scale, for example, a group of small "patches" of seagrass will be mapped together as a single habitat area if the total area exceeds 0.04 hectares, and if the distribution and appearance of the patches indicates continuity within that habitat, and if intervening unvegetated areas are not large relative to the minimum mapping unit. Similarly, a small patch of seagrass adjacent to a large area of seagrass may be ignored, delineated independently, or delineated as continuous with that larger habitat. The outcome is based on its size and proximity to the large area of seagrass relative to the minimum mapping unit.

Delineated seagrass habitat will tend to be conservative. The degree of underestimation of seagrass habitat will depend upon the quality of photography and physical complexity of the area being photographed. Photography obtained under guidelines given in Table 2 will optimize delineations and minimize underestimations of seagrass habitat. Edges of seagrass habitat may or may not be clearly distinct in the photography (e.g. dense growth of seagrass along a clear water channel with a steep bank of light-colored sediment is relatively simple to delineate compared to sparse, patchy growth along a shallow depth gradient in turbid water over dark-colored sediment).

The accuracy of delineating seagrass habitat will depend upon quality of both photography and surface level information for verification. In order to delineate seagrass habitat, recognizable and verified seagrass signatures must be present in the photography. Seagrass (and other benthic features) in a given area will present a variety of signatures depending upon species present, bottom sediment, depth, season and local conditions. Seagrass with unrecognized signatures due to poor photography or incomplete surface level information will not be mapped. Benthic signatures derived from features other

Figure 1. Aerial photograph (scale 1:24.000) from the Chesapeake Bay
    showing seagrass beds (arrows point to several representative
    beds).

than seagrass must not be confused and mapped with those of seagrass.

More detailed information on seagrass species, bed form, and index of cover, or information on other benthic features may be obtained from aerial photography depending on region, scale, and local conditions. Under certain conditions in some tropical areas, species distributions and photographic signatures may be sufficiently distinct to discriminate seagrass habitats by species. This generally is not possible in east coast temperate areas.

Bed form (e.g. circular, doughnut-shaped, or irregular patches and continuous cover) can be observed under certain localized and ideal conditions (Fig. 2). Contrast and resolution in the photography are affected by some factors which are likely to co-vary with bed form, however, which limits this application. These factors include water depth, color of bottom sediments, density of seagrass, and presence of other benthic features co-occuring with seagrass (e.g. dense beds of macroalgae).

An index of cover can be obtained from aerial photography. An example of this is an adaptation of the crown density scale originally developed to categorize tree crown cover from aerial photography (Orth and Moore, 1983b; Orth and Nowak, 1990). This approach can provide an index of percent cover of seagrass habitat but is vulnerable to those factors which affect quality of photography. Analysis of change over time in this index requires consistent conditions and photography. Scale changes, for example, will lead to inconsistent index assessment. Contrast and resolution of fine features within seagrass habitat will affect the ability to discriminate edges of seagrass patches within the habitat. Less than optimal photographic conditions will tend to bias estimates up or down depending upon conditions. Overestimation of percent cover may result if adjacent patches appear to blend into one another. Underestimation may result if sparse seagrass between thicker patches cannot be distinguished from, and is interpreted as, background sediment. Delineation among different categories of cover, based on the crown scale, within a seagrass bed showing different degrees of cover can be particularly difficult.

A number of benthic features, in addition to seagrass, can be distinguished and identified in aerial photography. The signatures of these features must be distinct from each other and from those of seagrass. Potential features include macroalgae, coral reefs, oyster rock, and microalgal (diatom or cyanobacteria) mats at the sediment surface. Determination of

signatures for those submergent features requires careful study
of the photography and extensive surface level verification.

Surface Level Surveys

        Surface level information must be acquired before,
during, and after the photographic mission. Stations are located
in the field by LORAN C or the Global Positioning System (GPS).
Prior to the photographic mission, surface level surveys,
nautical charts, aerial reconnaissance, pertinent literature, and
local seagrass experts define general location, water depth, and
seasonality of seagrass and water clarity. The photographic
mission is planned with this information, combined with data
obtained from aeronautical charts.

        Surface level surveys concurrent with the photographic
mission are systematic and extensive for seagrass species (e.g.
species composition, relative growth state, and relative
abundance), sediments, and ancillary water quality data. An
appropriate statistical design should be incorporated to insure
validity of data, both in terms of location of stations, and
intensity, frequency, and type of sampling at any one location.
A stratified sampling regime should be incorporated to account
for seagrass growth in restricted depth zones, dependent on water
clarity in the particular region. The maximum water depth for
occurrence of seagrasses ranges from two meters in the Chesapeake
Bay and eastern North Carolina to nine meters for the northwest
coast of Florida.

        Surface level surveys conducted subsequent to acquisition
of photography are directed to clarify and verify
photointerpretation. Areas with new or questionable signatures
in the photography are visited to determine the features present
and/or verify the presence of seagrass habitat.

        The level of detail of samples collected at each station
will depend on the specific questions being addressed by the
survey, and can range from presence/absence and relative
abundance data, to measurements of standing crop and phenological
parameters, such as leaf length, width, shoot density, biomass,
and leaf area. Ancillary data should include water depth,
sediment composition (grain size, organic content, water
content), water clarity, salinity, and general descriptive
observations of the site.

Base Maps, Compilation and Cartographic Products

The map for compilation for seagrass habitat polygons
should be on stable base film, cartographically calibrated to an
external grid system (i.e. latitude and longitude), contempory
with the film and possess cultural and natural features
sufficient to horizontally control and, if necessary, to scale
and rectify the photography. United States Geological Survey
(USGS) 7.5 minute topographic quadrangles often are the best,
although they may not be current maps. Their scale (1:24000)
meets the requirements of a seagrass mapping program. In some
areas these maps are out of date which can cause problems in
application of local horizontal control and scaling accuracy of
the seagrass polygon. This is particularly true for those
regions with limited permanent cultural features such as road
intersections and buildings. Natural features such as shoreline
can change over time and reduce the positional and scaling
accuracy of seagrass polygons. Alternative base maps are NOAA's
National Ocean Survey shoreline manuscripts, and, if available,
should be considered in the mapping program.

Seagrass beds are represented as polygons traced on
stable base film at the photographic scale. Two approaches can
be taken. In both approaches the photography is positioned
according to local horizontal control points present in the
photography and base maps. The first approach is monoscopic and
the tracing film is the base map. In the second approach, the
tracing film is not the base map and monoscopic or stereoscopic
devices can be used to assist visualization. Seagrass polygons
are traced at the photographic scale and then compiled on the
base map using a zoom transfer scope if necessary. Local
horizontal control is achieved by superimposing the combined
image of the source photography and the interpreted seagrass
polygons on the base map. Each approach has its advantages. The
first avoids error associated with the additional transfer of
polygons. The second can employ stereo optical techniques and
improves the accuracy of horizontal positioning of the photograph
since the scale of photography may be affected by altitude
changes, and the deviation from true vertical warps the
photographic image. Both approaches to mapping require
digitization of polygons subsequent to compilation. New digital
image processing systems (analytical plotters) scan photographs
into a computer. Photointerpretation occurs on the computer
screen once the photograph is rectified to a base map. This
instrumentation is very costly but offers the advantage that
habitat polygons are interpreted and digitized in a single step
(Kiley, et al., 1990).

Seagrass habitat must be related via local horizontal controls to an external grid system in order to convey positional information and to allow for trends analysis. Seagrass habitat should be mapped relative to cultural or natural features on a USGS or NOAA map to provide this spatial orientation. Accurate mapping will require rectification of the photography and horizontal control, the accuracy in many cases being dictated by the accuracy of the base map. These errors can be minimized by precise control of aircraft altitude and vertical orientation and use of map bases most contempory with the photography.

Digitization of compiled seagrass polygons can be achieved by manual digitization or optical scanning methods. In areas where seagrass habitat is extensive and extends across one or more base maps, seagrass habitat is edge-mapped (Orth and Nowak, 1990). Digital data should be stored in a Geographic Information System (GIS) package, e.g. ARC-INFO. GIS systems should be able to resolve regionally specific problems (e.g. extensive seagrass beds of Florida's Gulf coast and eastern North Carolina) or discontinuities in habitat (e.g. seagrass habitat surrounding substantial unvegetated shallow or deep areas). Technology is shifting toward digital cartography. Seagrass habitat mapping should adopt these cost-effective, accurate, and precise mapping techniques as they are developed and proven reliable.

Historical Trends Analysis

Analysis of historical photography for seagrass habitat trends analysis is important in order to understand those factors governing distribution and abundance of seagrasses, as well as to effectively manage, protect, and potentially restore lost or damaged seagrass habitat. Since most aerial photographic missions in the past have been flown for purposes other than seagrass habitat mapping, those mission guidelines required for optimal seagrass mapping (Table 2 from Orth, et al., in press a) were generally not considered. Serious potential limitations in historical photography are the season and time of day of the photographic mission, water transparency, and spatial coverage. In most cases dates of photography coincide with periods of seasonal clear air and not during peak standing crop of the seagrass in that region. Poor water transparency and sun glare can obscure much of the seagrass habitat, in particular near the photographic edges. Also, the spatial coverage of seagrass habitat is often incomplete since beds may extend kilometers from shore beyond flight line coverage. Historical photography cannot be verified by surface level surveys. Because of the above limitations, interpretation of historical photographs for seagrass habitat should be attempted only by photointerpreters with substantial experience in the study area. In general, historiacal photography will permit visualization of only those seagrass habitats that are relatively easy to see (e.g. dense seagrass in shallow, clear water areas with light colored bottoms). Maps based on historical photography, however, will tend to be conservative in term of extent of seagrass habitat. Historcal photography has been used successfully to examine long term trends in the Chesapeake Bay (Orth and Moore, 1983b) (Fig. 2).

Historical photography is available from the Earth Resource Observation System (EROS) Data Center in Sioux Falls, South Dakota, the National Aeronautical and Space Administration, the U. S. Fish and Wildlife Service, the Soil Conservation Service, the U. S. Geological Survey, the U. S. Forest Service, the Agricultural Stabilization and Conservation Service, and many state highway departments. The earliest photography for consideration in seagrass habitat mapping trends analysis is the mid 1930's.

Conclusions

Seagrasses are an important living resource that are of national and worldwide concern because of serious declines in their distribution and abundance in many areas. As with emergent wetlands, human activities have resulted in either permanent loss of seagrass habitat or have caused serious stress on remaining areas. Effective management for preservation and protection of existing seagrass habitat and restoration of lost habitat will require detailed knowledge on the distribution and abundance of seagrass habitat. A national program to map seagrass habitat and conduct trends analysis is crucial, not only to fully understand the extent of existing and past seagrass habitat, but also to help develop improved water quality standards that must take into consideration levels of water clarity critical to seagrass survival (Kenworthy and Haunert, in prep; Orth, et al., in press b).

Acknowledgements

We would like to thank those state and federal (EPA, NOAA, USFWS, Army Corps of Engineers) agencies who have supported our seagrass mapping efforts and provided the background for the workshop. We would like to thank J. Nowak, B. Pawlik, and L. Wood who have contributed substantially to the seagrass mapping efforts. We would like to thank NOAA for the financial support for the seagrass habitat mapping workshop held in Tampa, Florida,

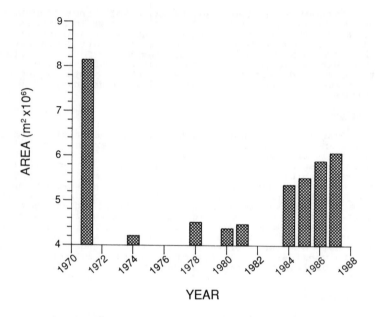

Figure 2. Historical abundance of seagrass in the York River, Virginia,
as determined from aerial photography.

July, 1990. We would especially like to express our appreciation
to all the workshop participants, in particular L. Handley and J.
C. Zieman, who provided additional comments subsequent to the
workshop. This is contribution number 1644 from the Virginia
Institute of Marine Science, School of Marine Science, College of
William and Mary, Gloucester Point, Virginia.

References

Cambridge, M. and A. J. McComb. 1984 The loss of seagrass from
    Cockburn Sound, Western Australia. I. The time course and
    magnitude of seagrass decline in relation to industrial
    development. Aq. Bot. 20:229-243.

Den Hartog, C. 1970. The seagrasses of the world. Verh. K. Ned.
    Akad. Wetensch. afd. Natuurk. Ser. No. 2 59:1-275.

Ferguson, R. L. in press. Mapping submerged aquatic vegetation
    in North Carolina with conventional aerial photography. Proc.
    Workshop on Status of Coastal Wetlands Mapping Programs.
    Slidell, LA. U. S. Fish and Wildlife Serv. Biol. Rep. 90
    (18).

Ferguson, R. L., R. J. Orth, and K.D. Haddad. in prep. Mapping
    submerged aquatic vegetation for National Oceanic and
    Atmospheric Administration's Coastal Ocean Program. Proc.
    Workshop Tampa, Florida, July 13-15, 1990.

Ferguson, R. L., J. A. Rivera and L. L. Wood. 1989a. Submerged
    aquatic vegetation in the Albemarle-Pamlico estuarine system.
    Albemarle-Pamlico Estuarine Study, Project No. 88-10, North
    Carolina Dept. of Nat. Res. and Com. Develop. Raleigh, N.C.
    and U. S. Environmental Protection Agency, National Estuary
    Program. 68pp.

Ferguson, R. L., J. A. Rivera and L. L. Wood. 1989b. Seagrasses
    in southern Core Sound, North Carolina. NOAA-Fisheries
    submerged aquatic vegetation study, southern Core Sound, North
    Carolina, NOAA-Fisheries, Beaufort Lab., Beaufort, N.C. 3'x4'
    chart, with text and illustrations.

Ferguson, R. L., L. L. Wood, and B. Pawlak. in press. SAV habitat
    from Drum Inlet to Ocracoke Inlet, North Carolina. NOAA-
    Fisheries submerged aquatic vegetation study, southern Core
    Sound, North Carolina, NOAA-Fisheries, Beaufort Lab.,
    Beaufort, N.C. 3'x4' chart, with text and illustrations.

Field, D. W., C. E. Alexander and M. Broutman. 1988. Toward developing an inventory of U. S. coastal wetlands. Mar. Fish. Rev. 50:40-46.

Haddad, K. D. and B. Harris. 1985. Use of remote sensing to assess estuarine habitats. pp. 662-675. In: O. T. Magoon, H. Converse, D. Minor, D. Clark, and L. T. Tobin (eds.), Coastal Zone 85, Vol. 1, Proc. Fourth Symp. on Coastal and Ocean Management. 1294 pp.

Iverson, R. L. and H. F. Bittaker. 1986. Seagrass distribution and abundance in eastern Gulf of Mexico coastal waters. Est. Coast. Shelf Sci. 22:577-602.

Kenworthy, W. J. and D. Haunert. in prep. Results and recommendations of a workshop convened to examine the capability of water quality criteria, standards and monitoring programs to protect seagrasses from deteriorating water transparency. Workshop Proc. South Florida Water Management District, West Palm Beach, Florida. Nov. 7-8, 1990.

Kiley, K.P., R.J. Orth, and K. A. Moore. 1989. Feasibility of the Map and Image Processing System (MIPS) as a device for the collection and display of information on the distribution and abundance of submerged aquatic vegetation from aerial photographs. Final Report. Virginia Council on the Environment, Richmond, VA. and Maryland Dept. of Natural Resources, Annapolis, MD. 54 pp.

Macomber, R. T. and D. Allen. 1979. The New Jersey submerged aquatic vegetation distribution atlas. Final Report New Jersey Department of Environmental Protection. 25 pp.

Orth, R. J. and K. A. Moore. 1983a. Chesapeake Bay: An unprecedented decline in submerged aquatic vegetation. Sci. 222:51-53.

Orth, R. J. and K. A. Moore. 1983b. Submersed vascular plants: Techniques for analyzing their distribution and abundance. Mar. Tech. Soc. J. 17:38-52.

Orth, R. J. and J. F. Nowak. 1990. Distribution of submerged aquatic vegetation in the Chesapeake Bay and tributaries and Chincoteague Bay- 1989. Final Report. U. S. Environmental Protection Agency. Chesapeake Bay Liaison Office. Annapolis, MD. 249 pp.

Orth, R. J., A. A. Frisch, J. F. Nowak and K. A. Moore. 1989.
Distribution of submerged aquatic vegetation in the Chesapeake
Bay and tributaries and Chincoteague Bay- 1987. Final Report.
U. S. Environmental Protection Agency. Chesapeake Bay Liaison
Office. Annapolis, MD. 247 pp.

Orth, R. J., K. A. Moore, and J. F. Nowak. in press a.
Monitoring seagrass distribution and abundance patterns: a
case study from the Chesapeake Bay. Proc. Workshop on Status
of Coastal Wetlands Mapping Programs. Slidell, LA. U. S. Fish
and Wildlife Serv. Biol. Rep. 90 (18).

Orth, R. J., et al., in press b. Development of water quality
standards based on species' habitat requirements: a case study
from the Chesapeake Bay using submerged aquatic vegetation.
Sym. Proc. Water Qual. Standards for the 21st. Century. Dec.
10-12, Arilington, VA.

Orth, R. J. and J. van Montfrans. 1990. Utilization of marsh and
seagrass habitats by early stages of Callinectes sapidus: a
latitudinal perspective. Bull. Mar. Sci. 46:126-144.

Perry, H. M. (ed.) 1984. A profile of the blue crab fishery of
the Gulf of Mexico. Gulf States Mar. Fish. Comm. No. 9, 80
pp. Ocean Springs, MS. 80 pp.

Phillips, R. C. 1984. The ecology of eelgrass meadows in the
Pacific Northwest: A community profile. U. S. Fish Wildl. Ser.
FWS/OBS-84/24. 85 pp.

Reyer, A. J., D. W. Field, J. E. Cassells, C. E. Alexander and C.
L. Holland. 1988. The distribution and areal extent of coastal
wetlands in estuaries of the Gulf of Mexico. Strategic
Assessment Branch, Rockville, MD, National Oceanic and
Atmospheric Administration. 18 pp.

Roach, E. R., M. C. Watzin, J. D. Scurry and J. B. Johnston.
1987. Wetland changes in Mobile Bay. Pages 92-101 in T. A.
Lowery, ed. Symp. on the Natural Resources of the Mobile Bay
estuary, Alabama SeaGrant Ext. Ser. MASGP-87-007. Mobile,
Alabama.

Thayer, G. W., W. J. Kenworthy, and M. S. Fonseca. 1984. The
ecology of eelgrass meadows of the Atlantic coast: A community
profile. U. S. Fish Wildl. Ser. FWS/OBS-84/02. 147 pp.

Thomas, J. P. and R. L. Ferguson. in press. National Oceanic and
Atmospheric Administration's habitat mapping under the Coastal

Ocean Program. Proc. Workshop on Status of Coastal Wetlands
Mapping Programs. Slidell, LA. U. S. Fish and Wildlife Serv.
Biol. Rep. 90 (18).

Tiner, R. W., Jr. 1977, An inventory of South Carolina's coastal
marshes. South Carolina Marine Res. Cntr. Tech. Rep. No. 23.

Tiner, R. W., 1985a. Wetlands of New Jersey. U. S. Fish and
Wildlife Service, National Wetlands Inventory, Newton Corner,
MA. 117 pp.

Tiner, R. W., 1985b. Wetlands of the Chesapeake Bay watershed: An
overview. Pages 16-24 in D. M. Burke, H. A. Groman, T. R.
Henderson, J. A. Kusler and E. J. Meyers, eds. Wetlands of the
Chesapeake Bay. Env. Law Inst. Washington, D. C. 389 pp.

Zieman, J. C. 1982. The ecology of the seagrasses of south
Florida: A community profile. U. S. Fish Wildl. Serv. FWS/OBS-
82/25. 158 pp.

Zieman, J. C. and R. T. Zieman. 1989. The ecology of seagrass
meadows of the west coast of Florida: A community profile. U.
S. Fish Wildl. Serv. Biol. Rep. 85(7.25). 155 pp.

Defining Wetlands--A Balance of Policy,
Practicality, and Science

David Ivester[1]

Abstract

Defining wetlands for regulatory purposes calls for
decisions on policy and practicality, as well as science.
One must first decide what is the resource to be
protected and then go about describing that resource in
a practical and scientifically meaningful fashion. The
modest aim of this paper is to establish this seemingly
obvious, but sometimes disputed, proposition and to
suggest that recognition of this fact will facilitate the
development of an appropriate wetland definition by
focusing needed attention on all three aspects--policy,
practicality, and science.

Introduction

Inherent in any program to regulate wetlands is the
need to define what a wetland is. After all, unless a
wetland can be found and recognized as such, it cannot be
regulated.

Choosing how to define wetlands for regulatory
purposes calls for decisions on policy and practicality,
as well as science. The government must first decide
what is the resource that it wants to protect and then go
about describing that resource in a practical and
scientifically meaningful fashion. The modest aim of
this paper is to establish this seemingly obvious
proposition and to suggest that recognition of this fact
will facilitate the development of an appropriate wetland
definition by focusing needed attention on all three
aspects--policy, practicality, and science.

[1]Partner, Washburn, Briscoe & McCarthy, a law firm
specializing in land-use and resource issues with
offices at 55 Francisco Street, San Francisco,
California, 94133.

Why bother proving something obvious? Because the experience of the United States over the last 15 years reveals that politics and bureaucratic processes sometimes obscure even the self-evident. Federal regulatory agencies recently issued a wetland delineation manual developed almost exclusively by technical personnel who ostensibly focused on scientific matters and eschewed any thought of making policy choices. The manual touched off a political firestorm as landowners across the country found the regulatory agencies reaching into their fields and forests and backyards to dub as "wetlands" areas that most had never dreamed could be worthy of the title. As a consequence, intense criticism threatens the entire wetland regulatory program and the agencies have announced corrective measures to scale back the expansive wetland claims made under the new manual. In other words, the policy considerations ignored (or hidden) during development of the manual have come back to haunt the government, and after-the-fact adjustments reflecting those considerations are now being made.

It has sometimes been said that defining wetlands is a purely technical matter, and so we should allow wetland scientists to tell us what they are. When defining wetlands for regulatory purposes, however, the whole point is to draw a legal boundary distinguishing a resource to be protected from other areas. The legal boundary reflects a policy decision that society values the delineated resource sufficiently to set it apart from other areas and treat it differently under the law. Science serves two functions in this process: (1) to advise the policymakers of the resource values of various types of areas so the policymakers can make informed choices and (2), once the policymakers have identified the resource they want to protect, to develop technical criteria and methods that faithfully describe that resource so it can be located on the ground.

Analysis

The primary consideration in fashioning a wetland definition is, not surprisingly, its purpose. Someone engaging in scientific research on mosquitos may be interested in areas with certain characteristics, while someone intent on preserving (and perhaps paying for) a valued natural resource may have areas with very different characteristics in mind.

In the United States, federal, state, and local agencies have developed many different wetland definitions designed to serve various purposes. These varying definitions have caused confusion and controversy

as wetland programs of one sort or another have
multiplied and expanded in recent decades. Efforts have
been made to achieve some uniformity of definition. With
the publication of the report of the National Wetlands
Policy Forum in 1988 and the Federal Manual for
Identifying and Delineating Jurisdictional Wetlands in
1989 some steps toward uniformity have already been
taken. The outcome, however, remains uncertain, and
controversy rages on.

Depending on whether a wetland definition is
intended for regulatory or non-regulatory purposes, the
controversy generally is shaped by the sometimes
competing, sometimes complimentary, pressures of policy,
practicality, and science. Science, for example, may
call for a detailed, broad definition suiting the needs
of biologists. Policy, on the other hand, may suggest
some limitations on the scope of the definition in order
to reflect society's view of what is interesting or
valuable or worth the expenditure of limited resources to
protect.[2] When a wetland definition establishes the
scope of a regulatory program, environmentalists
generally argue for a broad definition, while landowners
and others affected by the wetland regulations seek a
narrower one. Practicality may point to a definition
that is simple and workable.

When defining wetlands for regulatory purposes, the
basic policy question is: What is the natural resource
that we intend to protect? The technical question is:
How best can we describe that resource and distinguish it
on the ground? And the practical question is: How can
we locate and delineate that resource on the ground with
reasonable expenditures of time, money, and expertise.

Ideally, a government seeking to define wetlands
will consider all three questions. Conceptually, they

---

[2]Illustrations of the different influences of
science and policy in regulatory and non-regulatory
programs may be drawn from the United States'
experience. The U.S. Fish and Wildlife Service,
focusing on the biological functions and values of
wetlands with respect to many and various plant and
animal species, has developed a rather broad definition
for use in its non-regulatory mapping and other
programs. The Environmental Protection Agency and the
U.S. Army Corps of Engineers, on the other hand,
focusing on the prohibition of fill activities in
wetlands, have developed a somewhat narrower definition
for use in their regulatory program.

should be considered in the order listed above. It only
makes sense first to decide what is the resource that you
want to protect, then develop technical criteria that
faithfully describe that resource, and finally work out
practical, reasonable methods of applying those criteria
in order to identify and delineate the resource in the
field.

As a practical matter, however, it is difficult to
draw a clean line between the three issues and discuss
one without discussing the others. And, moreover, it is
not necessary. They can just as well, probably better,
be considered in some integrated fashion. The important
point is that all three should be recognized and
consciously addressed.

The necessity and desirability of considering policy
and practicality, as well as science, in developing a
regulatory definition of wetlands can perhaps best be
illustrated by examining this country's failure to do so.

In the United States, wetlands are regulated under
the Clean Water Act.[3] That Act prohibits "discharges" of
"dredged or fill material" into "waters of the United
States" without a permit.[4] The Environmental Protection
Agency ("EPA") and the U.S. Army Corps of Engineers
("Corps") administer the Act. In 1975, those agencies
defined "waters of the United States" to include not only
rivers, lakes, and other readily recognizable
waterbodies, but also "wetlands."[5] Thus was born one of
the most controversial regulatory programs in the
country.

Since 1977, the Corps and the EPA have defined
wetlands by regulation as:

> those areas that are inundated or saturated by
> surface or groundwater at a frequency and
> duration sufficient to support, and that under
> normal circumstances do support, a prevalence
> of vegetation typically adapted for life in
> saturated soil conditions. Wetlands generally

---

[3]33 U.S.C. § 1251 et seq.

[4]33 U.S.C. §§ 1311, 1344.

[5]40 Fed. Reg. 31,324-31,325 (1975)(codified at 33
C.F.R. § 209.120(d)(2)(1975)); 40 Fed. Reg. 41,293,
41,297 (1975)(codified at 40 C.F.R. § 230.2(b)(1975)).

include swamps, marshes, bogs, and similar areas."[6]

Caught in the political and legal crossfire between those who see wetlands as natural resources and those who view them as private property, the EPA and the Corps have gradually extended the meaning of "wetlands" beyond the original notion of swamps, marshes, and bogs to encompass much drier areas, including some hardwood forests, fields, and cultivated farmland, many of which may be saturated with rainwater for as little as a few weeks during the course of a year. Throughout this period, the language of the regulatory definition has remained unchanged. The extensions have been accomplished by incremental adjustments in the technical criteria used to identify and delineate wetlands in the field.

The Corps began developing a technical manual to identify and delineate wetlands in the early 1980s. The first "draft" manual was issued in 1981.[7] The Corps recognized at the outset that development of the manual involved both policy and technical issues. Perhaps most telling of the various documents confirming this is a memorandum from Forrest T. Gay, III, Brigadier General, Deputy Director of Civil Works (for the Chief of Engineers) to the Assistant Secretary of the Army (Civil Works) dated March 12, 1982. In that memorandum, the deputy director discussed the jurisdictional limits in wetlands, which he characterized as "one of the most difficult and controversial problems facing the Corps regulatory program." Noting that exclusive dependence on vegetation as the primary indicator of wetlands had led to errors in determining jurisdictional limits because vegetation adapted or tolerant to wet conditions is generally quite capable of growth and reproduction in

----

[6]42 Fed. Reg. 37,144 (1977)(codified at 33 C.F.R. § 323.2(a) & (c) (1977); 45 Fed. Reg. 85,346 (1980)(codified at 40 C.F.R. § 230.3(s) & (t) (1980)); 47 Fed. Reg. 31,810 (1982)(codified at 33 C.F.R. § 323.2(a) & (c) (1982)); 51 Fed. Reg. 41,250-41,251 (1986)(codified at 33 C.F.R. § 328.3(b) (1986)).

[7]From 1981 to 1989, the Corps issued at least six different versions of a wetland delineation manual. All were "drafts" circulated to Corps personnel ostensibly for field testing. None were formally adopted by the Corps, although all were used to a greater or lesser extent by Corps district offices to determine the extent of their jurisdiction over wetlands.

areas that are rarely or never inundated, the deputy
director pointed out that "the Corps has been working to
develop an improved delineation system, the so-called
multiple parameter approach." He then explained:

> The multiple-parameter approach is nothing more
> than a tool--a yardstick if you will--for
> technically measuring where to draw a line in
> the transition zone continuum between "always
> wet" and "always dry." . . . .   [¶] In
> developing the multiple-parameter approach, the
> Corps has recognized that there are two basic
> parts of the jurisdictional extent problem:
> the policy problem of deciding how far to
> regulate and the technical problem of providing
> a tool to clearly establish, in a repeatable,
> defensible way, a line in the field dictated by
> the policy decision. . . . The jurisdictional
> limits can be as restrictive or expansive as
> policy considerations determine. The technical
> delineation methodology is simply the tool to
> help implement the policy decision.

In order to tackle the jurisdictional limits
problem, the Corps developed a four-step plan of action:

a.   Develop a technical report (so-called
      matrix study) which would array resource
      information in a matrix displaying
      development potential, potential water
      quality effects, and administrative
      requirements against wetness regimes;

b.   Use this information as an aid in reaching
      an internal Corps policy decision on
      jurisdictional limits;

c.   Conduct inter-agency policy discussions
      (especially with EPA) on limits coupled
      with proposed rule making to announce the
      decision; and

d.   Proceed with final rule making and
      implementation.[8]

---

[8]Memorandum from E.R. Heiberg, III, Major General,
Director of Civil Works (for the Chief of Engineers)
for the Assistant Secretary of the Army (Civil Works)
dated April 5, 1982; see also a Corps document entitled
"A Review of the Adequacy of the Corps Definition of
Wetlands," dated May 22, 1981.

Based on the report of the matrix study,[9] the Corps developed a series of options. As explained in the Corps' April 5, 1982 memorandum, the matrix report and the options "refer to a series of wetness zones. . . . They represent natural geographic boundaries established by varying hydrologic factors. These zones make up the natural transition from 'always wet' to 'always dry' and the zones can be readily explained to the public." The five zones are described in the memorandum as follows:

Zone I: Intermittently   exposed—soil inundation or saturation by surface water or groundwater typically exists on a nearly permanent basis throughout the growing season of the prevalent vegetation, except during extreme drought periods.

Zone II: Semi-permanently   inundated   or saturated—soil inundation or saturation by surface water typically occurs during the spring and summer months. Frequency of occurrence ranges from 51 to 100 years per 100 years, and the total duration of time for the seasonal event(s) typically exceeds 25 percent of the growing season of the prevalent vegetation.

Zone III: Seasonally inundated or saturated— soil inundation or saturation by surface water or groundwater typically occurs until the beginning of the summer season. Frequency of occurrence ranges from 51 to 100 years per 100 years and the total duration of time for the seasonal event(s) typically ranges from 12.5 to 25 percent of the growing season of the prevalent vegetation.

Zone IV: Temporarily inundated or saturated— soil inundation or saturation by surface water or groundwater typically occurs at a frequency ranging from 11 to 50 years per 100 years. The total duration of time

---

[9]Corps, Wetland Jurisdiction Decision Matrices, February 15, 1982.

for the seasonal event(s) typically
ranges from 2 to 12.5 percent of the
growing season of the prevalent
vegetation.

Zone V: Intermittently   inundated   or
           saturated--soil inundation   or
           saturation by surface water or
           groundwater occurs with a typical
           frequency of 1 to 10 years per 100
           years. The total duration of time
           for these seasonal event(s) is
           typically less than 2 percent of the
           growing season of the prevalent
           vegetation.

The Corps posed a series of options, including
regulating to the end Zone I, Zone II, Zone III, and
Zone IV, respectively. The Corps' assessment of the
advantages and disadvantages of regulating to the end of
Zone III and Zone IV, respectively, is interesting:

Option 1.c. Regulate to the end of Zone III.
Characterized by inundation typically limited
to spring, the vegetation in this zone includes
drier bottomland hardwoods, wet meadows, and
tundra. Advantages of regulating this zone
include:

-    Majority of the medium risk to water
     quality areas identified are afforded
     regulatory management;

-    Reasonable probability of interagency
     agreement on jurisdictional limits;

-    Good probability of agency cooperation in
     developing additional general and
     nationwide permits; and

-    Reduced risk of environmental special
     interest legal challenge because of
     interagency agreement.

Disadvantages include:

-    Increased Corps manpower and resources
     required to effectively administer the
     program, even with increased use of
     general permits; and

-       Does not alter, and in some regions
        expands, the present geographic scope of
        the program.

Option I.d. Regulate to the End of Zone IV.
Characterized by periodic and typically brief
inundation, the vegetation in this zone
includes driest bottomland hardwoods, playa
lakes, and vernal pools.   Advantages of
regulating this zone include:

-       All pro-environmental interests are
        satisfied.

Disadvantages include:

-       Would extend regulatory coverage to
        low-risk to water quality areas;

-       Expansion of the already burdensome 404
        regulatory program;

-       Reasonable probability of legal challenge
        from development interests (most likely on
        taking issues); and

-       Need for doubling of Corps manpower and
        resources to administer the program even
        with increased dependence on general
        permits.

As acknowledged by the Corps in its April 5, 1982,
memorandum, development of a manual asserting
jurisdiction over Zone IV would expand the Corps'
jurisdiction to new areas that the Corps had not before
claimed were wetlands. The Corps' acknowledgement of
this fact is based on a survey that it conducted of its
various districts to ascertain the limits of the
jurisdictional claims actually being made by the
districts. That survey "showed that most districts were
regulating somewhere within Zone III."[10]

---

[10]Memorandum from C. G. Goad, Chief,
Construction-Operations Division, Directorate of Civil
Works (for the Commander) regarding Section 404
Jurisdictional Limits, dated December 21, 1982; see
also Memorandum from C. G. Goad, Chief,
Construction-Operations Division, Directorate of Civil
Works (for the Commander) regarding Current Limits of
Individual Permit Jurisdiction Exerted by Districts,
dated August 11, 1982.

Throughout the 1980s, the Corps issued a series of draft wetland delineation manuals. Each new version of the draft manual appeared to expand the Corps' jurisdiction more than its predecessor.

The Corps acknowledged, however, during this period that the manual was merely a draft and that the policy issue of the proper jurisdictional limits remained to be decided. The Chief of Engineers' office noted in 1983 that the manual's "status will remain unclear until the policy issues surrounding how and at what level we should establish jurisdictional limits are settled."[11] In much the same vein, on April 3, 1985, Acting Assistant Secretary of the Army (Civil Works) Robert K. Dawson instructed the Director of Civil Works:

> Please ensure that Corps field offices understand that the manual is purely technical in nature and has not been reviewed from a legal or policy perspective. As such, it may not represent proper limits of jurisdiction under Section 404.[12]

Why the policymakers remained mute for so long while their technical personnel developed a manual is not apparent. In any event, the technical personnel (largely biologists and soil scientists) filled the policy vacuum. They could not help but do so since any line they drew distinguishing wetlands from uplands necessarily reflected some policy. With little or no guidance from agency policymakers, therefore, the technical personnel drew the line where they saw fit. As one might expect, wetland scientists (and particularly those serving in public agencies) typically exhibit a special appreciation of wetlands as a natural resource; that, after all, is the focus of their professional lives. It should come as no surprise then that they chose a line reflecting their special perspective.

Throughout this time, many in the regulated community complained about perceived flaws in the draft manuals, but found no established process or forum in

<hr/>

[11]Memorandum from C. G. Goad, Chief, Operations and Readiness Division, Directorate of Civil Works, regarding Wetlands R & D Program, dated August 31, 1983.

[12]Memorandum from Acting Assistant Secretary of the Army (Civil Works) Robert K. Dawson to the Director of Civil Works, dated April 3, 1985.

which to press such complaints.[13]   Their comments were typically shrugged aside with a statement that the manual was merely a draft that was still undergoing revisions. They awaited with increasing anticipation the day they could voice their concerns.

So what finally happened?   The Corps continued to develop its manual.   In the late 1980s, the EPA started developing its own separate manual.   The agencies then got together to resolve their differences and, in discussions between themselves (and two other federal agencies, the U.S. Fish and Wildlife Service and the Soil Conservation Service), developed the Federal Manual for Identifying and Delineating Jurisdictional Wetlands, which they "adopted" and "implemented" on March 20, 1989. That manual, by all appearances, extends the Corps' jurisdiction further than ever--well into and perhaps through Zone IV and maybe even into Zone V.   And, notwithstanding the Corps' earlier plans, the agencies never held a public hearing or otherwise sought public input on the policy issues inherent in the manual.

It was not until one year after issuance of the manual that the agencies opened the door to public debate.   And then, they tried to restrict discussion to the "technical" issues.   The announced purpose of a series of public meetings was merely to "obtain comments on the technical aspects of the Manual" to assist the inter-agency committee (composed of representatives from

---

[13]The primary complaint of those in the regulated community was that the manuals simply extended the agencies' jurisdiction too far--far beyond the Congress' intent in passing the Clean Water Act, far beyond the Corps' intent in promulgating the regulatory definition of wetlands in 1977, far beyond the Corps' actual claims of jurisdiction before issuance of that version of the manual, and far beyond anything necessary or reasonable for protection of aquatic resources.   Many technical provisions of the manual troubled landowners.   For example, the use of "facultative" plants, which can grow equally well in uplands, as indicators of wetlands struck them as inappropriate and unjustified.   The further notion that a few weeks of wetness suffices to qualify an area as a wetland left many dumbfounded.   Exacerbating these perceived flaws was a concern that many plants on the U.S. Fish and Wildlife Service's wetland inventory plant list appeared to be improperly categorized, skewing the list to produce wetlands where properly none existed.

the Corps, EPA, U.S. Fish and Wildlife Service, and Soil
Conservation Service) in assessing "the need for
clarification and/or technical changes" "[n]ow that the
Manual has been in use for a full year."[14] Matters of
policy were not to be discussed:

> It is important to understand that the purpose
> of these meetings is to obtain comments on the
> technical aspects of the Manual. The meetings
> are not for the purpose of discussing broader
> policy issues such as "no net loss" of
> wetlands. The President's Domestic Policy
> Council Interagency Task Force on Wetlands will
> be holding public meetings to discuss policy
> issues related to wetlands later on this
> year.[15]

Notwithstanding such pronouncements, the agencies
got an earful of policy advice at the meetings on the
manual and at the later meetings by the Wetlands Task
Force. They heard the regulated community condemn the
manual--and the entire federal wetland regulatory
program--for intruding too far into the province of
private property.

Members of Congress heard as well, and some
initiated efforts to trim back the agencies' wetlands
jurisdiction.

Responding to the groundswell of criticism (and,
perhaps, to the threat of legislative dictates), the
Corps and the EPA have begun adjusting the manual to
reflect policy decisions resulting from the public outcry
over the manual's perceived excesses. On September 26,
1990, the Corps (with the EPA's approval) issued a
regulatory guidance letter to "clarify" the wetland
definition with respect to farmland. The Corps announced
that it would curtail its jurisdictional claims to
conform to the Soil Conservation Service's program under
the so-called Swampbuster provisions of the Food Security
Act of 1985. (Under that act, generally any person who
farms a wetland converted after December 23, 1985, the
day the act was passed, will be ineligible for various
federal benefits.) The Corps declared it would not
regulate "prior converted croplands," i.e. wetlands that
"were both manipulated (drained or otherwise physically
altered to remove excess water from the land) and cropped

---

[14]Corps Public Notice No. CECW-OR-1, June 22, 1990.

[15]Id.

before 23 December 1985, to the extent that they no
longer exhibit wetland values."[16] According to various
reports, 20 to 60 million acres of land were affected by
this clarification. Representatives of the Corps and the
EPA explained that the intent was to enable the agencies
to concentrate on protecting truly valuable wetlands
without dissipating their limited resources on marginal
areas exhibiting little or no wetland value.

     Further changes are in the wind. There is, for
example, discussion of changing the minimum period of
inundation or saturation required to constitute a wetland
from one week (as the 1989 manual states) to two, three,
or four weeks. A change of this sort could shift
millions of acres from one side of the wetland-upland
boundary to the other.

     If these after-the-fact adjustments in the wetland
definition seem considerably grander than the technical
tweaking generally to be expected of new regulatory
provisions, the reason is that the agencies are only now
responding to the policy issues inherent in the
definition. They are, at last, deciding what is the
resource they want to protect. And they are finding that
the manual produced by their technical personnel (without
policy guidance) draws a line in the wrong place,
encompassing not only the resource to be protected, but
also many other areas (e.g. prior converted cropland and
areas experiencing only a few weeks of wetness in a year)
the agencies have chosen (or will choose) as a matter of
policy not to regulate.

Conclusion

     Controversy over the wetland regulatory program of
the Corps and the EPA, perhaps inevitable to some extent,
has been exacerbated by years of confusion and indecision
over what is and is not a wetland. For years, the
agencies shied away from deciding as a matter of policy
what is the resource they sought to protect. (Congress,
it should be noted, offered them little help.) Instead,
they busied themselves with preparing a technical manual.
Just what the manual was supposed to describe would be
decided later, they noted. Then, once the manual was in
hand, they seemingly blinded themselves to its policy
implications and simply published it with the happy
thought that it was merely a technical document. The

---

[16]Corps Regulatory Guidance Letter No. 90-7,
September 26, 1990.

COASTAL WETLANDS

critical policy questions, thus, were decided by default, rather than conscious deliberation. Now, confronted with strong objections to the manual's reach, the agencies are reconsidering (or, more accurately, considering for the first time) what it is they want to protect. They are finding it necessary to balance competing interests and make judgments about dissimilar factors, e.g. the natural resource values of various types of land and the interests of private property owners. In other words, they are making policy decisions.

Ultimately, in defining wetlands for regulatory purposes, a government must consider both policy and technical issues and should consider practicality as well. Development of an appropriate wetland definition can best be accomplished by recognizing each of these three aspects of the task and consciously deliberating with each of them in mind. It might also help to put the horse before the cart and identify the resource to be protected before undertaking to develop technical criteria to describe it.

APPENDIX

Codes and Statutes

Title 33 United States Code
    § 1251 et seq. 4
    § 1311 4
    § 1344 4

40 Fed. Reg. 31,324-31,325 (1975)
    (codified at
    33 C.F.R. § 209.120(d)(2)(1975)) 4

40 Fed. Reg. 41,293, 41,297 (1975)
    (codified at
    40 C.F.R. § 230.2(b)(1975)) 4

42 Fed. Reg. 37,144 (1977)
    (codified at
    33 C.F.R. § 323.2(a) & (c) (1977)).)).) 5

45 Fed. Reg. 85,346 (1980)
    (codified at
    40 C.F.R. § 230.3(s) & (t) (1980)) 5

47 Fed. Reg. 31,810 (1982)
    (codified at
    33 C.F.R. § 323.2(a) & (c) (1982)) 5

51 Fed. Reg. 41,250-41,251 (1986)
    (codified at
    33 C.F.R. § 328.3(b) (1986)) 5

U.S. Army Corps of Engineers Documents

"A Review of the Adequacy of the Corps
Definition of Wetlands,"
    May 22, 1981 6

Wetland Jurisdiction Decision Matrices
    February 15, 1982 7

Memorandum, Forrest T. Gay, III, Brigadier General,
Deputy Director of Civil Works (for the Chief of
Engineers) to the Assistant Secretary of the Army
(Civil Works)
    March 12, 1982 5

APPENDIX
(Cont'd)

Memorandum, E.R. Heiberg, III, Major General,
Director of Civil Works (for the Chief of
Engineers) for the Assistant Secretary of the
Army (Civil Works)
    April 5, 1982 6

Memorandum, C. G. Goad, Chief, Construction-
Operations Division, Directorate of Civil Works
(for the Commander) regarding Current Limits of
Individual Permit Jurisdiction Exerted by Districts,
dated
    August 11, 1982. 9

Memorandum, C. G. Goad, Chief, Construction-
Operations Division, Directorate of Civil
Works (for the Commander) regarding Section 404
Jurisdictional Limits
    December 21, 1982 9

Memorandum, C. G. Goad, Chief, Operations and
Readiness Division, Directorate of Civil Works,
regarding Wetlands R & D Program
    August 31, 1983. 10

Memorandum, Acting Assistant Secretary of
the Army (Civil Works) Robert K. Dawson to the
Director of Civil Works
    April 3, 1985. 10

Federal Manual for Identifying and
Delineating Jurisdictional Wetlands
    March 20, 1989 . . . . 3, 11

Public Notice No. CECW-OR-1
    June 22, 1990. 12

Regulatory Guidance Letter No. 90-7
    September 26, 1990 13

Coastal Louisiana:
Abundant Renewable Natural Resources - In Peril

William S. Perret[1]
Mark F. Chatry[2]

Abstract

Louisiana contains 40% of the nations coastal wetlands. These wetlands have been recognized as one of the
most productive fish and wildlife habitats in the world.
The renewable resources that inhabit and utilize Louisiana's coastal marshes and estuaries represent a combined
commercial and recreational value in excess of one-half
billion dollars annually. Louisiana has led the nation
in volume of total fisheries production, shrimp production, blue crabs, oysters, furbearers and alligators
harvested. Additionally, the Louisiana coastal zone
provides 4,000,000 acres of overwintering habitat for the
largest concentration of ducks and geese in the Nation.

These vast resources, however, are in peril. Over
the past 20 years, more than 800 square miles of Louisiana's coastal wetlands have been lost. This loss is
continuing at a rate of 50 square miles per year. While
the land loss along should be sufficient cause for alarm,
it is the associated and inevitable loss of fish and
wildlife resources that constitutes a hugh economic and
cultural loss to Louisiana and to this Nation.

The difficult questions remain unanswered, however,
Given the dynamic geologic and biologic processes involved, can the land loss be reversed or even stopped?
If so, can it be done without destroying the very thing
we are trying to protect - the economy and the quality of

[1]Louisiana Department of Wildlife & Fisheries, Marine
Fisheries Division, 400 Royal Street, New Orleans,
Louisiana 70130

[2]Fisheries Management, Inc., P.O. Box 537, Covington, LA
70434

life of the region. While the authors recognize that
change along the Louisiana coast can be preserved, and
possibly, even enhanced. Such an ambitious goal, however
will only be attainable if the agenda and objectives are
explicitly defined and carried out in a genuinely comprehensive fashion.

Introduction

        Louisiana contains the largest and most productive
coastal wetlands in the United States. These wetlands
have been recognized as one of our Nation's most biolog-
ically productive fish and wildlife habitats in terms of
recreational fishing, commercial fishing, hunting and
trapping. These productive coastal wetlands are one of
the major reasons Louisiana has been nicknamed "Sports-
man's Paradise." But, all is not well in our coastal
wetlands "paradise". Foremost, is the act of physical
loss of coastal wetlands, which is compounded by the
complexity of the land loss issues. Numerous governmental agencies, federal, state and local, as well as
various interest groups and individuals continue to be
concerned over the coastal wetland loss problem, however,
until 1989 little consensus has been brought forward to
address this complex issue. Now however, State, Federal
and local governments are attempting to develop a comprehensive plan to restore and enhance Louisiana's productive coastal wetlands (U.S. Army Corps of Engineers,
1989).

        Before we delve further it is important to provide
a general background into the processes responsible for
the formation of Louisiana's coastal wetlands.

Coastal Processes and Habitats

        Coastal landforms in the Mississippi Deltaic Plain
(Figure 1) are the results of delta constructional and
destructional process taking place over the last 7000
years. During the constructional phase of delta formation, sediment from the river is deposited into areas
adjacent to the river. The process has not been one of
continuous, regular growth. Instead, the Mississippi has
changed its course every 1000 or so years, building new
delta lobes with associated wetlands while older abandoned deltas eroded and deteriorated (Penland and Boyd,
1981). This process, referred to as deltaic switching,
results from the river seeking a shorter and slightly
steeper course to the Gulf of Mexico.

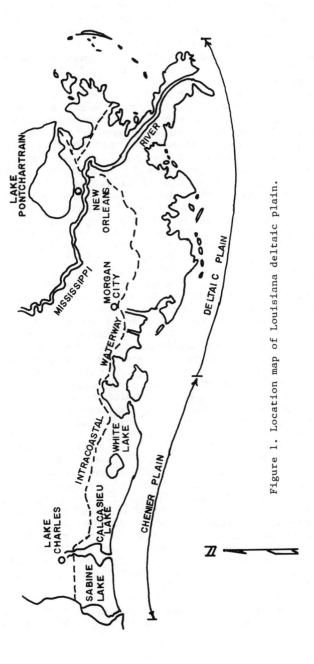

Figure 1. Location map of Louisiana deltaic plain.

During the construction phase of delta formation,
the river delivers its sediment load into a receiving
body of water. As deposition continues, a progradational
platform is developed upon which the delta plain is
established. Flood waters build this delta plain, and
since this aggravation is greatest adjacent to the river,
natural levees are formed. These natural levees are
composed of the coarser sediments which settle out first.
These sediments also build shoals vertically at the mouth
of the river, and freshwater stream-mouth marshes are
formed. As progradation continues, the delta plain is
enlarged and land adjacent to the natural levees forms
shallow interdistributary bays. The natural levees
continue to build vertically until only the highest
floodwaters overflow the river banks. As the sediment
adjacent to the levees become more stable, and the bays
fill in, they are invaded by freshwater and/or brackish
marsh plants depending on the topography. Continued
delta progradation leads to an overextension of the
distributary gradient, and a reduction in its slope and
hydraulic efficiency. These conditions will result in an
upstream distributary diversion. The channel will now
switch to a shorter, more efficient course with a steeper
gradient to the Gulf of Mexico. The abandoned delta,
which no longer receives the amount of sediments it did
when active, is not capable of prograding seaward or
building vertically to the same extent as before. Without this input of sediments, subsidence of the land
surface, due to compaction and dewatering of underlying
clays, becomes the dominant physical process controlling
the deltaic landforms. Freshwater marshes become
brackish as their elevation decreases relative to sea
level due to the aspects that control subsidence and, as
the salt water wedge of the Gulf of Mexico moves further
northward due to reduced freshwater outflow of the river
and lowered marsh elevation. The natural levees are also
subsiding and this results in the marsh also advancing
over these features. At the same time the brackish
waters push the distal boundary of the freshwater wetlands closer to the old distributary channel. With
continued subsidence, brackish marshes become salt
marshes as the seawater moves further landward. The old
distributary channel slowly fills from reworked sediments
as the shoreline transgresses and the brackish marshes
deteriorate initially to an open form which subsequently
coalesce into lakes, and finally open bays.
(Mendelssohn, et al, 1986).

Louisiana's coastal wetlands extend westward from
the eastern shore of the Pearl River (state boundary) to

the western shore of Sabine Lake. The northern boundary
is the Intracoastal Waterway except for Lake Pontchar-
train, Lake Maurepas and the Lake Charles areas, which
are north of the Intercostal Waterway. The southern
boundary was arbitrarily designed as the coastline.
Barrett (1970) reported that the total water area contained with this geographical description was 3,378,924
acres; the total land area was 3,695,700 acres.

In 1981, Gagliano, et al, reported that the
Louisiana wetlands in the deltaic plain were deteriorating at a rate of 25,204 acres per year as of 1980.
Using this rate, we calculate that Louisiana's deltaic
plan has experienced a loss of 504,040 (25,204 x 20)
acres of coastal vegetated wetlands to date in the last
20 years.

Another estimate of wetlands loss was reported by
Hawxhurst (1984); he found that more than 556,000 acres
of deltaic plans wetlands had been lost in the last 25
years, and that unless this activity is halted, coastal
Louisiana could lose one million more acres of wetlands
by the year 2040.

This is the issue that we would like to address in
this report - the extreme importance of this area to fish
and wildlife resources. We will also indicate one
approach that we feel would be successful in curtailing
the current rate of vegetated wetland loss, and enhance
our coastal fish and wildlife resources.

It is our contention that if the current processes
continue the loss of these wetland habitats will have
detrimental effects on the productivity of these areas
for future fish and wildlife resources.

It is hypothesized that deteriorating marshes
provide better access and more protection and food for
juvenile fishery species than do accreting marshes
(Zimmerman and Minello, 1984). Coastal marshes in the
northwestern Gulf are presently undergoing unusually long
periods of tidal inundation due to the combined area wide
affects of subsidence, loss of sediment nourishment, sea-
level rise and levee building by man. As a result, these
marshes are drowning, breaking up and creating ever
diminishing islands of marsh surface in expanding areas
of open water. In the process of deterioration, tidal
marshes offer a superior nursery environment that
stimulates production of penaeid shrimp and of at least
some economically important fishes. As marshes drown and

become progressively lower relative to sea level, the
duration of intertidal flooding increases, which makes
them more available for longer periods of exploration.
In theory, these conditions can account for recent
increases in fishery production. The gains, however,
come at the expense of losses in wetland habitat areas
that, if continued, assure a decline in future yield of
fisheries. Some indicators suggest that we may now be at
the peak of fisheries production and a decline will begin
within the next decade. (Browder, et.al, 1988).

Description of Resources

The renewable natural resources that inhabit and
utilize these coastal marshes and estuaries are extremely
diverse; they include numerous species of birds, mammals,
fish, crustaceans, mollusk, reptiles, and amphibians.
For the purpose of this paper we will concentrate on
those species of commercial and/or recreational importance.

The commercial production of fish, crustaceans,
mollusk, furbearing mammals, and reptiles (alligators) is
exceptionally high and furnishes a major percentage of
the total United States production. In fact, Louisiana
annually leads the nation in volume of commercial fish,
shrimp, blue crabs (Callinectes sapidus), oysters
(Crassostrea virginica) furbearers and alligators
(Alligator mississippiensis) harvested. Marine and
estuarine recreational fishing opportunities are un-
paralleled. The economic activity associated with marine
recreational fishing in 1985 was valued at over 173
million dollars (S.F.I., 1988). Additionally this
coastal zone provides over 4,000,000 acres of wintering
habitat for the largest concentration of waterfowl (ducks
and geese) in the United States.

The reported volume of commercial and recreational
fishery landings in Louisiana is tremendous. There is
also a significant amount that goes unreported (Perret,
1981). Additionally, it has been reported (N.M.F.S.,
1988) that recreational fisherman caught over 43 million
fish during 1988. This catch was comprised of over 50
species.

The commercial fisheries harvest for the U.S. and
Louisiana for the period 1980, 1985-1988 is presented in
Table 1. Highest production was in 1987 when over 1.8
billion pounds was harvested; these landings accounted
for 26% of the U.S. total that year. It is readily

apparent that the magnitude of the Louisiana catch constitutes a very significant portion of the entire U.S. commercial fisheries production.

It is extremely important to recognize that in discussing Louisiana's marine fisheries, we are with few exceptions, dealing with species that are estuarine dependent. Gunter (1967) reported that estuarine species made up 97.5% of the total commercial catch of the Gulf states. Estuarine dependent species were also the dominant fish harvested in the marine recreational fishery.

Therefore, any attempts at management of the state's marine fishery resources must recognize the importance of the estuarine environment and should include provisions for maintaining or even enhancing this habitat.

Fur & Alligator Production

Louisiana annually leads the nation in fur and alligator production. The major coastal furbearers are nutria (Myocastor coypus), muskrat (Ondatra zibethicus), mink (Mustela vison), otter (Lutra canadensis), and raccoon (Procyon lotor). Historically Louisiana's fur production ranking was due to phenomenal muskrat production. However in 1962 nutria took first place and is now the backbone of the industry.

As the continental leader in wild fur production, this state contributes 40 percent of the entire fur supply in the United States, and yields more pelts that all Canadian provinces combines (Murchison, 1978).

Table 1. U. S. Total and Louisiana commercial dockside landings (billions) and value (billions $).

Year	1980 Billion lbs. $	1985 Billion lbs. $	1986 Billion lbs. $	1987 Billion lbs. $	1988 Billion lbs. $
U.S.	6.5 2.2	6.3 2.3	6.0 2.8	6.9 3.1	7.2 3.5
LA	1.4 .18	1.7 .23	1.7 .32	1.8 .32	1.4 .32
LA%	21% 8%	27% 10%	28% 11%	26% 10%	19% 9%

Source: National Marine Fisheries Service

Louisiana's commercial fur harvest and value to the trapper for the years 1960, 1970, 1980, and 1986 is presented in Table 2. In 1980, the takers of furbearers had an economic value to the trapper of approximately $17 million dollars. During this 26 year time span a great deal of fluctuation has taken place, both in number of pelts harvested and value of this harvest. This is more a reflection of the economic value of the pelts, rather than the availability of animals to be harvested. The lowest number of pelts taken during this period was in 1986 (1,438,394), however, this was a result of low prices for the raw pelts.

The commercial harvest of furbearing animals is not only an economic venture, it is also an important marsh management tool. It helps to limit vegetation damage during periods of high animal population density.

This state is also the national leader in the harvest of alligators and has managed this species on a sustained harvest yield since 1972 (Joanen, et.al., 1984). While alligators occur throughout the state the largest concentrations are found in the lower third of the state (about 85% of the total) and the harvest from this coastal area accounts for 91% of the total alligators taken (Joanen and McNeese, 1987).

Table 2.   Louisiana's commercial fur harvest (millions of pelts) and value (millions $) for 1960, 1970, 1980, 1986.

1960 Millions # of Pelts	Value	1970 Millions # of Pelts	Value	1980 Millions # of Pelts	Value	1986 Millions # of Pelts	Value
2.38	$3.3	2.09	$4.5	2.25	$16.5	1.43	$6.8

Millions #'s of Meat	Value	Millions #'s of Meat	Value	Millions #'s of Meat	Value	Millions #'s of Meat	Value
		8.7	$ .75	1.63	$ .33	1.5	$ .49

TOTAL $ 5.25 $ 16.83   7.29
Source: Louisiana Department of Wildlife and Fisheries

The alligator harvest and value for the period 1972 through 1988 is presented in Table 3. From an initial harvest of 1,350 alligators taken in 1972 valued at $75,505, this activity has grown to a harvest of 23,526 alligators valued at nearly $8,000,000 in 1988. Additionally, some 600,000 pounds of meat valued at $3,000,000 was sold in 1988. If 91% of the value of this harvest is attributed to the coastal zone, this means that over $10,000 000 could be assigned to the area for 1988.

Waterfowl Resources

The coastal area is the most important wintering ground for migratory waterfowl in the nation. These coastal wetlands provide habitat for millions of waterfowl from September through March. It has been reported that more that two-thirds of the entire waterfowl wintering population utilize the Louisiana coastal zone (Bellrose, 1976). Over one-fourth of the North American dabbling duck population, such as mallards, (Anas platyrhnos), pintail (A. acuta), shovellor (A. cypeata), blue wing teal (A. discors), green wing teal (A. crecca), widgeon (A. americana), and gadwall (A. strepera) over-winter in the wetlands with a peak number of over 5.5 million in some years (Palmisano, 1973). An additional 5 million ducks and geese stop in Louisiana for a period of time before migrating further south to Mexico and South America. These include lesser scaup (Aythya affinis), ringneck (A. collaris), canvasback (A. balisinera), red head (A. americana), snow geese (Anser caerulescens), white fronted geese (A. albifrons), and Canada geese (Branta canadensis).

The economic impact of waterfowl hunting is of great significance to Louisiana. Byrd and Smith (1984), using the national average of money spent by duck hunters, estimated that Louisiana waterfowlers contribute over $37,000,000 annually to the economy of the state in pursuit of this sport. Approximately 65 percent or $24,050,000 of this total is attributed to the Louisiana coastal area.

TOWARD A SOLUTION

While it is generally perceived that concern for Louisiana's coastal fish and wildlife resources is only a recent phenomenon, a review of the literature indicates otherwise. The second Biennial Report of the Oyster Commission of Louisiana (1906) stated that the continuous

Table 3. Alligator harvest & value in Louisiana 1972-1988

Year	No. Taken	Value of Skins	Approx. Amount Meat Sold (lbs)	Approx. Value of Meat
1972	1,350	$ 75,505		
1973	2,921	268,994		
1975	4,420	258,791		
1976	4,389	512,240		
1977	5,474	488,499		
1979	16,300	1,711,500	100,000(bone-in)	$ 125,000
1980	17,692	1,609,972	100,000(bone-in)	125,000
1981	14,870	1,821,575	100,000(bone-in)	125,000
1982	17,142	1,621,633	100,000(bone-in)	125,000
1983	16,154	1,452,568	100,000(bone-in)	125,000
1984	17,389	2,556,183	100,000(bone-in)	125,000
1985	16,691	2,482,619	150,000(deboned)	675,000
1986	22,429	3,611,000	310,000(deboned)	1,395,000
1987	23,892	6,689,760	500,000(deboned)	2,250,000
1988	23,526	7,905,024	600,000(deboned)	3,000,000

Source: Louisiana Department of Wildlife & Fisheries

extension of the Mississippi River levee system below New Orleans had interfered with the natural oyster conditions. Further, the report recommended that "... if at appropriate intervals on the east bank gaps could be permitted through which the freshwater might flow and mingle with the saltwater of the Gulf, more suitable oyster producing conditions would exist."

Viosca (1928), in reference to the decline of Louisiana's wetlands and their wildlife and fishery resources, wrote of "the birth of the rather recent movement for the conservation of supply." Viosca claimed that it has indeed possible to reverse the decline. His suggested remedial measures included weirs and dams for water level control, siphons for freshwater from the Mississippi River and relief outlets along the Missississippi in connection with flood control. In summary, Viosca wrote,

"It seems that the time is ripe for an enormous development of the Louisiana wet lands along new and intelligent lines, the ideal conditions to be demonstrated by observation and research, and that this development should be included in a broad

program of conservation which has for its object
the restoration of those conditions best suited to
an abundant marsh and swamp fauna, but under some
degree of control at all times.

"It should be considered a state and national
problem equal in significance to agricultural
development, to the end that the state and
nation may enjoy a more balanced diet, more
healthful   recreation,   and   enduring
prosperity."

Unfortunately, Viosca's 60 year old vision of
"intelligent" development of Louisiana's wetlands remains
unfulfilled. In some respects, in fact, Viosca's remarkable insight seems to have been lost on many modern day
coastal managers and estuarine scientists. What is
lacking is a widespread, genuine understanding of the
value of the coast to society, i.e. its utility. Simply
phrased, how do we go about "saving" the coast if the
desired outcome of our efforts is undefined.

Mississippi River diversion projects will solve
major problems concerning marsh deterioration in southeast Louisiana. Installation of water control structures in southwest Louisiana will abate problems of
erosion, saltwater intrusion, and general marsh loss.

According to Chabreck and Condrey (1979), the
production of fish and wildlife in the marsh is definitely related to photosynthesis plants produced with the
marsh. These plants are a basic source of energy for
annual populations and conditions that enhance plant
growth and serve to benefit fish and wildlife within the
area. Installation of water control structures within a
wetland system is the technique utilized whereby habitat
productivity can be increased or stabilized.

Obviously, any plan with a prospect for success must
have its goals explicitly defined and be genuinely comprehensive. It must take into account the varied needs
of the people of south Louisiana. In addition to continued fish and wildlife productivity, these needs include flood control, navigation, freshwater, housing,
agriculture, hurricane protection and mineral extraction.
These needs in turn, must be considered in light of the
dynamic geologic and biologic processes at work along the
coast.

While this is certainly a formidable task against which little progress has been made over the past 60 years, there is cause of optimism. This optimism is best exemplified by three freshwater diversion projects in Southeast Louisiana which are in various stages of completion (Figure 2). In particular, the Caernarvon Project, located east of the river below New Orleans, is now under construction. After twenty-five years of planning, this project will divert 8,000 cfs from the Mississippi River, through the levee, and into the marshes of the Breton Sound Basin. The 25 million dollar project will slow the rate of land loss and allow the management of salinities for increased fish and wildlife production.

While this project alone is only a small step towards ensuring continued productivity and quality of life along the Louisiana coast, the lessons learned during planning, design and finally constructed and operation are essential to continued progress. The most important of these lessons undoubtedly concerns the generations of public and political support.

First, a project or plan will not proceed unless it has the substantive support of those directly affected. The most important among these groups are those who are economically tied to the coast. Their voices will be loud and they will be heard. The lesson here is that unless a project is demonstrably beneficial to those who live or work in the affected coastal area, the project will not proceed.

Second, public support must be actively sought. A worthwhile project will not be built simply because it is a "good idea." This is not to say, however, that a project should be oversold. Coastal resources managers have an obligation to provide a clear assessment of the costs of a given project as well as anticipated benefits.

Third, nothing will kill a project faster than discord among state and federal agencies or local governing bodies. The magnitude and complexity of the problems affecting the chances that such cooperation can be achieved will be greatest if all governmental entities involved collectively designed a genuinely comprehensive plan for the coast.

The Caernarvon Freshwater Diversion structure became operational in early 1991. The other two diversion

structures located at Davis Pond and Bonne Carre', should
be operational by 1994 (Figure 2).   Thus, the Comprehensive Southeast Louisiana Freshwater Diversion Plan may
become a reality, and can become a showcase for the world
in decreasing continuing wetlands deterioration.

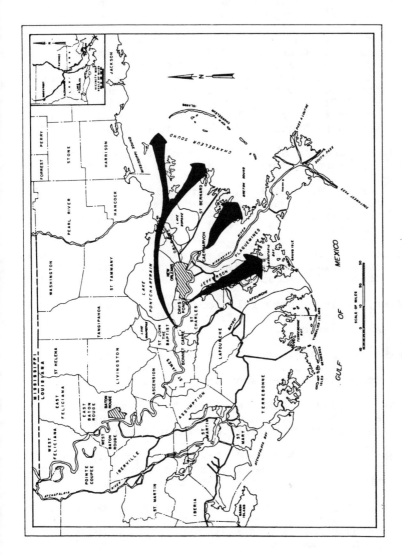

Figure 2. Proposed freshwater diversion sites and
areas of influence.

Literature Cited

Barrett, B. B. 1970. Water measurements of coastal
    Louisiana. Louisiana Wildlife & Fisheries
    Commission. New Orleans, LA 1-196.

Bellrose, F. C. 1976. Ducks, geese and swans of North
    America. Stockpole Books. Harrisburg, Pa. 543p.

Browder, J. A., L. N. May, A. Rosenthal, R. H. Baumann
    and J. G. Gosselink 1988. Utilizing remote sensing
    of thematic mapper data to improve our understand-
    ing of estuarine processes and their influence on
    the productivity of estuarine-dependent fisheries.
    Final Report to the National Aeronatics and Space
    Administration. Center for Wetland Resources,
    Louisiana State University, Baton Rouge, La. 178pp.

Byrd, W. and C. Smith. The good old days are now -
    Louisiana Conservationist. 1984 (36) 6:25-28.

Chabreck, R. H., and R. E. Condrey. 1979 Common vascular
    plants of the Louisiana marsh. Sea Grant
    Publication No. LSU T-79-003. Baton Rouge, LA.
    pp.116.

Gagliano, S. M., M. J. Meyer-Areudd, and K. M. Wicker.
    1981. Land loss in Mississippi River deltaic plain.
    Trans., Gulf Coast Assoc. of Geol. Soc. 31:295-300.

Gunter, G. 1967. Some relationships of estuaries to the
    fisheries of the Gulf of Mexico. In: Estuaries:
    Amer. Assoc. Adv. Sci. Publ. No. 83:621-638.

Hawxhurst, P. 1984. Marsh creation and barrier island
    stabilization. In, the Barataria Estuarine
    Complex:Past, Present and Future: 32-37.

Joanen, T. and L. McNeese. 1987. The management of
    alligators in Louisiana, U.S.A. In: Wildlife
    Management Crocodiles and Alligators: 33-42.

Joanen, T., L. McNeese, G. Perry, D. Richard, and D.
    Taylor. 1984. Louisiana's alligator management
    program -Proc. Annu. Cont. SEAFWA: 201-211.

Mendelssohn, I. A., J. G. Gosselink, W. Grip, and J. M.
    Hill. 1986. Report on the causes for wetland loss
    on the Harrison family property and critique of
    proposed management plan: 5-31.

Murchison, J. 1978. Bio-Scope Fur Louisiana Conservationist - 1st quarter 1978: 5-8.

Oyster Commission of Louisiana. 1906. Second biennial report 1904-1905, New Orleans, Louisiana.

Palmisano, A. W. 1973. Habitat preference of waterfowl and fur animals in the Northern Gulf Coast Marshes Proc. Coastal Marsh and Estuary Management Symposium. L.S.U., Baton Rouge, LA 163-190.

Penland, S. and R. Boyd. 1981. Shoreline changes on the Louisiana barrier coast. Oceans, 9/81: 209-219.

Perret, W. S. 1981. Management of Sciaenids in the Northern Gulf of Mexico. Mar. Rec. Fish No. 6:145-158. Sports Fish. Inst., Washington, D.C.

U. S. Army Corps of Engineers. 1989. Louisiana Comprehensive Coastal Wetlands Study, Reconnaissance Report, New Orleans District; 1-28.

Viosca, P. 1928. Louisiana wetlands and the value of their wildlife and fishery resources. Ecol. 9(2):216-229.

Zimmerman, R. J. and T. J. Minello. 1984. Fishery habitat requirements: utilization of nursery habitats by juvenile penaeid shrimp in a Gulf of Mexico salt marsh. pp. 371-383. In: B. J. Copeland, K. Hart, N. Davis and S.

Incorporating Global Positioning System Technology Into Coastal Mapping and Research Efforts

William K. Michener[1]

William H. Jefferson[2]

David A. Karinshak[3]

Charles Gilbert[4]

Abstract

The United States Department of Defense developed the global positioning system (GPS) to provide accurate positional data for any location on the earth's surface. The GPS utilizes a ground control segment and a constellation of 21 NAVSTAR satellites (plus three spares) orbiting at an altitude of 20,200 kilometers to triangulate positions on earth. Commercially available GPS receivers differ in position accuracy, cost, ease of transportability, and susceptibility to the adverse environmental conditions (cold, heat, moisture, sand, saltwater, dust, etc.) found in various coastal habitats.

Global positioning system technology coupled with a geographic information system can greatly facilitate coastal mapping and research efforts. In the past these and similar activities have been hindered by necessary reliance on existing technology (conventional costly and time-consuming surveying methods and two-dimensional database management and analytical tools).

- - - - - - - - -
[1]Research Assistant Professor, Baruch Institute, University of South Carolina, Columbia, SC 29208.

[2]Research Specialist, Baruch Institute, University of South Carolina, Columbia, SC 29208.

[3]Senior Cartographer, Baruch Institute, University of South Carolina, Columbia, SC 29208.

[4]Technical Specialist, Trimble Navigation Limited, 645 North Mary Avenue, P.O. Box 3642, Sunnyvale, CA 94088-3642

In this paper, we discuss global positioning system technology, current and potential applications for GPS, and design considerations of a GPS survey. Furthermore, we present specific examples of how GPS technology has facilitated coastal mapping efforts and been utilized to locate intertidal oyster reefs and other sampling sites for long-term research and resource management.

Introduction

As research attention focuses on long-term and large-scale spatial phenomenon (Callahan, 1984; IGBP, 1988), accurate knowledge of our position on the earth's surface becomes critical. Geographic information system (GIS) technology has become an integral component of coastal mapping and research efforts (Michener, et al., 1989). The most costly aspect in development and maintenance of a GIS is acquisition of high quality data layers. Data input to a GIS requires that the coordinates of all sampling sites (points), boundaries (lines), and areas (polygons) be known. Also, as the scale of remotely sensed imagery increases, highly accurate position fixes of ground truth sites become necessary.

In the past, coastal mapping and research projects have been hindered by necessary reliance on existing technology (conventional costly and time-consuming surveying methods and two-dimensional database management and analytical tools). Global positioning system (GPS) technology coupled with a geographic information system can greatly facilitate coastal mapping and research. Collection of ground truth data with GPS is essential in poorly mapped regions and where features suitable for ground-based navigation by triangulation are absent. GPS does not require intervisibility between sites, fewer personnel are required for GPS surveys, and GPS surveys can be performed under inclement weather conditions which may hinder conventional surveys.

In this paper, we discuss global positioning system technology, current and potential applications for GPS, and design considerations of a GPS survey. Furthermore, we present specific examples of how GPS technology has facilitated mapping salt marsh vegetation communities and been utilized to locate intertidal oyster reefs and other sampling sites for long-term research and resource management efforts.

The Global Positioning System

The United States Department of Defense developed the global positioning system to provide timing information, accurate positional data for any location on the earth's surface, and to aid in navigation (Leick, 1990). The GPS utilizes a ground control segment and a constellation of 21 NAVSTAR satellites (plus three spares) orbiting at an altitude of 20,200 kilometers to triangulate positions on earth.

The orbital period for each satellite is approximately 12 hours. The satellites emit coded radio signals which contain timing information, satellite ephemeris information, and information regarding the "health" of the satellites. Ground-based GPS receivers decode the timing and ephemeris information to calculate ground positions. Position fixes can be received at a rate of 1/second. GPS receivers calculate 2-dimensional (latitude and longitude) or 3-dimensional (latitude, longitude, altitude) positions based on data emitted simultaneously by the satellites. The unknown receiver clock bias requires that 3 satellites be utilized to triangulate 2-dimensional positions. Similarly, 4 satellites are necessary for 3-dimensional positions.

Presently (January 1991), 16 satellites are in orbit and additional deployments are scheduled for summer 1991. Satellite 23 was launched on November 26, 1990 and was set healthy on December 10. Satellite 23 was the tenth block II satellite successfully launched by the Department of Defense. The ten block II satellites combined with the six working block I satellites provided worldwide GPS availability averaging 18.4 hours for 3D coverage (10°, PDOP <8) and 23.32 hours for 2D coverage (10°, PDOP <12). The complete constellation (24 satellites) is projected to be in place by mid 1992 providing continuous world-wide 3-dimensional coverage. The satellites will be deployed in six orbital planes inclined 55° to the equator. Eventually, the GPS will replace both LORAN and TRANSIT which currently provide 2-dimensional position fixes.

The accuracy of position fixes obtained by GPS receivers depends primarily on the magnitude of error introduced by the United States Department of Defense. On March 25, 1990 the Department of Defense introduced selective availability (SA) which is a method of denying unauthorized GPS users access to high position accuracy. When SA is not activated the accuracy of autonomous positions is on the order of 10-15 meters. With SA switched on, the accuracy of autonomous position fixes may vary up to 100 m (Georgiadou and Doucet, 1990). Nevertheless, accurate position fixes can be obtained when SA is enabled by using differential GPS techniques. Differential GPS entails establishing one stationary GPS

receiver at a known location as a "base" station. One or
more GPS receivers may then be used to collect data at
remote locations in conjunction with the data being
acquired at the base station. SA errors can then be
corrected since the position of one receiver is known and
any introduced errors can be determined and subtracted
from the remote position fixes. Some GPS systems support
differential correction as a real-time operation, whereas
others compute corrected fixes as a post-processing
operation.

The NAVSTAR satellites transmit two carrier signals:
Link One (L1) and Link Two (L2) which differ in broadcast
frequency (L1 - 1575.42 MHz; L2 - 1227.60 MHz) and
wavelength (L1 - 19 cm; L2 - 25 cm). Both single
frequency (L1) and dual frequency (L1 and L2) GPS
receivers are available commercially. Both the L1 and L2
carrier signals are modulated with pseudo-random ranging
codes which allow time delays to be measured. There are
two situations where dual frequency receivers can be
beneficial: working in high latitudes where ionospheric
disturbances might affect the propagation of radio
signals, or when there is a particularly long base line
between the base receiver and the remote receiver.
Having a dual frequency receiver allows you to calculate
an "ionospheric free" solution.

GPS receivers differ in accuracy of position fixes,
cost, ease of transportability, durability, and
susceptibility to the adverse environmental conditions
(cold, heat, moisture, sand, saltwater, dust, etc.) found
in various coastal habitats. Based on the precision that
can be delivered, GPS receivers can be grouped into two
broad categories, "code phase receivers" and "carrier
phase receivers". Code phase receivers are able to
utilize the code that is being transmitted by the
satellites. Carrier phase receivers are capable of
utilizing both the code and the carrier wave upon which
the code is modulated. Several relatively inexpensive
code phase GPS receivers are commercially available which
can provide 12 meter autonomous accuracy and 2-5 meter
differential accuracy. Carrier phase receivers are more
costly and generally less easily transported, but are
necessary for use in high precision surveys where
latitude, longitude, and altitude position fixes accurate
to approximately 0.5-1 cm (plus 1-2 mm/kilometer baseline
length) are required. Carrier phase receivers may be
either single or dual frequency.

When evaluating GPS receivers for purchase it is
advisable to fully assess user needs for accuracy (i.e.
single versus dual frequency, code phase versus carrier
phase, plus number of channels), planned and potential

applications, durability, and portability. Other
important factors which should be carefully considered
include availability and power of existing software,
accessibility of software and technical support, quality
of documentation, and training.

GPS Applications

GPS technology represents a cost-effective and
efficient method for establishing primary and secondary
control points (Crossfield, 1989; Wood, 1989; Newcomer,
1990). GPS technology has been successfully utilized to
provide first-order control points for use in development
of a geographic information system (Amenda, 1990). GPS
is also a useful tool for resolving differences in
horizontal datums and adjusting different data sets to a
common datum (Henderson and Quirion, 1988; Shrestha, et
al., 1990). It is often imperative in a GIS environment
to have data that are tied to a common reference frame.

GPS surveys can provide the accurate vertical
control data necessary for monitoring land subsidence
(Blodgett, et al., 1990). GPS has been successfully used
to accurately locate ground truth sites for integration
of field data with Landsat TM multispectral imagery.
Wilkie (1989,1990) found that three-dimensional data
could be obtained in forest openings > 0.125 ha where the
angle to the horizon did not exceed 50° and canopy
closure was less than 30 per cent.

The Office of Charting and Geodetic Services
(National Oceanic and Atmospheric Administration)
currently utilizes GPS technology for the high-precision
North American geoid, to develop a national absolute
gravity reference system in conjunction with the Defense
Mapping Agency, to determine precise satellite orbits, to
monitor and analyze crustal motion, and to precisely
determine global and regional sea level changes (Hull,
1990).

Whereas most of the aforementioned applications
required the use of high precision receivers, there are
numerous coastal mapping and research needs which could
be satisfied with a lower level of precision (Table 1).
GPS receivers are indispensable for use in offshore and
coastal mapping and research efforts where the 300 m
accuracy of LORAN units is inadequate. Code phase GPS
receivers are particularly useful in situations where 2-5
meter accuracy is required, but transportability and ease
of setup are primary concerns.

Table 1

Potential GPS Applications for Coastal Mapping and Research

A. Surveying point features
   1. study sites
   2. developing digital elevations
   3. ground control points for registering aerial photos and satellite imagery
   4. relate hand-held radiometer spectral measurements to satellite data
   5. deployment and relocations of buoys or scientific instruments
   6. locating historical sampling sites

B. Mapping linear features
   1. streams, channels, creeks
   2. sand dune lines
   3. beach transects
   4. habitat boundaries
   5. bulkheads
   6. disturbance boundaries

C. Mapping areas
   1. coastal plumes, oil spills, etc.
   2. permanent plots
   3. community mapping (oyster reefs, etc.)
   4. hurricane and storm damage
   5. wetland demarcation
   6. fishing grounds

Designing a GPS Survey

Blair (1989) outlined the factors that should be considered when evaluating the GPS equipment available on the market and additional items that are necessary for implementing a GPS survey. Minkel (1989) and Craymer et al. (1990) discuss error sources in GPS surveys and make recommendations regarding equipment and field procedures which can minimize errors. Prior to a survey, necessary GPS and field equipment should be inventoried and tested. The planning stage should include estimation of setup time, travel time between stations, and satellite visibility windows. Contingency plans should be formulated in the event of "unhealthy" satellites, equipment malfunctions, or unanticipated obstructions to signal reception.

Specific Examples of GPS Applications

A GPS survey was planned for January 10, 1991 in order to accomplish two primary objectives: (1) collect accurate position fixes for 30 vegetation ground truth stations within a North Inlet salt marsh basin and (2) obtain position fixes at 31 oyster reefs located throughout the North Inlet Estuary. The vegetation sites were sampled for aid in interpreting low altitude color infrared aerial photography which had been digitized, spatially registered, and classified into potentially different vegetation communities (polygons). The oyster reefs were flagged and have been sampled for accumulation of heavy metals, pesticides, herbicides, and coliform bacteria. At sites within the creek channel and adjacent to those oyster reefs, water samples were also collected for nutrient analyses.

A satellite visibility chart was obtained using Trimble GPS PATHFINDER System software to determine the optimum schedule for field surveying (Table 2). Results suggested that a 5 hour period (1000 through 1505 offered the longest, uninterrupted period for data collection during the day. Three GPS PATHFINDER receivers were utilized in the field work. One receiver was established at a known control point. A second receiver was used for the boat survey and the third was carried as a backpack unit for collecting ground truth data in the marsh.

Position fixes at the base station had a standard deviation of 11.5 meters, with a maximum offset from the mean of over 40 meters (Figure 1). Position fixes varied up to 88 meters at the base station (diagonal line connecting two most distant points). This relatively high degree of scatter can be attributed to one of two sources. Selective availability (SA) could possibly account for the high variance, or a period of poor satellite geometry (high PDOP, position dilution of precision) could account for degraded positions.

The first possible error source (SA) can be entirely eliminated by performing a differential correction in the post-processing. For example, the position fixes collected at one of the oyster reefs had a standard deviation of 9.8 meters (Figure 2). However, after differential post-processing the standard deviation decreased to 1.9 m (Figure 3). All point data from both the oyster reefs and vegetation communities were processed in a similar manner. In addition, several "walk" files were generated by continually receiving position fixes while walking through specific vegetation communities (Figure 4). These files were also differentially post-processed.

Table 2

Satellite Visibility Chart for North Inlet Base Station,
January 10, 1990

User location: 33° 20' N, 79° 12' W North Inlet
Elevation mask: 10° PDOP switch: 8 PDOP mask: 12
Satellites to consider: 2 3 6 9 11 12 13 14 15 16 17 18
19 20 21 23

GPS Availability from Thu Jan 10 06:00:00 1991 to Fri Jan
11 06:00:00 1991 LCL

LCL		Solutions Possible
06:00	3D	solutions available
06:45	2D	solutions available
07:35	No	solutions available
07:40	3D	solutions available
08:20	2D	solutions available
09:10	3D	solutions available
09:25	No	solutions available
09:30	2D	solutions available
10:00	3D	solutions available
15:05	2D	solutions available
15:30	3D	solutions available
17:00	2D	solutions available
17:40	3D	solutions available
17:45	2D	solutions available
18:25	No	solutions available
19:10	2D	solutions available
20:10	No	solutions available
20:15	2D	solutions available
21:05	3D	solutions available
21:15	No	solutions available
21:30	2D	solutions available
22:45	3D	solutions available
00:40	2D	solutions available
01:00	3D	solutions available

2D for 22:50   3D for 15:30

The second error source, poor satellite geometry,
cannot be eliminated during post-processing operations.
However, satellite geometry is an absolutely predictable
phenomena. Thus, the occasional brief period of poor
geometry can be anticipated well in advance and work can
be planned around these periods.

After completion of the post-processing operation,
position fixes were exported to the ARC/INFO GIS software

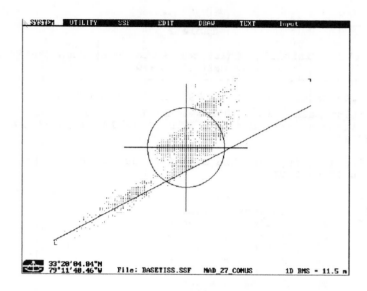

Figure 1.   North Inlet, SC Base File.

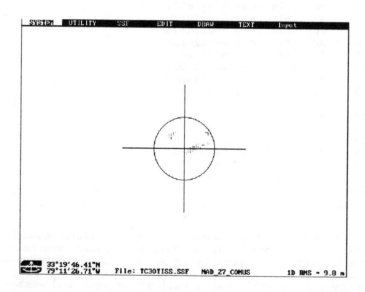

Figure 2.   Uncorrected Point File for Oyster Reef Sampling Site.

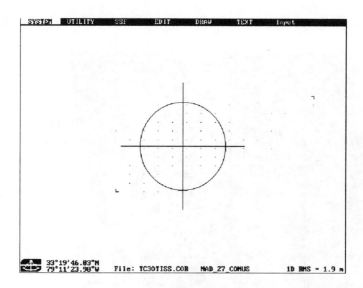

Figure 3.   Corrected Point File for Oyster Reef Sampling
Site.

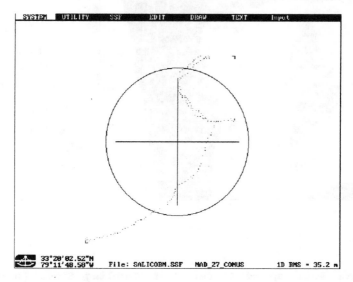

Figure 4.   Continuous Point File Collected During "Walk"
Through a Vegetation Community Dominated by <u>Salicornia</u>.

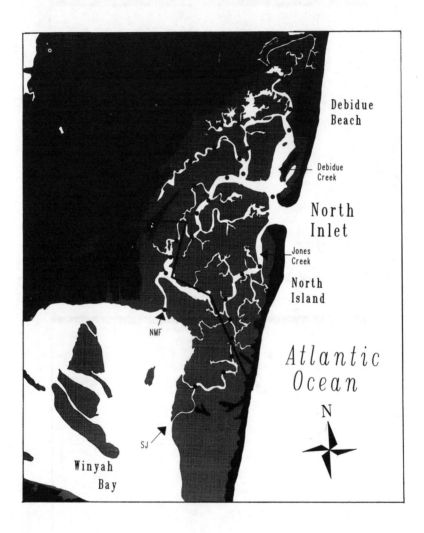

Figure 5. Map of Oyster Reef Sampling Sites Located Throughout the North Inlet Estuary, SC.

BLY CREEK
VEGETATION PATTERNS

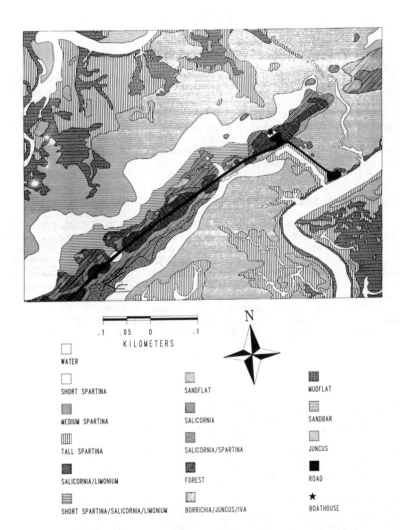

Figure 6. Map of Vegetation Communities Found Within the Bly Creek Basin in the North Inlet Estuary, SC.

package and used in demarcating the oyster reefs (Figure
5) and classifying the vegetation communities identified
in the color infrared photography (Figure 6). The
position fixes necessary for these two activities were
collected by two individuals in less than five hours.
Post-processing operations were completed in another 2-3
hours.

Conclusions

As spatial resolution of remotely sensed imagery and
the need for development of coastal geographic
information systems increase, accurate geographical
positions will be essential for registration of spatial
data layers. GPS technology will likely serve an
integral role in meeting these needs and facilitating
sophisticated spatial analyses. Miniaturization of
components and reduced costs should make GPS technology
accessible to scientists and resource managers for many
diverse applications.

Acknowledgments

Funds for this study were provided by NSF grant BSR-
8514326 and NOAA grant NA90AA-D-SG672. Catherine Coleman
identified the vegetation samples, edited the Arc/Info
coverages, and provided editorial assistance. Sharon
Lawrie and Jennifer Osterman digitized and edited the
registered aerial photography. GPS PATHFINDER is a
trademark of Trimble Navigation, Limited. ARC/INFO is a
trademark of Environmental Systems Research Institute,
Inc. This paper is Contribution Number 836 of the Belle
W. Baruch Institute for Marine Biology and Coastal
Research.

References

Amenda, L.B. 1990. Developing a geodetic framework:
    the Fresno GPS project. Surveying and Land
    Information Systems 50(3):185-188.

Blair, B.R. 1989. Practical applications of global
    positioning system. Journal of Surveying
    Engineering 115(2):218-222.

Blodgett, J.C., M.E. Ikehara, and G.E. Williams.
    1990. Monitoring land subsidence in Sacramento
    Valley, California, using GPS. Journal of
    Surveying Engineering 116(2):112-130.

Callahan, J.T. 1984. Long-term ecological research. Bioscience 34:363-367.

Craymer, M.R., D.E. Wells, P. Vanicek, and R.L. Devlin. 1990. Specifications for urban GPS surveys. Surveying and Land Information Systems 50(4):251-259.

Crossfield, J.K. 1989. Evaluating efficient surveying technology for the land information system environment. Surveying and Mapping 49(1):21-24.

Georgiadou, Y. and K.D. Doucet. 1990. The issue of selective availability. GPS World 1(5):53-56.

Henderson, T.E. and C.A. Quirion. 1988. Use of GPS-derived coordinates in GIS environment. Journal of Surveying Engineering 114(4):202-208.

Hull, W.V. 1990. Geodesy. Surveying and Land Information Systems 50(2):73-76.

IGBP. 1988. The international geosphere-biosphere programme: a study of global change. IGBP Secretariat, Stockholm, Sweden. 200 pp.

Leick, A. 1990. GPS Satellite Surveying. John Wiley & Sons. New York, 352 pp.

Michener, W.K., D.J. Cowen, and W.L. Shirley. 1989. Geographic information systems for coastal research. Proceedings of Sixth Symposium on Coastal and Ocean Management/ASCE. pp. 4791-4805.

Minkel, D.H. 1989. GPS antenna set-up procedures and error sources. Journal of Surveying Engineering 115(3):297-303.

Newcomer, J.D. 1990. GPS as fast surveying tool. Journal of Surveying Engineering 116(2):75-81.

Shrestha, R., R. Taylor, S. Smith, and L.Stanislawski. 1990. Preliminary results for the Florida high-precision GPS network. Surveying and Land Information Systems 50(4):299-302.

Wilkie, D.S. 1989. Performance of a backpack GPS in a tropical rain forest. Photogrammetric Engineering and Remote Sensing 55(12):1747-1749.

Wilkie, D.S.   1990.   GPS location data: an aid to
   satellite image analyses of poorly mapped regions.
   International Journal of Remote Sensing 11(4):653-
   658.

Wood, L.P.   1989.   Combined photogeodesy and GPS network.
   Journal of Surveying Engineering 115(1):160-163.

MARINE RESOURCE MAPPING AND MONITORING IN FLORIDA

Kenneth D. Haddad[1]
Gail A. McGarry[1]

INTRODUCTION

The State of Florida has one of the most extensive coastlines in the United States and climatically ranges from tropical and sub-tropical to temperate. This results in a very complex and diverse assemblage of species and habitats that are often unique and fragile. Florida's population growth is one of the highest in the nation with over 80% of state inhabitants living within 16km of the coastline. The resultant impacts on our marine and estuarine resources, although at times obvious, have been poorly understood, rarely quantified, and assumed to be far reaching.

Most marine resource management strategies and actions in Florida have been oriented to single species. As technical data on the status and trends of our coastal and marine resources have become available, it has become evident that this targeted approach to management is inadequate over the long-term. Habitat has been lost, species abundance has declined, polluted waters have reduced shellfish harvest areas, and fisheries have been closed. This realization has stimulated the evolution of an ecosystem approach to resource management. This approach is based on the fact that without an understanding of species interactions, communities, community interactions, and cumulative environmental impacts (natural and man-induced), our management actions will often be reactive rather than preventive or corrective. Major obstacles to a comprehensive management approach are synthesis of the large volume of existing data, data inconsistencies, and lack of data.

A first step in building a digital ecosystem database is the determination of the extent and location of critical habitat. In

----

[1]Florida Marine Research Institute, Department of Natural Resources, 100 8th Ave. SE, St. Petersburg, FL

1983 the Department of Natural Resources, through the NOAA Office
of Ocean and Coastal Resource Management and Department of
Environmental Regulation, initiated a program to map and monitor
coastal wetlands (including saltmarsh and mangroves) and submerged
habitat (including aquatic vegetation, oyster reefs, and
unconsolidated bottom).

HABITAT MAPPING

Initially, mapping techniques were evaluated to determine cost
and production time comparisons between digital image processing of
Landsat Thematic Mapper data (TM) and cartographic aerial
photography methods. A 69% cost saving and 83% production time
reduction was realized with TM data (Haddad and Harris, 1985a). It
was also determined that aerial photography was often needed for
photointerpretation and digitization into the resource map when
submerged habitats were being mapped (Haddad and Harris, 1985b).
Based on these results, a systematic mapping of Florida's estuarine
and marine wetlands, excluding the Everglades National Park and
Biscayne Bay, was initiated. The primary data source was LANDSAT
TM.

Base Map and Geographic Referencing

A decision on a base map was required early in the program.
The base map is the digital map to which all data are referenced.
As is common for many areas, a digital base map was not available
on a statewide basis and the cost of digitization was prohibitive.
It was determined that the only reasonable approach was to make the
TM data the base map and digitally rectify any additional habitat
types (i.e., seagrass, oysters) would be digitally rectified to that
base.

TM data consist of 6 spectral layers of information for each
1/4 acre (30m x 30m) on the ground and a thermal layer with 4-acre
resolution. Each spectral band was rectified to 7.5 minute USGS
quadrangles in a Universal Transverse Mercator (UTM) projection
using a bilinear interpolation technique (Junkin et al, 1981). This
type of process can achieve accuracy standards for 1:50,000 scale
maps and approach standards for 1:24,000 scale maps. Rectification
of the individual spectral bands, rather than the finished product,
is standard because of our need to continually return to the
spectral data for additional analyses.

Wetlands Image Analyses

We have not developed a rigid process for statistical analysis
of the satellite image data, but workable techniques have been
standardized. Numerous types of statistics have been tested for
their ability to classify marine and estuarine wetlands and for
their relative computer processing times. Standard classifiers such
as the maximum likelihood, which can use either supervised or

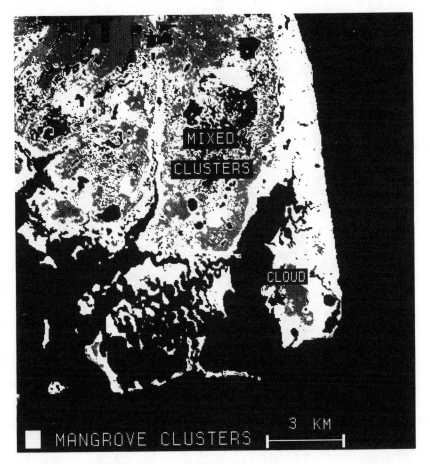

Figure 1.   Statistical clustering of TM green, red and near-infrared spectral bands into 256 clusters. Forty clusters make up the majority of the mangrove but many of those clusters are mixed with other habitat categories.

unsupervised approaches to generate statistical clusters, have been found to be processing intensive and cumbersome in a production operation.   This observation is based on our specific needs, relative to coastal wetlands, and does not consider the use of this approach for general mapping needs. With this type of algorithm, and with most algorithms in use, the higher the spatial resolution the more difficult it is to resolve confusion within and among clusters.   At   some   point   human   intervention   with   a photointerpretive-like process is necessary.

Our approach has been to use a very rapid parallelepiped type of classifier (Junkin et al., 1981) to initially process the data

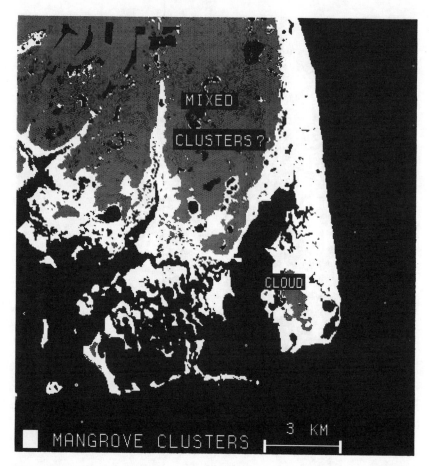

Figure 2. Statistical clustering of TM red, near-infrared, and mid-infrared spectral bands into 256 clusters. A close clustering relationship is evident with 15 clusters making up the majority of mangroves. Confusion between mangrove and other habitat types is minimal. Analysis emphasis on the infrared bands allows this enhanced differentiation.

into 256 clusters. The classifier is applied to the green, red, and near-infrared TM bands, and the red, near-infrared, and mid-infrared TM bands, respectively, to generate two statistical images. The first image, pictorially similar to a color-infrared photograph, can be image-interpreted by identifying those clusters that represent the categories of interest. It is often advantageous to use the second image because of its accentuation of the infrared bands. In particular, we have found that the mid-infrared band enhances our ability to differentiate wetlands. An analysis of mangrove distribution for a 7.5 minute USGS quadrange (Punta Gorda SW), in

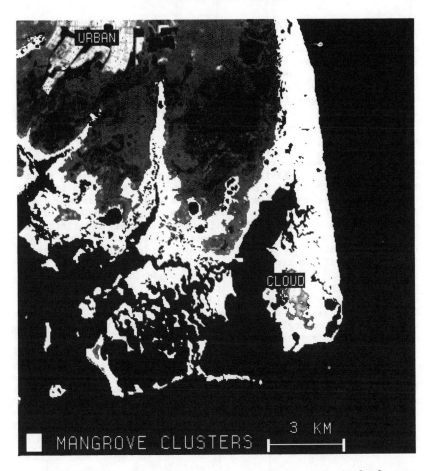

URBAN

CLOUD

3 KM

■ MANGROVE CLUSTERS

Figure 3. Composite image based on the best statistical clusters from figure 1 and figure 2 and interactive interpretation to delineate mangove habitat.

the Charlotte Harbor area of Florida, depicts the differences in the analysis of different spectral band combinations. In Figure 1, the green, red, and near-infrared bands were clustered and then grouped to delineate mangrove distribution. In this case mangroves were represented in 48 statistical clusters that also included freshwater wetlands, pine flatwoods, and other habitat types. The same analysis using the red, near-infrared, and mid-infrared bands (Figure 2) resulted in 15 clusters delineating mangroves, with minimal confusion between mangroves and other habitat types. In many cases, we use both images to selectively differentiate categories of interest, with the results being a third image comprised of the best clusters from each of the two images (Figure 3).

Although this approach is rapid and effective, it still does not meet land cover identification accuracies expected for wetlands mapping when compared to interpretation of photographs at similar spatial resolutions. The associative and subjective analyses performed by a photointerpreter are not yet reproducible statistically. On the other hand, use of the TM mid-infrared band can have advantages in certain analyses where identification of different levels of moisture content enhance the ability to differentiate wetlands types beyond those observable in an infrared photograph.

Once the images are statistically clustered, available aerial photographs, existing National Wetlands Inventory (NWI) maps, ground truthing, and a myriad of other data sources are used to identify or confirm clusters that are not pure to a given wetlands type. For example, some of the clusters representing mangroves may be confused with freshwater wetland resulting in an unacceptable identification accuracy (Figures 1 and 2). The remote sensing literature has many examples of this type of confusion and reports the statistical inaccuracies of this type of analysis (Dottavio and Dottavio, 1984; Hopkins et al., 1988). This reflects academic studies for image analysis and not a production approach. We routinely "fix" the confused clusters by using simple digital manipulations based on the interpreter's assessment of the data (Figure 3). Freshwater wetlands, for example, would be "manually" reclassed into appropriate categories that would increase identification accuracies.

This flexible and rapid approach to wetlands mapping results in a high accuracy product, but only for wetlands. We standardly produce a final map product which merges the wetland types with the original color infrared-like image (Figure 3). By providing this pictorial image for the background data, the user is able to orient to the image and eliminate the need for a summary presentation of the data not classified as wetlands.

HABITAT TREND ANALYSES

Habitat trend analyses have also been completed for selected areas around the State from the 1940s to the present. A major conclusion from the trend analyses has been that submerged aquatic vegetation has often undergone the greatest loss, and this loss is no longer due to mechanical impacts but, rather, changes in water quality. This is supported by the fact that losses often occur in deeper waters within the estuaries, suggesting insufficient light penetration as a causative factor. Compared to the 1940s-1970s, the rate of loss of marsh and mangrove has substantially decreased in Florida, and where sufficient protective measures have been established, increases in aerial extent have been observed (Haddad and Harris, 1985b).

Trend analyses for coastal wetlands can be conducted with numerous techniques. The development of the multi-temporal databases to be used for comparative analyses must be done with caution. In most cases, it is difficult to separate errors in the mapping process from actual habitat changes. Trend analyses cannot be conducted on data that use different classification systems and categories that have not been normalized. In fact, it is very difficult to compare data that have been interpreted by different investigators, even when using the same classification system, if tedious interpretive calibrations are not conducted. However, if done properly, habitat trend analyses can provide valuable insights on impacts of habitat management regulations and changes in marine resources which utilize those habitats.

Historical Trend Analyses

Historical analyses have been accomplished for a number of areas in Florida by photo-interpreting archived photographs from the 1930s to the 1970s and 1980s. We rectified the photo-interpreted data to the Landsat base map and table digitized the results as a separate data layer. When using aerial photography, the interpretations often must be transferred to a USGS quadrangle, using standard transfer techniques, to geo-correct the data for spatial and scale inconsistancies prior to digitizing. We can occasionally by-pass this step by using a three-point triangulation method and then digitizing directly from the photographs into the TM base image. This method can only be used if the photographs were well controlled during flight acquisition. During the digitizing process, if positional deviations are observed relative to the base map, new points are selected and the digitizing is continued. If the interpretation of the historical photographs is compatible to the TM analyses, trend analyses can be conducted.

When building a database for trend analyses it is important to create an accurate habitat data layer to which historical and future data will be compared. If an acceptable habitat map does not already exist then the most current data should be used to develop that layer. The recent data can be ground truthed and corrected for errors in classification which cannot be accomplished for historical photographs. This also gives the investigator a "feel" for the area and increases the potential for accurate interpretation of the historic photos. By expending initial efforts in the development of the most current data, a considerable reduction in effort is realized when developing the historical database and conducting future map updates.

Database Updates

One approach to updating a habitat database for trend analysis is to remap an entire area for comparison with the original maps. That process is time consuming. It may not be necessary to remap an entire area when, if updating is done on a 1-5 yr basis, change would be expected to involve a small percentage of the total area. Since TM data are digital, we use a technique that takes advantage of that attribute. When working with a focused database, such as

coastal wetlands, we process the new TM data into 256 classes as
previously described. This process results in an image that can be
manipulated to update the original map. The original data are used
to mask a given habitat that is then compared in a very rudimentary
fashion to the new TM image. For example, when updating mangroves,
we would use the original coverage of mangroves (Figure 4) to locate

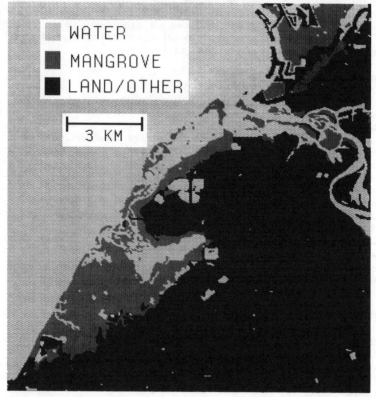

Figure 4.   1982 digital NWI map depicting the distribution of
mangroves used to mask the 1987 TM data for the map update process.

those areas in a new TM image which should contain mangroves (Figure
5).   Mangroves, in the new image, can be expected to fall within a
specific range of statistical clusters, and those clusters that fall
outside that range are identified as potential areas of change
(Figure 6).   These areas can then be assessed to confirm the
changes.   An inverse process can be used to identify areas of
mangrove growth.   We have not tested this approach due to
insignificant amounts of growth in wetlands since our initial
mapping effort utilizing TM data. The final result of this process
would be an update of the original map with the change depicted in
Figure 6.

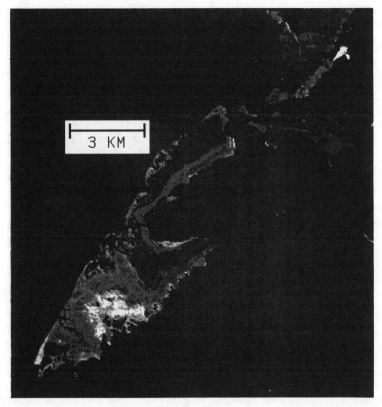

Figure 5.   Statistically clustered 1987 TM data for the areas defined as mangrove in 1982.

Update Cautions

Although Figures 4-6 depict the updating process, we have used mangroves digitized from a 1:24,000 scale 1982 U.S. Fish and Wildlife Service National Wetlands Inventory aerial photograph mapping effort, instead of TM classified data, as the mask to a 1987 TM image. This was done to show both the process and, if using disparate data sources, the problems. Areas of change can represent differences in original and final product resolution, spatial misregistration, habitat classifications, and real changes in habitat. Figure 4 is a general map of a coastal area of Tampa Bay, Florida. The data have been consolidated to 3 classes and are a digital representation of the 1982 NWI map. Figure 5 represents the statistically clustered 1987 TM data for the area of mangroves delineated in the 1982 data. Figure 6 shows those areas that were labelled as mangrove in 1982 but that were not classified as mangrove in 1987. Quantitatively the area was reduced from 2,435 ha of mangrove to 2,156 ha, an 11% (279 ha) loss. However, when

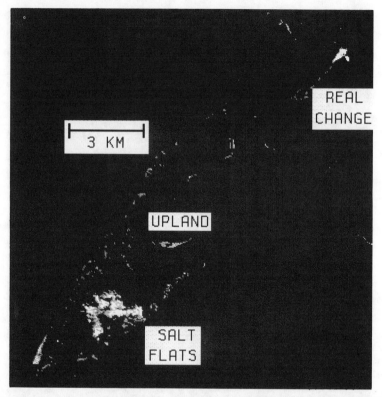

Figure 6. Areas depicted as mangrove in 1982 but not within the statistical clusters defining mangrove in the 1987 TM data. These clusters would represent areas of potential change.

investigating the changes it becomes obvious that a large portion of that change is not real and represents differences in interpretation techniques and classification systems. Actual loss is < 1%. Many of the smaller areas of change are actually uplands within the mangrove complex (Figure 6). These types of features are averaged by the photointerpreter to become mangroves. In the photointerpretation and digitization process it becomes impractical and costly to try to delineate less than one acre features. The photointerpreter makes a conscious decision to delineate them, or lump them into the mangrove classes; digital processing automatically maintains their separation.

The utilization of different classification systems and remote sensing techniques contributes to the greatest discrepancies in data updating and trend analyses. The NWI maps are based on the Cowardin system (Cowardin, 1979), while the State of Florida uses the Anderson (Anderson, 1976) system modified to state needs. Use of TM data often precluded effective application of either of these two common classification schemes. Figures 7-9 depict a subset of the test area, labeled as salt flats in Figure 6, interpreted for mangrove and flats using the NWI Cowardin System from 1982 photography, the Florida modified Anderson System from 1982 photography, and spectral analyses and clustering from a 1987 TM image, respectively.

The NWI interpretation (Figure 7) depicts 1050 ha of mangroves and no tidal or salt flats. The Anderson interpretation (Figure 8) depicts 816 ha of mangroves and 164 ha of tidal flats. The TM interpretation (Figure 9) depicts 919 ha of mangroves and 131 ha of salt flats. For mangrove, between 1982 and 1987, this represents a 10-22% difference in mangroves area and a 20-100% difference in tidal or salt flat area even though little change, if any, has occurred. The differences can be explained by the following:

1. The NWI Cowardin interpretation actually identified 198 ha of the 1050 ha of mangroves as E2SS/F1 indicating the area to be mangrove predominant with irregularly flooded tidal flat observable but not delineated. This forced us, for comparison, to label those polygons as mangrove. This interpretation was a result of the classification system used then and subjective interpretation by the photo-interpreter not to delineate the flats.

2. The Anderson interpretation resulted in the separation of the mangroves from tidal flats, and thus, a significant difference between the NWI and Andereson interpretations were observed.

3. The TM analysis was based on spectral reflectance, and thus, the highly reflective surface of the exposed sand allowed easy differentiation between the flats and mangroves. The difference between the Anderson and TM analysis was due to subjective lumping, by the photo-interpreter, of smaller areas of mangroves into the salt flats.

Finally, neither the NWI or Anderson classifications adequately describe the flats. NWI would define them as intertidal, irregularly flooded flats, while Anderson would call them tidal flats (protected from wave action, lacking vegetation, and intertidal). In reality, they are a unique land cover called salt flats or salterns and are characterized by hypersaline soils and sparse distributions of salt-tolerant vegetation that can support endemic species.

Although we have pointed out some of the discrepancies in our attempt to compare NWI or other photointerpreted maps with TM data, there is inherent value in integrating these data and other digital data in the mapping process. This can range from using masking techniques for selecting TM image analysis training areas, to verification of TM analysis results or the generation of a hybrid NWI/TM map, that can be updated with digital imagery. It may be that, to conduct trend analyses for multiple habitats, TM to TM are

Figure 7.   NWI classification depicting mangrove distribution and no salt or tidal flats for 1982.   This is a subset of the area depicted in figure 4-6.

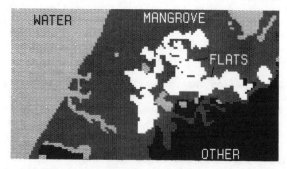

Figure 8.   Anderson classification depicting mangroves and tidal flat distribution for 1982.

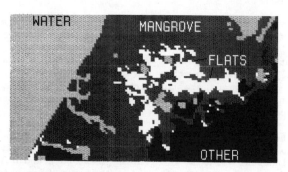

Figure 9.   LANDSAT TM analyses depicting mangrove and salt flat distribution for 1987.

the only comparable data, and comparisons must be made at the spectral level, not with the final map product. These types of applications need further investigation.

The point to be made is that trend analyses must be conducted with caution and with a full evaluation and understanding of the data being compared. Despite the data discrepancies depicted in Figure 4-9, there are areas of actual change delineated. If the original 1982 image used was TM based, rather than NWI, then the data updating would not have the problems that have been identified. This does not indicate that one process is better than the other, just that they are different and require special considerations if they are to be compared.

The importance of the classification system cannot be under estimated when using satellite image processing for habitat delineation; this is something that must be addressed in the initial stages of the mapping program. Because of inherent differences, it is best to determine the limits of the existing classification systems relative to TM processing and develop a hybrid system. If this is not done, much effort can be spent attempting to force a classification of the image data, which reduces the ability to efficiently conduct trend analyses.

SEAGRASS MAPPING AND TREND ANALYSIS

Seagrass mapping presents special problems for satellite image analyses. Landsat only collects an image over a given area once every 16 days. This means that conditions conducive to mapping must all coincide on that given day. If the water is clear, clouds do not obscure the area, and the tides are appropriate, there is good potential for using imagery for seagrass mapping (Haddad and Harris, 1985b). We have not found any statistical analyses that adequately define seagrasses. Variations in water clarity, water depth, and sediment type preclude the efficient use of standard spectral analyses. The image must be manually "photointerpreted" in either the blue, green, or red spectral bands. Because of these current obstacles, we commonly use aerial photography, either existing or contractually flown, to map seagrasses. The photographs are photointerpreted, and the seagrass coverages are digitized as wetlands types into the wetlands database.

There are currently no standard classification systems for seagrass mapping and trend analysis. There are no remote sensing techniques, including aerial photography, that can be used to consistently delineate seagrass species, so the most common approaches for mapping have been based on varying levels of density. A standard approach is to use some variation of percentage crown closure in discreet ranges between 0% and 100% (Orth, 1989). The more density categories defined (i.e., 0-10%, 10-20%, 20-30%,....) the less likely that valid trend analyses can be conducted without lumping the categories. In Florida, with over 1.5 million acres of seagrass, we, the Department of Natural Resources, have simplified

the mapping process and conducted trend analyses based on seagrass-
bed morphology, which is not as susceptible to subjective mapping
or seasonal variations in seagrass densities. We now suggest 3
categories: (1) medium/dense; seagrass present in contiguous
distribution patterns (2) patchy; medium/dense seagrass interspersed
with sand/mud patches or sand/mud interspersed with medium/dense
seagrass patches and (3) sparse; not observable by remote sensing
techniques because the density is too low to change the character
of the bottom substrate. This is only mapped by in situ
observation.

If trend analyses and updating are to be conducted for
seagrasses, the development of the base map, resolution of the
photography, and the mapping classification categories must be
considered prior to initiation. There is a need, on a national
basis, to develop a flexible process for seagrass mapping and
conducting trend analysis (Orth et al., in prep.).

CONCLUSIONS

The Florida Department of Natural Resources has developed and
implemented a coastal mapping effort designed for efficient and cost
effective mapping and monitoring of Florida's geographically
expansive coastal wetlands. A combination of Landsat satellite
imagery, aerial photography, ground truthing, and ancillary map data
are used to produce digital maps from a Landsat TM map base.
Aerial photography is used for seagrass mapping and trend analyses.
We have described, generally, the techniques and concepts we employ
in the map making and habitat trend analyses. The success of this
effort has been based on the flexibility built into the
standardization of the mapping process.

Many issues, such as ground truthing and digital and hardcopy
data distribution, have not been discussed. All require substantial
planning and can become major operational components of an effective
program. We have also evaluated SPOT satellite data for mapping
efficiency and use it when higher resolution mapping is required.
But the spectral superiority (particularly the mid-infrared bands)
and lower costs of Landsat TM data make its use more advantageous
for large geographic areas.

Although our habitat mapping effort is substantial, it has
little long-term meaning if the habitat is not considered as part
of an ecosystem. The wetlands are just one layer of information out
of many that we are compiling in the Marine Resources Geographic
Information System. Linkage to dredge-and-fill permits and other
types of permits, which allow us to reconstruct permitted habitat
losses which cannot be mapped, is being investigated. Concurrent
with the habitat mapping efforts, field research is being conducted
to quantitatively assess species utilization and production within
different habitats. All of these efforts will eventually provide
the information necessary to implement an ecosystem approach to
coastal resource management.

ACKNOWLEDGMENTS

Funds for this project were provided by the Department of
Environmental Regulation, Office of Coastal Management using funds
made available through the National Oceanic and Atmospheric
Administration under the Coastal Zone Management Act of 1972, as
amended.

LITERATURE CITED

Anderson, R., E. Hardy, T. Roach, and E. Witmer. 1976. A land use
and cover classification system for use with remote sensor
data. U.S. Geol. Surv. Prof. Paper 964. 22 pp.

Cowardin, L. M., V. Carter, F. C. Golet, and E. T. LaRoe. 1979.
Classification of wetlands and deepwater habitats of the
United States. U.S. Fish Wildl. Serv. 103 pp.

Dottavio, C. L., and F. D. Dottavio. 1984. Potential benefits of
new satellite sensors to wetland mapping. Photogram. Eng. and
Remote Sensing 50(5):599-606.

Haddad, K. D., and Harris, B. 1985a. Use of remote sensing to
assess estuarine habitats. Pp. 662-675 in: O. T. Magoon, H.
Converse, D. Minor, D. Clark, and L. T. Tobin (eds.), Coastal
Zone 85, Vol. 1, Proc. Fourth Symp. on Coastal and Ocean
Management. 1294 pp.

Haddad, K. D., and Harris, B. 1985b. Assessment and trends of
Florida's marine fisheries habitat: An integration of aerial
photography and Thematic Mapper Imagery. In: Eleventh
International Symposium, Machine Processing of Remotely Sensed
Data, LARS/Perdue University, Indiana, pp. 130-138.

Hopkins, P. F., MacLean, A. L., and T. Lillesand. 1988. Assessment
of thematic mapper imagery for forestry applications and lake
states conditions. Photogrammetric Engineering and Remote
Sensing, Vol. 54, No. 7, pp. 61-68.

Junkin, B., R. Pearson, R. Seyfarth, M. Kalcic, and M. Graham.
1981. ELAS Earth Resources Laboratory Applications Software.
NASA/NSTL Earth Resources Laboratory Rep. No. 183. NSTL
Station, MS

Orth, R. J., A. A. Frisch, J. F. Nowak, and K. A. Moore. 1989.
Distribution of submerged aquatic vegetation in the Chesapeake
Bay and tributaries and Chincoteague Bay 1987. Chesapeake Bay
Program, U. S. Environmental Protection Agency, Annapolis,
MD. 247 pp.

Orth, R., Ferqusen, R., and K. Haddad, in prep. CZ91 Proceedings.

THURSTON REGIONAL WETLAND MAPPING METHODOLOGY

Steven W. Morrison, Senior Planner[1]

ABSTRACT

In Washington State and particularly within the Puget Sound region, local governments find themselves facing the challenge to protect wetlands without adequate resources. Unfortunately the federal government and state agencies have not yet developed a standardized wetland inventory methodology. Aerial photographic interpretation (similar to the National Wetland Inventory) is an inexpensive means of mapping a large area, however it lacks adequate detail for many uses. Field inventories generate adequate detail, but are generally limited to smaller areas because of their higher cost. However, the most cost effective option may be to combine the advantages of both techniques through the use of a geographic information system (GIS).

The Thurston Regional Planning Council (TRPC) is a 15-member intergovernmental board made up of local governmental jurisdictions within Thurston County including Olympia, the state capital. TRPC has developed a methodology to inventory wetlands using large-scale color infrared aerial photographs with a limited amount of field boundary reconnaissance. The aerial and field data were collected and analyzed using a GIS. The TRPC Pilot Project mapped 21 square miles of wetlands and stream corridors in northern Thurston County during 1990. The primary objective was to develop an aerial survey methodology which balanced cost with accuracy.

SETTING

Thurston County, Washington is a rapidly urbanizing county at the southern terminus of Puget Sound, 60 miles south of Seattle. (See Figure 1) Olympia, the state capitol, lies at the heart of this urbanizing area which also includes the neighboring cities of Lacey and Tumwater. In the southern part of the county lie four small towns. This

---

[1]   Thurston Regional Planning Council, 2000 Lakeridge Drive SW, Olympia, WA 98502

717 square miles county includes a diverse geographic mixture of low foothills, peninsulas, prairies and river valleys much as a result of the pleistocene glaciation. Wetlands are most common in northern Thurston County and can be found along the 90 miles of Puget Sound shoreline, adjacent to the county's 107 lakes, or in association with the three major rivers systems. The TRPC Profile has estimated that approximately 8.0% of the County is covered by wetlands with another 1.4% being cover by lakes or ponds, based upon the existing local wetland maps.

THURSTON COUNTY, WASHINGTON
Figure 1

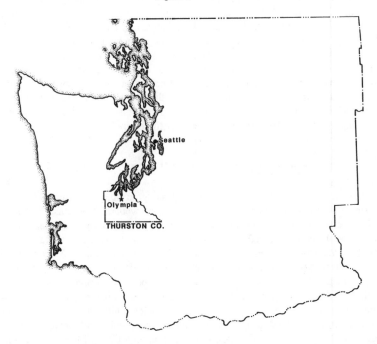

WETLAND PROGRAMS IN THE THURSTON REGION

Over the past decade TRPC has been involved in wetland and stream corridor issues including those projects listed in Table 1. All except for the last have been funded by grants from the Washington State Department of Ecology through the Coastal Zone Management program. As a result TRPC has come to recognize the increasing significance of preserving wetlands to maintain and improve water quality within the Thurston Region.

TABLE 1 -- TRPC WETLAND AND STREAM CORRIDOR PROGRAMS

1983   Stormwater Management in North Thurston County (Vols. I & II)
1986   Wetland and Stream Corridor: Phase I
1987   Wetland and Stream Corridor: Phase II (Vols. I and II)
1989   Wetland and Stream Corridor: Pilot Project
1991   Wetland and Stream Corridor: Demonstration Project

There are a number of reasons why the Thurston Region needs updated wetland and stream corridor mapping. Briefly, several local jurisdictions have adopted or are updating their land use regulations which require more accurate maps. The new state Growth Management Act requires local jurisdictions within Thurston County to adopt "Critical Areas" regulations for wetlands and streams. The 1991 Puget Sound Water Quality Management Plan will also require those same jurisdictions to adopt wetlands and stormwater programs to control water quality pollution. Better wetland and stream corridor maps are also needed to support several stormwater, lake management, and groundwater programs. Because of these extensive regional and intergovernmental activities more accurate wetland maps are needed by both planners and stormwater engineers. Since the needs are region-wide, a cooperative regional mapping and data collection program was instituted by TRPC.

WETLAND MAPPING IN WASHINGTON STATE

Wetland and stream corridor mapping in Washington State is currently in transition as the state begins to require the adoption of local wetland programs. However, the State does not have an established protocol to map or inventory wetlands and stream corridors. Historically, wetland maps were prepared as a part of a wetland inventory survey which placed a heavy emphasis on field reconnaissance. While this method create reasonably accurate wetland maps, its application on a regional basis is limited by its relatively high cost which then in turn possibly limits its scope. In this technique field biologists rely upon aerial photos and other maps of varying scales and accuracies. Over the next decade, State agencies could require local governments to implement wetland programs without an economic and accurate means of accomplishing this task. Thurston Regional Planning Council has made a first step at bridging this gap.

DESIGNING THE TRPC WETLAND PILOT PROJECT

In 1989 Thurston Regional Planning Council initiated a wetland Pilot Project to address its regional needs. Simply stated, the goal of TRPC was to produce reasonably accurate wetland and stream corridor maps of a large geographic area, for both urban and rural areas, at an affordable price. After researching the previous wetland inventories in Washington State, it was discovered that several focused on the collection of field data to support a wetland ranking system. A similar focus in Thurston County would not meet the local needs and appeared to be a waste of money without a statewide ranking system. Therefore, TRPC initiated a wetland mapping program as a first step, while recognizing that wetland ranking system requiring detailed field information could be added at a later date.

The staff evaluated different types of aerial photographic techniques. Aerial scanning, which uses video tape, was discarded as an experimental technique and remote sensing, which relies on satellite images, would not provide adequate resolution, particularly in the urbanizing areas. While the staff had documented significant inaccuracies in the preliminary National Wetlands Inventory (NWI) maps from which the local wetland maps were produced, the fundamental technique of an aerial survey approach appeared to be promising.

The interpretation of standard low level aerial photography was a known and reliable technique which could be flown by a number of contractors. Since the NWI program used color infrared (CIR) film to delineate wetlands, its seemed the most appropriate even though CIR is more expensive than black and white or true color. From discussions with numerous wetland and aerial photography experts, it was determined that the best "window" to take the aerial photographs is in early spring prior to leaf out of deciduous trees.

The scale of the aerial photographs was also evaluated in terms of cost, accuracy and ease of handling the photographs. Registration of the photos to the ground could be accomplished through surveying coordinate system, which require at an absolute minimum 3 points per photo. Photo scales of from 1" = 500' to 1" = 2,000' were evaluated and resulted in a compromise scale of 1" = 1,000'. At this scale an entire square mile fits onto a 9" x 9" print. By comparison the NWI scale for aerial photo declination of wetlands is almost 5 times smaller at a scale of 1" = 4,800'. It was therefore assumed that an accuracy greater than that of the NWI could be obtained by using a similar technique but with larger scale CIR photos.

Aerial photographic interpretation require some ground truthing which can involve a brief trip to the field to match the photographic "signatures" with the habitat types and review questionable signatures. However, it was believed by some that aerial photography _alone_ could not provide the desired level of accuracy and therefore additional field verification would be necessary. TRPC needed wetland maps which had enough detail that they could be used for various planning purposes, but which would not substitute for a wetland boundary evaluation for a specific development permit.

Concern about how the data would be managed and updated lead TRPC to investigate the use of geographic information system (GIS). The Thurston County Public Works

Department operated a regional computer mapping system, called the Thurston Geographic Information Facility (TGIF), which could perform several GIS functions such as the electronic comparison of polygons and the creation of variable scale color maps. The system was also capable of a technique called "rubber sheeting" which allows the computer to enlarge or reduce an image in one axis and at a different ratio from the other axis. Although not as accurate as ortho-corrected photos, this technique could link the aerial photos with the assessor's base map, provided there are enough survey control points.

PILOT PROJECT METHODOLOGY

The Pilot Project was funded by two Coastal Zone Management grants totaling $48,600 administered by the Washington State Department of Ecology (ECOLOGY). Originally designed to cover twenty square miles of northern Thurston County in and around the cities of Olympia, Lacey and Tumwater, an additional square mile was added along Puget Sound. This 21 square mile area was divided into three flight lines (A, B, & C) centered along a north-to south axis of section boundaries. (Refer to Figure 2).

PILOT PROJECT FLIGHT LINES
Figure 2

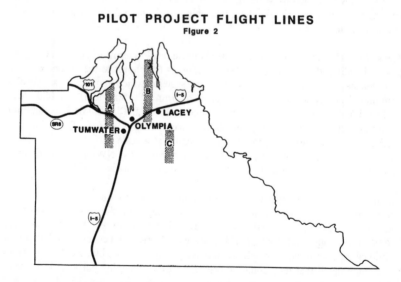

To complete the Pilot Project, TRPC contracted with three consultants to form a Pilot Project Team. The Washington Department of Transportation (WDOT) was hired to take the aerial photographs and Shapiro Associates Inc. (SHAPIRO) of Seattle was engaged to interpret the aerial photographs and undertake the field reconnaissance. The Thurston Geographic Information Facility (TGIF) provided computer mapping and data analysis services. Due to limited budget and time constraints, TRPC bypassed the standard request for proposal process in lieu of consultants with whom it had worked or had proven expertise.

The TRPC Pilot Project used large scale (1" = 1,000') color infrared (CIR) aerial photographs taken during the spring. SHAPIRO used a stereoscope to interpret the CIR aerial photos and marked the survey coordinate points on a mylar overlay. They used United States Geological Survey (USGS) quads, NWI, soils, and local wetland maps as references during the interpretation process. Wetlands were classified by using the United States Fish and Wildlife Service (USFWS) classification system (Cowardin, 1979).

A key part of the Pilot Project design was to designate the wetland boundaries as either "certain" or "uncertain" based upon the interpreter's level of confidence. Certain boundaries had greater than a 90% confidence, with uncertain boundaries being any level less than that. Solid lines were used to delineate "certain" wetland boundaries and dashed lines for "uncertain" boundaries. Both certain and uncertain boundaries were field surveyed in order to test the accuracy of this technique. Although TRPC wanted a wetland threshold of 0.25 acres, given the scale of the photos SHAPIRO was only able to distinguish a 1/8 inch square, which corresponds to 0.35 acre wetland. Stream corridors less than 125 feet in width were represented by a line.

Before the field reconnaissance began, it became necessary to resolve a difference of opinion between TRPC and Ecology on the level of field documentation required. Two field protocol methods and two soil analysis techniques were designed which SHAPIRO tested in a selected square mile of flight line C. The results indicated an equal accuracy in the mapping of wetland boundaries, but the method which required a more detailed data collection sheet and the use of soil test pits resulted in 40 % fewer observation and data points. All the parties meet and evaluated the data in a collaborative process which resulted in the redesign of field data forms and the soil testing protocol which meet the needs of TRPC, budget limitations of the Pilot Project, and Ecology grant requirements.

The SHAPIRO Aerial data was digitized by TGIF. Colored maps comparing the SHAPIRO Aerial and NWI boundaries were used by SHAPIRO to select locations for field verification. SHAPIRO visited both certain and uncertain boundaries and all types of wetland classes. In some locations an observation point was noted on the map, whereas at other data points a field data sheet was filled out. In the field SHAPIRO also attempted to identify areas which were incorrectly identified from the aerial photo interpretation, here referred to as "over coverage" or "under coverage". Over coverage occurred when a site was identified as a wetland from the aerial photograph which was not a wetland. Under coverage was when a site identified as an upland area from the aerial photograph, but was found to be a wetland.

A letter from TRPC was sent to all potential wetland property owners in each flight line before SHAPIRO did any field reconnaissance. The letter explaining the Pilot Project and purpose for the field visit. In all, TRPC mailed 550 letters which represented 4.2% of the property owners in those areas.

All the SHAPIRO aerial and SHAPIRO field data was digitized into the TGIF computer system. The TGIF system consists of a county-wide database stored on a VAX main frame. The data base consists of section maps at the parcel level. Each section map was exported from the VAX in DXF format for data entry and updating with an AutoCad software package. Once data entry was complete, the files were transferred to the GIS system where the layers were electronically overlaid and compared. These comparisons were then input to an RBase base file for tabular output.

Eventually, six wetland mapping layers which were digitized into the TGIF system for the purpose of comparing accuracy. These included the National Wetland Inventory (NWI), Washington Department of Natural Resources (WDNR) soil maps showing hydric soils, and existing wetland maps from Olympia and Thurston County. The local wetland maps were created from preliminary NWI and WDNR soil maps in the mid 1980's. The current NWI and WDNR maps were then combined to create a composite layer.   The mapping from SHAPIRO was digitized after aerial photography interpretation and after the field reconnaissance. These layers are referred to as SHAPIRO Aerial and SHAPIRO Field respectively. All the other layers were compared against the SHAPIRO Field data, which was assumed to be the most accurate by TGIF, using the calculation in Table 2 and described in Figure 3.

TABLE 2 -- ACCURACY CALCULATION FORMULA

Percent mutual coverage (A_m / A_{std}) X 100%
Percent under coverage (A_u / A_{std}) X 100%
Percent erroneous (over) coverage $((A_e - A_m) / A_{std})$ X 100%

A_{std} = The SHAPIRO Field
A_e = Area given by method to be evaluated
A_m = Area mutually covered by both methods
A_u = Area not covered by method being evaluated

WETLAND LAYERS COMPARATIVE ANALYSIS
Figure 3
SHAPIRO FIELD

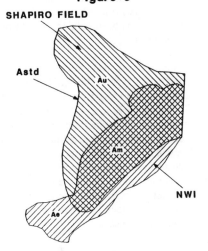

PILOT PROJECT RESULTS

TRPC confronted a normal range of issues in completing the Pilot Project. Some delays were encountered, but none were critical. The only annoying delay was with TGIF which was unable to work on the analysis of the wetland data due to a previous commitment on a major project which was also delayed. TRPC work well with all the consultants, and to its credit, SHAPIRO was especially adaptive and responsive to necessary changes in the scope and products. Because much of the wetland data analysis is yet to be completed, this report will contain very little detailed analysis on the various project costs.

There do not appear to be any fatal flaws in Pilot Project approach, but the cost effectiveness of the methodology still needs to be fully evaluated. The Pilot Project approach seems to be an appropriate design for the TRPC Demonstration Project which will map 260 square miles of northern Thurston County in 1991. For that project, consultants or additional expertise will again be needed to take the aerial photographs, interpret the wetland boundaries, undertake the field reconnaissance, and digitize all the data.

The SHAPIRO Field maps closely matched TRPC's expectations and needs. Figure 4 indicates the type of map produced by TGIF. In their colored version, these wetland maps have proved to be an extremely persuasive public relations tool. It appears that SHAPIRO found few wetlands under 1 acre in size, and the minimum threshold for wetlands and stream corridors was larger than desired, particularly in the urban area. The ratio of observation points to data points was roughly 6 to 1 and averaged 17.5 to 3 per square mile. Further evaluation will be necessary to determine if this is a reasonable ratio for the Demonstration Project. Field data sheets were collected for 64 sites, and TRPC has been seeking ways to put this required part of the ECOLOGY grant to a useful purpose.

One of the most common corrections made to the SHAPIRO Aerial maps was the misidentification of wetland habitat types. The field reconnaissance resulted in a 66% increase in wetland diversity which ranged from 33% to 140%. This was a mean increase of 4 wetland classes per flight line. The field reconnaissance added a total of 162 wetland acres which represents an increase of 7.5% over the SHAPIRO Aerial data. An evaluation of all the additions and deletions found that 45% were emergent wetlands, 32% were forested wetlands and the remaining 23% were scrub/shrub wetlands. In the SHAPIRO Field data, the ratio of certain to uncertain wetland boundaries was 3 to 1. A comparison between the wetland Aerial and Field maps is needed to determine if the field work actually decreased the length of the uncertain boundary, hence an improved accuracy. A more comprehensive analysis of the wetlands in each flight line is also needed.

Public relations required more time than either TRPC or SHAPIRO expected. The TRPC staff spoke with 29 wetland property owners (5.2%) over the phone, 9 of which (1.5%) did not want the survey team to visit their property. SHAPIRO experienced only a few situations where the residents had not received a letter. The most common question was to inquire about the "true" nature of the pilot project, and the ultimate use of the data for regulatory purposes. Whereas, SHAPIRO spent the most time at flight line C, their initial flight line, TRPC received twice the number of phone calls on flight line A which also included 90% of the prohibited access.

The accuracy of the various wetland mapping layers is summarized in Table 3. These results are based upon the calculations described in Table 2 and shown in Figure 3. The totals in Table 3 were calculated on a proportionate basis by the area in each flight line. The local wetland maps were shown to be the least accurate source when compared to the SHAPIRO Field maps. A possible reason for these low numbers may be the lack of wetland mapping by some jurisdictions and mapping from less accurate preliminary NWI maps. Another factor contributing to the low accuracy of both the local wetland and WDNR soils maps may be the exclusion of lakes and marine waters as mapped "wetlands" according to the USFWS classification system. In addition to highlighting areas of "Mutual Coverage", Table 3 also identifies over and under coverage, which in some cases can be greater than the area of mutual coverage. The accuracy of the SHAPIRO Aerial was found to be 93.4% as the field reconnaissance with only a 10.5 % total of over and under coverage. For flight line C, the mutual coverage between the SHAPIRO Aerial and Field data was 99%!

TABLE 3 -- ACCURACY OF VARIOUS WETLAND MAPPING SOURCES*

Flight Lines →	A	B	C	TOTAL
1. LOCAL WETLAND				
Mutual Coverage	51.49	53.80	26.45	**46.41**
Under Coverage	48.51	46.20	73.35	53.55
Over Coverage	45.64	57.47	4.08	40.26
2. NWI				
Mutual Coverage	44.88	62.32	79.10	**59.68**
Under Coverage	55.12	37.68	20.90	40.33
Over Coverage	7.22	13.59	3.38	8.73
3. SOILS				
Mutual Coverage	77.50	69.72	64.42	**71.43**
Under Coverage	22.50	30.28	35.58	28.58
Over Coverage	52.19	50.35	6.27	40.56
4. SOILS & NWI				
Mutual Coverage	82.07	80.78	82.28	**81.64**
Under Coverage	-37.98	19.22	17.72	-2.93
Over Coverage	55.91	58.31	8.60	45.57
5. SHAPIRO AERIAL				
Mutual Coverage	91.24	91.91	99.00	**93.35**
Under Coverage	8.76	8.08	0.90	6.63
Over Coverage	2.27	2.09	9.90	4.02

NOTE: All Numbers in Percent (%)

* Compared to SHAPIRO field

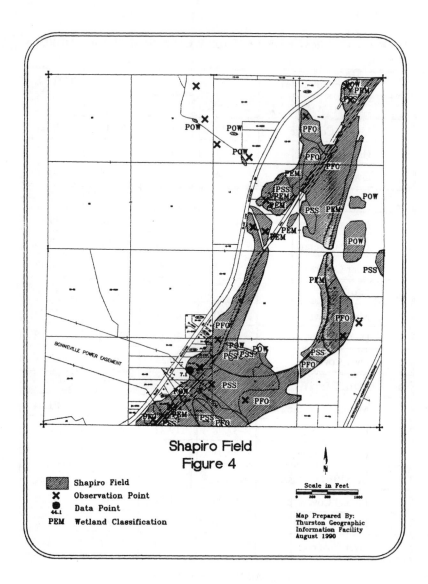

Shapiro Field
Figure 4

Shapiro Field
X Observation Point
● Data Point
44.1
PEM Wetland Classification

Scale in Feet
0   250   500   1000

Map Prepared By:
Thurston Geographic
Information Facility
August 1990

These results will need to be further evaluated to determine if the last 10% of accuracy of a Pilot Project map is worth the substantial cost of the field reconnaissance.

The Pilot Project budget was adequate even with several additions which included: printing extra CIR photos taken earlier in the year by WDOT; the addition of 1 square mile in Flight Line A; evaluating the field protocol and data sheets in a test section, and increased time for field crews. The actual time for tasks by SHAPIRO closely paralleled the project estimates. The analysis of other consultant costs will be forthcoming. A desire to increase accuracy or data collection for the purposes of rating wetlands will increase the costs per section.

The cost of consultant services for the Pilot Project represented the "going rate" during 1990. It is believed that the limited budget for field reconnaissance may have minimized the time spent searching for unmapped wetlands. Even though hiring a wetland field consultant may cost more than if the project is done in-house, their expertise can minimized the field time on questionable boundaries, provide valuable guidance, and otherwise relieve the trauma often associated with managing a first-of-its-kind effort.

PILOT PROJECT RECOMMENDATIONS

The following section combines the evaluations of all the Pilot Project consultants and TRPC staff. It includes recommendations for potential modifications which would enhance the quality and effectiveness of this methodology.

Scope and Budget. The scope and budget for the TRPC Demonstration Project should further test the finding of the Pilot Project, accommodate unanticipated changes in scope during the project and provide for financial contingencies.

Consultants. A request for proposal process should be considered for all consultant work. It is important for all consultants meet each other and work together as a TEAM. Encourage the in-house staff to use the opportunity to get into the field with consultant.

Aerial Photography. TGIF recommends that the number of photo registration points per photo needs to be increased. To maintain the registration between photos and to provide for a wetland threshold of 0.25 acre in the urban areas, the photo scale should remain at 1" = 1,000' with enlargements of the urban area at 1" = 500'.

Photo Interpretation. SHAPIRO recommends completing the photo interpretation and field verification of areas in small blocks.

Wetland Sources. The creation of a NWI & Soils map appears to be an appropriate interim product until a more detailed wetland inventory can be completed.

Field Reconnaissance. The field data forms should help the jurisdictions rate the wetland using the ECOLOGY four tier system. ECOLOGY recommends the use of a soil bore rather than an auger for quick soil investigations. SHAPIRO recommends concentrating field time on uncertain boundaries and problem areas including:

- Forested wetland adjacent to forested upland
- Scot's Broom shrub areas
- Aquatic beds and emergent communities adjacent to open-water wetlands
- Boundaries of emergent versus shrub adjacent to large, open-water wetlands
- Tall shrubs versus saplings
- Riparian wetland boundaries
- Agricultural wetland fields and
- Wetlands of one acre or larger

GIS Mapping. Additional time needs to be allocated to proof the wetland maps and data after is has been digitized. Wetland maps from TGIF need to be user friendly, easily accessible and segmented by section. To make use of this data TRPC will need to have a menu driven, stand alone, PC system with a color printer available to the counter staff and other users of this data.

Public Relations. Good wetlands public relations does not just happen and more time needs to allocated to a public relations effort. It's to be expected that people would be concerned about "government" or their consultants on their property. Therefore, this situation needs to be viewed as a valuable opportunity to provide residents information on the project and about the benefits of wetlands. As much public relations information as possible should proceed the field reconnaissance phase. Mail letters to all potential wetland owners and copies should be provided to the field crew. It will be important to work with residents to gain access to wetland boundaries on their terms. And mail out follow up notices when maps or report are available.

LITERATURE CITED

Cowardin, L. M.; V. Carter, F. C. Golet; and E. T. LaRoe. Classification of Wetlands and Deepwater Habitats of the United States. U.S. Department of Interior, Fish and Wildlife Service, US Government Printing Office, Washington D.C. 1979.

Shapiro and Associates, Inc. Thurston Regional Planning Council Phase I Pilot Project Wetland Inventory, Seattle, WA. 1990.

Shapiro and Associates, Inc. Thurston Regional Planning Council Phase II Pilot Project Wetland Inventory, Seattle, WA. 1990.

Thurston Regional Planning Council. The Profile, Thurston Regional Planning Council, Olympia, WA. 1988.

U.S. Environmental Protection Agency. Office of Ground-Water Protection. Guidelines for Wellhead Protection Areas. US Government Printing Office, Washington D.C., 1987.

20:mm\0191f22.181

Mapping at the National Wetlands Research Center

James B. Johnston and Lawrence R. Handley[1]

Abstract

Over the past 10 years, the U.S. Fish and Wildlife
Service's (FWS) National Wetlands Research Center (Center;
formerly the National Coastal Ecosystems Team) has been
continuously involved in the production of maps for use by
coastal decisionmakers. The types of maps produced by the
Center have been national, regional, or local in scope,
depending on user needs. Map scales have ranged from
1:24,000 to 1:250,000. Themes depicted have included
biological resources, including wetlands and seagrasses;
upland habitat or land use; water resources such as water
quality, bathymetry, and salinity; cultural features such
as ownership, archaeological sites, and dredge-spoil
disposal areas; and soils and landforms. This paper
presents overviews of the various mapping programs of the
Center. Highlighted are efforts such as the ecological
inventories of the Atlantic, Gulf, and Pacific coasts; the
ecological characterization atlases of the Gulf of Mexico;
and the large-scale (1:24,000) habitat maps of various
coastal regions of the United States. Center methods and
techniques are discussed, including the collaborative
efforts between the Center and the FWS's National Wetlands
Inventory for updating wetland maps and adding upland and
seagrass bed delineations to Inventory maps. We also make
recommendations for future coastal ecosystem mapping
programs that use conventional and automated mapping
methodologies, such as geographic information systems and
image processing.

Introduction

The National Wetlands Research Center (Center) of
the U.S. Fish and Wildlife Service (Service) has an ongoing

---

[1]U.S. Fish and Wildlife Service, National Wetlands Research
Center, 1010 Gause Boulevard, Slidell, LA 70458.

program in habitat mapping of wetlands and uplands. We cooperate with the National Wetlands Inventory (NWI), using their processes of photointerpretation, quality control and assurance, and distribution. We differ from NWI in mapping biological data and resources at other scales of 1:100,000 and 1:250,000, in adding upland habitat to the wetland maps, and in developing time-sequenced mapping for habitat trend analysis.

Habitat Mapping

All of our habitat mapping projects are developed as special interest programs (e.g., Louisiana land loss or seagrass mapping in cooperation with other Federal and State agencies such as the U.S. Army Corps of Engineers (USACE), the U.S. Environmental Protection Agency (EPA), and the Louisiana Department of Natural Resources (LDNR), or as technical assistance within the Service). We have added uplands to wetland maps and developed the criteria for the incorporation of the additional upland categories. We have provided updates of habitat maps that the Center completed previously. From the sequential dates of mapping we can look at a trend analysis of loss and habitat gain. We use the Classification of Wetlands and Deepwater Habitats of the United States (Cowardin et al. 1979) as the primary criteria for wetland delineation through the various systems, subsystems, classes, and subclasses. If, however, additional information is available, we are able to add modifiers. For example, for coastal Louisiana, we are adding a salinity modifier onto the habitat maps. Through additional coordination at the local level, we are able to gather this type of information for many of our special projects. We have completed a number of projects related to habitat mapping at the 1:24,000 scale. Coastal Louisiana was mapped for 1956, 1978, and 1988 for the entire coast, and in 1983 for the lower Mississippi River Delta and the Terrebonne Marsh area. In 1985, we coordinated with the National Aeronautics Space Administration (NASA) to fly aerial photography of coastal Louisiana. Only 11 habitat maps were developed at that time for the State of Louisiana to compare data with pre-Hurricane Juan satellite imagery. We mapped coastal Texas for two time periods, 1956 and 1979, and mapped 10 quadrangles (quads) for 1983. For coastal Mississippi, we mapped habitats for 1956 and 1979, and for Alabama, we completed 32 quads for the Mobile Bay area for 1956 and 1979. For the west coast of Florida, we completed selected quads of the panhandle area; for example, Tampa Bay had 26 quads done for three dates--mid-1950's, 1962, and 1982. For San Francisco Bay, we completed 20 quads of the south bay for the mid-1950's, 1976, and 1985. For the north bay, an additional 24 quads were done for 1976 and 1985, and another 63 quads were done for only 1985 for a surrounding

area of the bay. These are most of the habitat mapping
projects that we have completed with multiple year updates.

The upland classification that we use is patterned
after Anderson's <u>A Land Use and Land Cover Classification
System for Use with Remote Sensor Data</u> (1976), used by the
U.S. Geological Survey (USGS) in their land-use mapping.
Upland classification, however, is gradually evolving just
as the wetland classification has changed over time, and we
have added more and more identifiers to the uplands, as the
special projects dictate greater detail and varying needs.

Examples of identifiers include urban forest,
rangeland, agricultural, and barren lands, plus additional
subcategories, such as rice fields, parks, cemeteries, golf
courses, spoil areas, and transportation corridors. The
North American Waterfowl Management Plan is interested in
rice fields as habitat for wintering ducks, and the Service
is interested in the breakdown of forested land into scrub-
shrub or evergreen and deciduous forests as habitats
suitable for nesting red-cockaded woodpeckers (<u>Picoides
borealis</u>).

One of the great needs in habitat mapping is the
addition of the upland classification to properly assess
and analyze habitat changes and processes. In assessing
the habitat changes for an area, it is very difficult to
explain "where did the wetlands go?" if the only category
is uplands. To understand the processes of change at work
in the landscape, it is necessary to know what particular
type of uplands replaced the wetlands. Were the wetlands
filled in or drained for urban development or did they go
into the upland agricultural, rangeland, or forest? Around
San Francisco Bay, with its wholesale development, it is
assumed that any loss of wetlands went into the uplands
category, but this is not always the case. Lastly,
wetlands categories lose acreages to other wetland
categories as the water regimes change.

The addition of the upland categories aids in
understanding the overall picture of habitat change for a
particular area. In the San Francisco Bay area, wetlands
certainly lost large acreages, but the greatest loser of
all was upland agricultural land. The land of the market
gardens, truck farming, and alfalfa around the San
Francisco Bay lost almost four times as many acres as the
wetlands. The upland categories are very important, and
the need is certainly there for the development of a
comprehensive and systematic upland classification system
that will tie into the existing wetland classification.

We have developed projects to analyze habitat
trends and changes. For example, in San Francisco Bay, we

compiled a change map for the mid-1950's, 1976, and 1985. We have also done this for the lower Mississippi River Delta with three dates, and we are presently working on a fourth date for 1987-88. In addition, for several of our mapping projects we are developing wetland maps that include selected indicator species. For example, in the Lower Mississippi River Delta, we are adding a habitat modifier for two particular species--Spartina alterniflora (smooth cordgrass) and Australis phragmites (common reed). Spartina is primarily an indicator of salinity and is an indicator of fresh and brackish water. We are also interested in seeing how much loss or gain has taken place for each species between 1978 and 1983, and between 1983 and 1987.

The mapping of the Chandeleur Islands is primarily seagrass mapping. In this project, we have photography for three dates (1978, 1982, and 1987) that has been interpreted. We have photography for an additional four dates (April 1969, October 1969, November 1988, and June 1989) that we will interpret this year. Also, we are acquiring aerial photography of the Chandeleur Islands on a quarterly basis to examine seasonal variations in the seagrass coverage. This project was originally undertaken in conjunction with other Center studies on the redhead ducks winter at the Chandeleur Islands, but it has evolved into a seagrass photointerpretation study of its own.

The Louisiana coastal zone project was performed for the State of Louisiana to update data following Hurricane Juan. The State is using these habitat maps to compare with Landsat Thematic Mapper Simulator digital data acquired prior to Hurricane Juan to analyze the hurricane impact for marsh breakup.

For coastal Louisiana, we are also photointerpreting and mapping uplands and wetlands in 330 quads using 1988 photography. This project will take approximately 3 years to complete. At present, we are in Phase I, which is the photointerpretation of 110 quads. Phase II and Phase III will complete photointerpretation of the remaining 220 quads over the next 2 years. This project will provide an update using 1988 photography to add to our existing 1956 and 1978 data bases of coastal Louisiana.

In the San Joaquin Valley, we are mapping 83 quads of which 26 are focusing on uplands and the other 57 are an update of wetlands and uplands. This project is for the San Joaquin Valley Drainage Program and will be used in analyzing the Kesterson National Wildlife Refuge selenium problem.

In Mobile Bay, we are mapping 26 quads to update the 1956 and 1979 wetland maps.

To show the overall aspects of some of these projects, we not only produced habitat maps for San Francisco Bay for several dates, but we also developed two reports that provide an analysis of the habitat trends in the south bay and the north bay, a report on the comparison of a natural marsh usage by fish and wildlife with an artificial marsh's usage, and two large format habitat maps of the bay area. The information for this trend analysis is being extensively used in the bay's waterfowl management plan, by EPA's Estuarine Program assessments, in two court cases, by the California Attorney General's office, and in at least a dozen other projects, programs, and studies.

Specialized Mapping

Special projects completed or ongoing at the Center are generally performed for the Service. In particular, we provide technical assistance to the Service's Fish and Wildlife Enhancement (FWE) Field Offices, Regional Offices, or National Wildlife Refuges. At present, we are mapping seagrasses in Perdido Bay, in Florida and Alabama, for 1940, 1978, and 1987, for the Panama City FWE Field Office. On Eglin Air Force Base in Florida, we are developing ecological community maps to be used by the Air Force and the Service in surveying red cockaded woodpecker habitats for active colonies.

The information we have collected has been used to develop digital data bases that can be entered into the Center's geographic information system (GIS) to implement natural resource inventories, habitat trend analyses, and cartographic modeling projects. We work with other Federal and State agencies in need of the habitat maps and the digital data to conduct their work, in particular, with the National Park Service, USACE, EPA, and LDNA. We have developed a digital data base of the habitat maps that cover coastal Louisiana, Texas, Mississippi, Alabama, and portions of the Gulf Coast of Florida. In addition, we have digital data for other selected areas of the country (e.g., New Jersey, lower Chesapeake Bay, the St. Lawrence Seaway, and San Francisco Bay).

Ecological Atlases and Inventories

We have developed maps for atlases at 1:100,000 scale and inventories at 1:250,000. In 1978, we began developing the first of four ecological atlases for regions of the Gulf Coast: the Mississippi Deltaic Plain Atlas, the Texas Barrier Islands Regional Atlas, the Coastal Alabama Ecological Atlas, and the Florida Ecological Atlas.

A fifth atlas, the Chenier Plain Ecological Atlas, is in progress; it will fill in the gap along the Gulf Coast. The mapping for each of these atlases is completed on 1:100,000 USGS base maps, and five topics per map are displayed: biological resources, socioeconomic features, soils and landforms, oil and gas infrastructure, and climatology and hydrology. Each topic includes the accumulation of much information from many resources in mapped form, text format, site visits, meetings with regional experts, and reviewers' comments. Reports produced as part of these projects include bibliographies of biological and socioeconomic literature, syntheses of information, map narratives, and some special reports on modeling, ecological community profiles, and seagrass atlases. The Minerals Management Service, EPA, and various State agencies were instrumental in funding, collecting data, writing reports, and reviewing the atlases and reports.

The ecological inventories were completed by the Center in 1984 and covered the Atlantic, Gulf of Mexico, and Pacific Coasts, and the lower Mississippi River Valley. The scale of maps used was 1:250,000, and they included an inventory of only biological resources. Examples of items mapped are fish spawning areas, bird rookeries, bird nesting areas, locations of endangered species habitats, major natural waterways, turtle nesting areas, and State and Federal wildlife refuges and management areas. The ecological inventory maps were conceived as aids to site planning of thermal power plants along the Atlantic Coast; their scale, however, made specific site planning difficult. Overall use has far overshadowed the deficit, and they have become aids at regional impact assessment and environmental analysis, oil spill risk assessment, oil spill sensitivity, and oil spill clean-up planning.

The Resource Planning Institute of Columbia, SC, an oil industry consultant, has developed a set of maps of the coastal United States dealing with the particular sensitivity of segments of the coast to oil spill clean-up activities. MERG, an oil industry consortium working through consultants such as Coastal Environmental, Inc., Baton Rouge, LA, has developed sets of maps that provide segments of the coastal United States that should be protected from oil spill impact on a priority basis.

Several groups have developed products based on NWRC mapping projects. S.L. Ross of Canada has developed a computerized data base that many of the oil companies are using on microcomputers for oil spill risk assessment and oil spill clean-up. All of the products have one major flaw: the data used to develop the maps. Particularly, the biological resource information taken from the

Ecological Inventory Maps and Ecological Atlases we have
produced is outdated and in some cases was highly
generalized. For example, the priority resources to be
protected or cleaned up may not be in those locations any
longer. For instance, 35% of the gulf coast bird rookeries
and nesting sites have either disappeared or changed
locations.

A group of representatives from oil companies and
State and Federal agencies met on 7 December 1989 to
discuss the need for a comprehensive updated biological
resources mapping program. Nationwide, the greatest need
in thematic mapping is to update the Service's ecological
inventory. One of the suggestions being made by the Center
is that the 1:250,000 scale maps do not lend themselves
well to detail for site-specific analysis, oil spill risk
assessment, or to digitizing. We propose that the
ecological inventory be updated using the 1:100,000 scale
USGS maps as the mapping base. When the USGS 1:100,000
Digital Line Graphs are completed for the country, they
will provide additional theme overlays of political
boundaries, hydrology, and transportation networks. This
scale of maps would provide a manageable and usable product
that would be more meaningful to planners and environmental
consultants and analysts and would be USGS more specific
and detailed for oil spill clean-up, risk assessment, site
planning, and permit analysis.

Aerial Photography Acquisition

Another aspect of the mapping that the Center
provides is the coordination and organization of flights to
acquire aerial photography and digital data over many
areas. We organized a consortium of Federal and State
agencies to provide the funding for a flight of coastal
Louisiana, Mississippi, Alabama, and a portion of the
western Florida Panhandle. The groups included the
Service, EPA (Atlanta Region and Dallas Region), USACE (New
Orleans District and Mobile District), the State of
Alabama, and the State of Mississippi. Nearly 3,000 line-
miles were flown, over which 100 coloring-infrared
photographs at 1:65,000 scale, and airborne Thematic Mapper
Simulator digital data were collected. The coastal
Louisiana, Mississippi, and Alabama flights were completed
between 6 November 1988 and 30 March 1989. Due to the
success of this flight, we were asked to organize a similar
group to fund a flight of coastal Texas for the fall of
1989. The consortium included the Service, EPA (Dallas
Region), COE (Galveston District), and the Soil
Conservation Service. The coastal Texas flight, flown by
NASA out of Ames Research Center at Moffett Field,

California, encompassed about 3,000 flight-line miles, resulting in a thousand 1:65,000 scale color-infrared photographs and TMS digital data. The Texas coast was flown between 27 November and 15 December 1989.

References

Anderson, J.R., E.E. Hardy, J.T. Roach, and R.E. Witmer. 1976. A land use and land cover classification system for use with remote sensor data. U.S. Geol. Surv. Prop. Pap. 964. 28 pp.

Cowardin, L.M., V. Carter, F.C. Golet, and E.T. LaRoe. 1979. Classifications of wetlands and deepwater habitats of the United States. U.S. Fish and Wildl. Serv. Biol. Serv. Program FWS/OBS-79/31. 103 pp.

REQUIEM FOR A WATERMEADOW - THE KAISER SITE, GRAYS HARBOR, WA

David E. Ortman*

SUMMARY

In August of 1976, the Seattle District Corps of Engineers issued a Section 404 Department of the Army permit to the Port of Grays Harbor to construct a berm, fill, pier and marine ways in the Grays Harbor estuary in Washington State, at the mouth of the Hoquiam River for the manufacturing and assemblying of off-shore drilling platforms by the Kaiser Steel Corporation. Approximately 36 acres (14.6 ha) of saltmarsh and tideland were filled, including at least 16 acres (6.5 ha) of sedge marsh. This represented a 12.5 percent loss of total Grays Harbor sedge habitat. However, no off-shore drilling platforms were ever built on this site.

Fifteen years later, the site remains undeveloped and, except for the initial watermeadow filling, no jobs or economic benefits to the Grays Harbor community have occured. No compensation for the loss of the saltmarsh, tidelands or sedge marsh functions or values took place.

The timespan of the Kaiser site fill, beginning in 1975 with the Port's application, covers the evolution of the Corps Section 404 regulatory program. The Port's initial application to fill this site also set in motion, in 1976, the events leading to the ongoing development of a Grays Harbor Estuary Management Plan (partially funded through Section 306 of the Coastal Zone Management Act). Plan adoption by the U.S. Dept. of Commerce's Office of Ocean and Coastal Resource Management was still pending in early 1991, fifteen years later.

This paper will review the Corps Section 404 permitting decision on the Kaiser Site application, the development of an environmental impact statement, public participation, permit issuance, mitigation, subsequent permit revisions, and letters of agreement signed between the Port of Grays Harbor and the U.S. Fish and Wildlife Service which have been violated by the Port. In addition, an analysis of the interrelationship between the Kaiser fill, the Grays Harbor Estuary

---
*Northwest Representative, N.W. Friends of the Earth,
4512 University Way N.E., Seattle, WA 98105

Management Plan (32) and Grays Harbor National Wildlife Refuge will
also be presented.

GRAYS HARBOR

Grays Harbor is a large coastal marine estuary on the Washington
Coast, which extends from the mouth of the Chehalis River to the
Pacific Ocean, 14 miles to the west. The Grays Harbor salmon run,
which includes chum, chinook, and coho salmon, as well as steelhead
and sea-run cutthroat trout, produces over 20 million small
downstream migrants each year. Approximately 210 species of birds
are found in the area, which is located within the Pacific flyway.
The Harbor is heavily utilized as a resting and feeding habitat for
waterfowl, particularly black brant and canvasback ducks. The
tideflats and marshes along the margin of the upper harbor are
extremely important to the biological productivity of Grays Harbor.
The shallow intertidal flats are the site of much of the primary
productivity in Grays Harbor. Salt-marsh vegetation, including
grasses, rushes, and submerged plants, provides food and habitat for
various invertebrates, shorebirds, and waterfowl. (38)

An area at the mouth of the Hoquiam River had been obtained by
the Port of Grays Harbor (the Port) in 1963 as an industrial
development district. (19) (See Figure 1.)

According to the U.S. Fish and Wildlife (U.S.F&WS)

This site has been inspected by Fish and Wildlife Service
biologists. It is known to support the following birds in the
wetlands and adjacent waters or upland fringes: mallard,
pintail, green-winged teal, wideon, canvasback, greater scaup,
lesser scaup, common goldeneye, bufflehead, whitewinged scoter,
surf scoter, common scoter, ruddy duck, hooded merganser, common
merganser, red breasted merganser, coot, common loon, Arctic
loon, red-necked grebe, horned grebe, western grebe, eared
grebe, double-crested cormorant, pelagic cormorant, great blue
heron, numerous sandpipers and plovers, several gulls and terns;
and several kinds of raptors and songbirds. Pheasants and deer
may be found in the wooded and brushy adjacent areas and will
visit this site. In addition, approximately 15 to 20 types of
small mammals ranging from shrews, to river otter and harbor
seal may be seen on or close to this site.

Mudflats and intertidal areas support small clams, polychaete
worms, and isopods and amphipod organisms which supply food to
birds and fish that feed in the areas. Species of fish known to
occur in waters adjacent to the site are as follows: American
shad, Pacific herring, Northern anchovy, chum salmon, coho
salmon, chinook salmon, searun cutthroat trout, steelhead trout,
Dolly Varden trout, surf smelt, three-spine stickleback, bay
pipefish, shiner perch, snake prickleback, crescent gunnel,
saddleback gunnel, Pacific staghorn sculpin, English sole,
starry flounder, green sturgeon, and Pacific tomcod. Many of

these species of fish and wildlife feed within the site to be
directly impacted by the proposed project and would suffer an
undetermined net population decline as a result of the
significant loss of food source and habitat provided by this
ecosystem segment in a critical estuarine zone.

The site in question is one of the last remaining undeveloped
areas along the North Channel within the city limits of Hoquiam
and Aberdeen. Past piecemeal development along the channel has
decimated fish and wildlife habitat and drastically limited
public access and fish and wildlife related uses. This is
essentially the last remaining area that has the potential for
becoming a waterfront park, providing public access to the
waterfront, or being dedicated as open space. We believe the
failure to consider this location for such uses is shortsighted
and does not constitute waterfront planning in a comprehensive
and coordinated manner (24).

CLEAN WATER ACT

The Federal Clean Water Act was passed by Congress in 1972.
Section 404 set out a permit program to regulate the disposal of
dredged or fill material into waters of the United States and was
phased in beginning in 1975. Regulatory guidelines (Section 404(b))
established by the Environmental Protection Agency must be met before
a permit may be issued by the Department of Army Corps of Engineers.

KAISER APPLICATION

The Port submitted a permit application dated 28 February 1975
to place fill for a log and lumber sorting and storage yard. This
application was the subject of Seattle District Corps of Engineers
(the Corps) Public Notice No. 071-OYB-1-002533. All Federal
resource agencies objected to the project. U.S. F&WS particularly
objected stating that the planned use was inappropriate because it
was non-water dependent (19).

In August, the Port submitted revisions changing the
configuration of the fill, increasing the quantity of fill from
381,000 to 457,500 cubic yards, adding a pier and marineways, noting
Kaiser Steel Corporation (Kaiser) as co-applicant and changing the
purpose of the proposed work to manufacturing and assembly yard for
off-shore drilling platforms (35).

The establishment of the Section 404 permit program had given
Federal resource agencies new leverage in opposing environmentally
damaging projects. U.S. F&WS and National Marine Fisheries Service
(NMFS) views were to be given "full" consideration and EPA could
prohibit the discharge of dredged or fill material into wetlands
under Section 404(c) of the Clean Water Act (a veto they have never
used in Region 10 in over fifteen years).

The Port was able to convince Senator Henry (Scoop) Jackson (D-

WA) to force the Federal resource agencies to back down.   Senator
Jackson's office summoned the U.S. F&WS, Region 10 Environmental
Protection Agency (EPA), the Port and Kaiser to a meeting in August
of 1975. At this meeting, U.S.F&WS asked Kaiser if they were certain
they would build on this site and the answer from Kaiser was "no they
were not sure." Kaiser would build only if enough oil leases are let
and orders received by them (7).

EPA summarized the entire project in an October 1975 letter to
the Corps

. . .there is no guarantee the site will actually be used for
manufacturing and assembling offshore drilling platforms as
proposed. To our knowledge there is only a lease-option on the
property. Depending on the availability of new offshore leases,
the proposed use may not occur. The possibility therefore
exists that this valuable wetland area will be destroyed for an
intended use that never occurs (39).

According to Kaiser, "The main reason for taking an option on any
property was to enable the permit application procedures to go ahead,
as we recognize the time required for such approvals and it is
imperative that we have such permits before we could finalize any
leases" (11).

Although the City of Hoquiam did not have an approved Shoreline
Master Program (33) until April of 1976, the Port of Grays Harbor
received a Shoreline Management Substantial Development Permit in
November 1975 from the City of Hoquiam with the concurrence of the
Washington State Department of Ecology (Ecology) after Ecology
appealed the project to the Washington State Shoreline Hearings Board
and signed a settlement favorable to the Port.

An environmental assessment was prepared under the State
Environmental Policy Act (SEPA). It was not circulated to the
public. In September 1975, the city of Hoquiam signed a "Declaration
of Non-Significance" and no SEPA environmental impact statement (EIS)
was prepared (47).

DEVELOPMENT OF A NEPA EIS

Because this was the first large development to enter the Corps'
404 permit program, the decision by the Corps on whether to prepare
an environmental impact statement under the National Environmental
Policy Act was the focus of intense pressure by the Port, Kaiser and
environmental organizations.

In January 1976, Kaiser wrote a letter to the Corps, stating
that as "co-applicant" it requested a permit be issued with no EIS.
The need for speed was to coordinate filling with maintenance dredging of the Corps' navigation channel planned for spring of 1976 (10).

On 15 January, a meeting was held by the Corps with

representatives of Friends of the Earth (FOE), the Sierra Club, the
Washington Environmental Council, and the Western Federation of
Outdoor Clubs. The environmental organizations requested that an EIS
be prepared. Patrick A. Parenteau, of the Northwestern School of
Law, Lewis and Clark College, represented FOE in making a legal case
for an EIS (34).

The Environmental Resources Section of the Corps determined that
the granting of the permit would be a Federal action which would have
a significant impact on the human environment and required an
Environmental Impact Statement (59). In January 1976, the Corps
District Engineer Col. Raymond Eineigl, informed the Port that an EIS
would be prepared. In his letter to the Port, Col. Eineigl noted the
delay will take some 10 months or

> potentially five to six months should it be necessary to obtain
> a Council on Environmental Quality waiver on full administrative
> processing requirements. . . Further, there are a number of
> indications that there is no immediate urgency to permit
> issuance in this situation since it is not apparent that the
> Kaiser Steel Corporation would have to exercise its option for
> development at Grays Harbor until approximately this time next
> year (17).

Major Gen. Wesley E. Peel, division engineer of the Corps' Northern
Pacific Division in Portland (OR), toured Grays Harbor in April 1976.
Gen. Peel stated that the Seattle district was working expeditiously
on the EIS and that the corps "will hand-carry it up the ladder
(through cognizant agencies) and hopefully we will be able to get
approval along the way" (1).

The Corps tentatively identified the scope of the EIS as
"Western Washington and Astoria, Oregon" (42). However, the Corps
subsequently limited the scope of the alternative sites study for the
draft EIS exclusively within Grays Harbor because other alternatives
were "beyond the jurisdiction and interest of the legal applicant"
(52).

The Corps prepared four EIS preparation schedules. All had a
"permit issued" milestone. None had a "permit denied" milestone.
Alt. A, the original schedule without an EIS, showed permit issuance
in January 1976. Alt. B, the routine schedule with EIS, showed
permit issuance in January 1977. Emergency Schedule C was to issue
the permit in November 1976, and Emergency Schedule D was to issue
the permit in July 1976 (54). Col. Eineigl also requested permission
of the North Pacific Division Engineer to seek a waiver of full CEQ
EIS processing requirements. The Corps then "proceeded on a 'crash
program' to produce the EIS and have it filed with CEQ, and were
successful in breaking some records" (18). The FEIS was filed with
CEQ on 30 July 1976 six months after the decision to prepare (35).

The speculative nature of the Kaiser proposal caused dissention
within the Corps. The Counsel for the Corps in late June of 1976,

noted that while Kaiser had in its application originally proposed to use the site for the "assembly of offshore oil drilling platforms", Kaiser backed off to a more general purpose of the "assembly of marine structures". The Counsel in a memo to the Engineering Division stated bluntly

> We believe that Kaiser has misrepresented the purpose of its project, just as it misrepresented the availability of alternative sites in Puget Sound. We believe it has done so in order to cash in on the current concern for expanding oil production. It has misled the Corps and, more seriously, it has made the Corps an unwitting partner in misleading the public and the agencies. Our Public Notice states flatly that the purpose of the project is a 'manufacturing and assembly yard for offshore drilling platforms. . . (58).

The memo also accused Kaiser of misrepresenting the availability of alternative sites in Puget Sound. At the time it was wooing the Port of Grays Harbor, Kaiser had also taken out an option on a 100 acre site at Everett, WA.

The office of Counsel recommended that

> A strong letter from the District Engineer to Kaiser and the Port should express his extreme disapproval of their behavior in connection with their application (58).

The Chief of the Engineering Division acknowledged that development by Kaiser hinged totally on contracts awarded from oil companies, that the number of jobs added to Grays Harbor was a key factor in priority of processing, and stated that the resource agencies are "little concerned about a Kaiser development" (45). No letter was sent out.

As a last note, the Final EIS p. 81 stated that

> In order to finalize the permit, the Port and Kaiser Steel Corporation must both sign the permit forms, thereby agreeing to limit the work to that shown on the drawings and in accordance with the terms and conditions of the permit (emphasis added) (43).

PUBLIC PARTICIPATION

The Corps 404 regulatory program was as new to environmental organizations as it was to the Corps. However, all parties recognized that significant precedents, such as the preparation of an environmental impact statement, would be set by the processing of this permit application.

Eight letters commenting on the public notice were received by the Corps. Six requested that a Federal EIS be prepared. Two opposed the proposed project. Commenting on the DEIS, FOE,

Washington Environmental Council, Sierra Club and Tahoma Audubon
objected to the issuance of a permit. A March 1976 letter from the
Sierra Club to the Corps stated "Kaiser has still not made a definite
commitment to the site; this proposed action must not degenerate into
a speculative fill" (26). No request for a public hearing was made
and no public hearing was held (35).

MITIGATION

By November 1975, the Region One U.S. F&WS office had already been
forced to accept the Grays Harbor estuary planning process as a
tradeoff for dropping opposition to the project

> There has been a memorandum of understanding developed and
> signed between the Fish and Wildlife Service and appropriate
> local governments with agreement on basic provisions of a
> comprehensive plan for Grays Harbor estuary and adjacent waters
> and shorelines with procedures and time frame for its
> development and completion as well as its coordination with
> other appropriate agencies or parties being incorporated (25).

However, the U.S. F&WS Olympia Field Office staff recommended
that "If we are forced to sacrifice this site then we should work
toward and achieve compensation for its loss. . .In essence, we
suggest that we start negotiations for compensating for losses now"
(22).

The Port made their views on the biological values of the Kaiser
site known early. The Port estimated that the Kaiser assembly plant
would employ approximately 175 to 250 persons (43). An August 1975
letter from the Port Engineer to the City Engineer at Hoquiam stated,
"As about 39 acres of marshland are involved, the maximum total
annual loss would be $162,000. This is negligible when compared to
the $2,500,000 payroll in direct wages alone, generated on the site.
." (29).

This amounts to $2,268,000 over the past fourteen years.

In a March 1976 letter to the Corps, FOE noted that

> Sec. 209.120(g)(4)(i) states that, "The applicant will be urged
> to modify his proposal to eliminate or mitigate any damage to
> such resources, and in appropriate cases, the permit may be
> conditioned to accomplish this purpose". We understand that
> Kaiser Steel was asked to mitigate damage done to their site in
> California. We would ask that mitigation be made a condition to
> the permit (30).

A Corps phone conversation in May of 1976 with the U.S. F&WS
records that Warren Baxter of the Corps had "discussed
[compensation] with Spearman [Seattle District Corps]. We believe
OCE [Office of the Chief of Engineers] policy prohibits mitigative
measures involving money or land tradeoffs" (27).

The Environmental Resources Section (ERS) of the Corps prepared the environmental impact statement which revealed substantial resource losses which would not be mitigated. Because the Corps was determined to grant a permit, the treat of litigation was seen by the Corps as the only means of stopping the project or mitigating for the loss of wetland resources. A statement by FOE that it would not appeal the issuance of a permit, signalled to the Corps that FOE would not challenge the project. As a result, mitigation was dropped from consideration by the Corps (23).

PERMIT ISSUANCE

A March 1976 Port letter stated that the initial lease with Kaiser Steel "will be for a period of ten years" (48). However, the issuance of the permit was complicated by the fact that the Port and Kaiser were listed on the application as co-applicants. An August 1976 meeting of the applicants with the Seattle Corps was held. Hank Soike, the Port Manager, inquired when the permit could be issued. He was told by the Corps that the permit could be approved by the end of August. Soike said that only the Port was the applicant and that Kaiser need not sign the permit. The Corps informed Soike that he was incorrect and that Kaiser and the Port were co-applicants and both must sign the permit. Carson (of Kaiser) was informed that Kaiser would have to sign the permit prior to issuance, but that work by the Port to fill the watermeadow could be delayed for up to one-year if desired. Carson revealed to the Corps that Kaiser might be unable to enter a lease agreement at such an early date. The Corps' position, if Kaiser could not the lease after the site had been filled, was that "the Corps, Kaiser, and the Port would suffer from some heat by the environmental group (sig.) The Corps would have to review any other potential user of the site to be sure that the use of the site was water and energy oriented" (53).

By this time, the Federal resource agencies had been forced by Senator Jackson into embarking on a multi-decade long planning effort in exchange for withdrawing the objections to the Kaiser fill. A 1976 Corps fact sheet on the project notes, the U.S. F&WS and NOAA objected and later withdrew objections. EPA, the State of Washington and the City of Hoquiam stated no objection. Environmental groups registered opposition to permit issuance (44).

Despite an October 1975 NMFS position that

> . . .we requested that this area not be filled for industrial purposes. The marsh area is covered with water during the high winter tides and therefore is a food production area for fish. We believe that this 45-acre fill and the manufacturing facility that is proposed for this site will ultimately damage fish and shellfish resources of Grays Harbor. . .We therefore request that this permit be denied (13).

NMFS withdrew their objection.

One reason that the U.S. F&WS could not maintain a position of
project opposition was given by Don Reese, Regional Supervisor of the
Division of Ecological Services in the Portland Regional Office,

> In the event we would object to the Kaiser Steel permit based on
> biological impact, you can bet it would be forwarded to
> secretarial level; and if I were to hazard a guess, I would bet
> that it would be approved because of other factors involved --
> water dependency, energy relationship, diversification of
> industry in a depressed area, etc. (40).

In a surprise eleventh hour move, Col. John Poteat made a
decision to issued the permit on 30 August 1976, not to the Port and
Kaiser as co-applicants, as the permit application and FEIS had
stated, but to the Port alone. The Corps noted that neither NOAA,
EPA or the Department of Interior had objections to the work (35).
No special conditions or mitigation were attached to the issued
permit (36).

The issuance of the permit to the Port alone was viewed as a
weaselly decision to allow the Port to fill the site and not be
encumbered in the likely event that Kaiser dropped the project. Col.
Poteat, in a September 1976 letter to FOE stated

> . . .you feel the Corps made a last-minute decision which
> changed an earlier position. I know of no such switch and,
> consequently, do not understand the context of your remarks
> (37).

SUBSEQUENT PERMIT REVISIONS

In April 1979, Kaiser notified the Port that they would let
their option expire (12). While informing the Corps, the Port
admitted that

> . . .it has always been recognized that without firm contracts
> to construct drilling platforms, Kaiser Steel would either find
> other water related metal fabrication uses of the site or allow
> the option to expire (49).

An article appeared in the Aberdeen Daily World soon after, which
revealed that Kaiser had formally withdrawn its interest in Grays
Harbor. Among the list of reasons Kaiser gave included

> Lease sales have been delayed.
> -Environmental concerns have caused large areas to be deleted
> from possible lease sales.
> -The few holes drilled have failed to disclose any significant
> quantities of gas or oil.
> -The process for obtaining permits to drill, develop, and
> transport are increasingly difficult, timeconsuming and costly.
> -A temporary glut of oil on the West Coast and the inability to
> refine and move it to other parts of the nation.

-The 'unreasonableness' of environmentalists and government
agencies (2).

As a result of Kaiser's announcement FOE wrote to the Corps
requesting that the permit be revoked and the site returned to its
original condition (31). The Corps refused.

LETTERS OF DISAGREEMENT

In May 1979, the Port requested from the Corps a one-year
extension of their 1976 permit and asked that all references to
Kaiser be deleted. Joe Blum, Area Manager of the U.S. F&WS readily
agreed provided that five conditions be attached to the permit (4).
The Corps refused to incorporate any of the special conditions
requested by U.S. F&WS (28).

In July of 1980, without public participation or review, Blum,
representing U.S. F&WS and Soike, the Port Manager, entered into a
Letter of Agreement to conditions on the proposed permit extension

The Service is cognizant of the fact that the Port of Grays
Harbor, ourselves, and several other entities have been working
together for over 2 years in developing a Grays Harbor Estuary
Management Plan. We recognize that this planning process was
undertaken as mitigation for signing off on the original
"Kaiser" permit and that the planning process entered into has
been based on the concept of balance between development and
non-development areas within the estuary.

In the event that a qualified tenant is not secured in the
allotted 2-year period, or a letter of agreement between the
U.S. Fish and Wildlife Service and the Port of Grays Harbor
agreeing to carry on normal functions within the framework of
the Grays Harbor Estuary Management Plan is signed, the
applicant will select, within 30 days, one of the following
measures:
    a) Removal of the entire fill of 457,5400 cubic yards and
restoration of the site to natural or as near the pre-existing
conditions as possible.
    b) Agree to work out a suitable mitigation, compensation,
or enhancement plan acceptable to the U.S. Fish and Wildlife
Service.

5. In the event a qualified tenant is not secure, or an
agreement has not been reached regarding the Grays Harbor
Estuary Management Plan, work on the selected alternative
mitigation measure in 4 (above) be commenced prior to September
1, 1983, and completed prior to September 1, 1984 (5).

The U.S. F&WS sent the agreement to the Corps which issued the permit
extension and changes.

Two years later, the Kasier site was still vacant. In August

1982, the Port convinced Blum to sign a supplemental agreement

> The Port and the Service agree to an additional two-year period,
> ending on September 1, 1984, to achieve one of the following:
>   a) approval of a final Grays Harbor Estuary Management
> Plan, or
>   b) Approval of a Letter of Agreement between the Port and
> the Service agreeing to carry on normal functions within the
> framework of a Grays Harbor Estuary Management Plan, or,
>   c) In the event that a) and b) above are not attainable,
> the Port and the Service will work out a suitable mitigation,
> compensation, or enhancement plan to be implemented prior to
> September 1, 1986 (6).

In July of 1986, when it became clear that the Port did not
intend to comply with any of the three conditions in the supplemental
letter of agreement, a representative of the U.S. F&WS's Olympia Area
Office met with the Port to discuss mitigation for the Kaiser site
fill. The Port bluntly told the U.S. F&WS that the Port never had
any intention of mitigating for the fill, and that they had always
assumed that the U.S. F&WS would continue to sign agreements (51).

With the departure of Blum from the U.S. F&WS Area Office,
signing worthless letters with the Port was abandoned. In October
1986, Charles Dunn, U.S. F&WS wrote to Col. Yankoupe of the Corps

> The terms of that agreement have yet to be met. . .We have
> indicated to the Port that they must therefore proceed
> immediately to begin preparation of a mitigation plan. The Port
> committed to do so in our joint agreement, but has resisted
> doing so to date. The Port has refused to begin development of
> a mitigation plan and instead has sought only to initiate
> another time extension (14).

Surprisingly, at the end of October, the Corps sent a letter to the
Port stating that the U.S. F&WS had informed the Corps that the Port
was not in compliance with the referenced agreement. According to
the Corps

> The amended Environmental Assessment (EA) and Finding of No
> Significant Impact (FONSI) prepared to evaluate the proposed
> revision referred to a "signed agreement" between the Port and
> U.S. Fish and Wildlife Service (U.S. FWS). The EA/FONSI
> document stated that the signed agreement adequately addressed
> concerns that were raised in response to our Public Notice. The
> Findings of Fact (FOF), which was the decision making document,
> specifically cited the agreement and referenced it to satisfy
> concerns raised by individuals and organized groups opposing the
> revision. This agreement between the Port and U.S. F&WS was an
> important factor in our environmental review and the final
> permit decision (21).

The Corps referred to general condition (K) of the permit and procedures of 33 CFR 325.7, "Modification suspension or revocations of authorization" (21).

Senator Jackson died in 1983 and was no longer available to the Port. But Ronald Reagan became President in 1981 on an anti-environmental platform and the Port began a strategy of fending off mitigation efforts by the U.S. F&WS by approaching Susan Recce, Asst. Sec. of Interior. The Olympia Field office expressed concern about the Port's efforts and pointed out

> The Agreement between the Port and the Service expired over two years ago, so any additional delays are unreasonable and will not change the fact that our expectations were not fulfilled within the agreed-to time period. . . .The Port freely admits that they always assumed that the Service would continue to sign agreements and so would never make them live up to the Agreement (15).

A December 1986 meeting was scheduled by the Corps. Five Corps representatives, two representatives from U.S. F&WS and two representatives of the Port were present. The Port's position was that because they continued to participate in the Grays Harbor Estuary Planning process, no further compensation was needed. U.S. F&WS stated that the Plan does not provide any extra measure of protection and does not mitigate for the Kaiser fill and that the deadline in the supplemental agreement had now passed. A follow-up meeting was scheduled for mid-January 1987. The Port was to establish where the Grays Harbor plan was modified to such an extent that it mitigated for the Kaiser fill. The U.S. F&WS was to prepare a report suggesting possible options and alternatives that might be acceptable mitigation or compensation for the Kaiser fill (8). The Port, seeking to undermine the local U.S. F&WS office, sent U.S. F&WS Director Dunkle a letter in April of 1987, requesting that the Service sign yet another Letter of Agreement (50).

Meanwhile, in June 1987, Stanley E. Senner of the U.S. Section of the International Council for Bird Preservation also wrote to Dunkle inquiring about the status of the two agreements the Service had with the Port (46). Even though the U.S. Fish and Wildlife had already gone on record indicating that the the Grays Harbor Estuary Management Plan was not in compliance with Department of Interior policy (16), based on the past representations to the Port by U.S. F&WS Area Manager Joe Blum, Asst. U.S. F&WS Director Steve Robinson replied to Senner that

> The Service entered into an agreement with the Port that stipulated that if an acceptable Plan could be developed and adopted by all of the various public and private interests, then the Service would accept this adopted plan as full mitigation for the resource impacts that resulted from the 45-acre fill on the "Kaiser" site. . . The Service intends to honor the past commitment to the Port of Grays Harbor to consider the Plan, when adopted as mitigation for the resource impacts that

resulted from the past filling on the Kaiser site (emphasis added)
(41).

EPA was no better in sticking with its conditions. When the
Port revised its permit to allow for the stockpiling of dredged
material on the Kaiser site, EPA wrote to the Corps in July of 1978,

> In a letter dated 10 December 1975, the Port of Grays Harbor
> entered into a written agreement with the Environmental
> Protection Agency to officially require water dependent use of
> the Kaiser fill site. The proposed use of this area as a
> stockpile site for dredged material is not water dependent and,
> therefore, inconsistent with our agreement with the Port of
> Grays Harbor (20)

However, EPA ultimately withdrew its opposition and the Corps issued
the permit amendment.

WILDLIFE REFUGE

Due to the nonresponse of the Grays Harbor estuary management
planning (GHEMP) (32) process to public participation** and
protection for critical habitat in Grays Harbor, FOE and Friends of
Bowerman Basin, a local citizen group, began circulating petitions in
1987 requesting that Congress protect the Bowerman Basin shorebird
migration area by adding it and an adjacent area, Point New, to the
National Wildlife Refuge System.

The local member of Congress, Rep. Don Bonker (D-WA) introduced
a bill (H.R. 3423 cosponsored by the entire Washington House
Delegation) in the House of Representatives in October 1987 to
protect the Bowerman Basin. An amended bill passed the House in
December of 1987, strengthening the relationship between
establishment of the Refuge and mitigation obligations of the Port of
Grays Harbor vis-a-vis Kaiser.

On the Senate side, Sen. Adams (D-WA) introduced S. 1755, a
companion bill to Bonkers', but Senator Dan. Evans (R-WA) introduced
his own version, S. 1758. Unfortunately, the Senators were persuaded
by the Port to introduce in December a new bill, S. 1979, which
included a provision sought by the Port that transfer of the Bowerman
Basin property to the U.S. F&WS be treated as "mitigation credits" as
meeting, in whole or in part, mitigation obligations of the Port of
Grays Harbor arising under section 404 of the Federal Water Pollution
Control Act for future watermeadow filling (55).

The Port acquired Bowerman Airport and approximately 50 acres of
tidelands north of the airfield in 1962 from Grays Harbor County for

---

(** FOE has documented yet another meeting of the Grays Harbor
Estuary Management Planning Task Force which took place on 15
November 1990 without notification of either the public or the Grays
Harbor Estuary Citizen's Advisory Committee.)

$1.00 and the approximately 1700 acres of tidelands west of Bowerman
Airfield in 1971 from Grayco Corp. for $8,660 (43).

FOE testified before the U.S. Senate Subcommittee on
Environmental Protection in February 1988 that "our preferred
alternative would be to compensate the Port $8,661 and condemn the
property under the U.S. F&WS's power of eminent domain. However, we
recognized that the House passed bill represents a carefully crafted
compromise between outright condemnation and the Port's scheme to
acquire mitigation credits for further wetland filling in the
estuary" (57). Despite the overwhelming evidence of the biological
importance of the Bowerman Basin to international shorebird
populations on the West Coast, the U.S. F&WS testified against the
bill, mainly due to budget constraints imposed by the Administration
(57).

The Seattle Corps took issue with three statements made by the
U.S. F&WS at the Senate hearing regarding the status of the Kaiser
fill

1) FWS/PGH agreement was special permit condition & allowed
issuance of permit. No. The agreement was discussed in Record
of Decision, but not a condition.
2) FWS/PGH agreement was modified to eliminate "energy
dependent" requirement on permit. Agreement, was probably
modified, but we modified permit, which is key.
3) Creation of National Wildlife Refuge will eliminate all work
in area - wrong 10/404 permits & balancing process still
applicable (3).

An amended version of S. 1979 was passed by Congress and signed
by the President on 19 August 1988. References to "mitigation
credits" was dropped and the Committee on Environment and Public
Works Committee Report specifically noted that "the Port of Grays
Harbor may have obligations to provide compensation for the impacts
resulting from past filling of wetlands in the Grays Harbor estuary.
Such obligations could be satisfied by the transfer authorized in
this legislation" (emphasis added) (56). The legislation also called
for the U.S. F&WS service to acquire the Bowerman Basin area from the
Port within three years of the bills enactment (by 17 August 1991).

As of January 1991, a final Grays Harbor Estuary Management Plan
has not been submitted to the Department of Commerce for approval as
an amendment to the Washington State Coastal Zone Program. However,
in 1990 the Federal Office of Oceans and Coastal Zone Management
threatened the State of Washington with cutoff of their Section 312
CZMA funding if the Grays Harbor Plan was not adopted despite
recognized inconsistencies between the Plan and current federal and
state wetland policies (9). The Bowerman Basin is still in Port
ownership and, nearly fifteen years after the Seattle Corps District
issued a permit to fill a critical habitat area at the mouth of the
Hoquiam River, the Kasier site remains undeveloped and unmitigated.

LESSONS TO BE LEARNED

(1) Resource agencies should not cut separate deals with applicants.

(2) Speculative fills should be opposed. Better to go down fighting than silently stand by.

(3) "Mitigation" proposals must be subject to public participation and review and ongoing monitoring.

(4) An Environmental Impact Statement does not always lead to correct decisions.

(5) When you negotiate with a Port, count your spoons.

(6) Never tell the Corps you've dropped plans to sue.

(7) Don't give up. Don't let them get away with it.

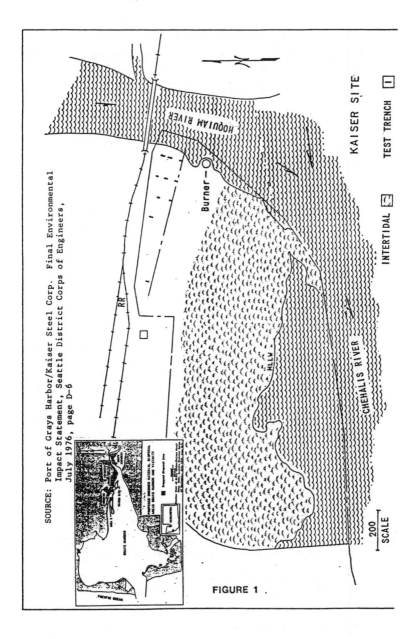

SOURCE: Port of Grays Harbor/Kaiser Steel Corp. Final Environmental Impact Statement, Seattle District Corps of Engineers, July 1976, page D-6

FIGURE 1 .

400   COASTAL WETLANDS

References

1. Aberdeen Daily World. 12 April 1979. "Corps is expediting Kaiser Project". Bill Steward.

2. --------------------. 11 April 1979. "Kaiser won't build plant on Port site".

3. Baxter, W. (Seattle District Corps of Engineers). 24 February 1988. Corps handwritten memo RE:G.H. NW Refuge Hearings. 1 page.

4. Blum, J.R. (U.S. Fish and Wildlife Service). 27 August 1979. Letter to Moraski, Col. L. (Seattle District Corps of Engineers). 3 pages.

5. Blum, J.R. (U.S. Fish and Wildlife Service) and Soike, H.E. (Port of Grays Harbor). 16 July 1980. Letter of Agreement. 2 pages.

6. --
----------------. 10 August 1982. Supplemental Letter of Agreement. 2 pages.

7. Brown, J.N. (U.S. Fish and Wildlife Service). 21 August 1975. Memorandum to files. 2 pages.

8. Bowlus, M.L. (Seattle District Corps of Engineers). 3 December 1986. MEMORANDUM FOR RECORD - SUBJECT: Coordination Meeting between Port of Grays Harbor, U.S. Fish and Wildlife Service (USFWS), and Seattle District Concerning Compliance with Letter of Agreement for the "Kaiser Fill" 071-OYB-2-002533. 3 pages.

9. Burgess, J.P. (Office of Ocean and Coastal Resource Management). 30 May 1990. Letter to Bloomquist, G. (Grays Harbor Regional Planning Commission). 2 pages.

10. Carson, W.C. (Kaiser Steel Corporation). 8 January 1976. Letter to Eineigl, Col. R. (Seattle District Corps of Engineers). 2 pages.

11. -------------------------------------. 12 January 1976. DRAFT #1. 3 pages.

12. -------------------------------------. 6 April 1979. Letter to Soike, H.E. (Port of Grays Harbor). 2 pages.

13. Cleaver, F. (National Marine Fisheries Service). 16 October 1975. Letter to District Engineer. (Seattle District Corps of Engineers). 1 page.

14. Dunn, C.A. (U.S. Fish and Wildlife Service). 16 October 1986. Letter to Yankoupe, Col. R.F. (Seattle District Corps of Engineers). 2 pages.

15. --. 24 November 1986.
     Memo to Director. (U.S. Fish and Wildlife Service). 4 pages.

16. --. 9 December 1986.
     Letter to Unsoeld, Hon. J. (U.S. House of Representatives). 4
     pages.

17. Eineigl, Col. R. (Seattle District Corps of Engineers). 23
     January 1976. Letter to Soike, H. E. (Port of Grays Harbor). 2
     pages.

18. --. 27 May
     1976. Memo to Division Engineer, North Pacific. (Corps of
     Engineers). 2 pages.

19. Evans, N. et al. October 1980. "The Search for Predictability".
     Washington Sea Grant 80-5. Institute for Marine Studies,
     University of Washington. 117 pages.

20. Geren, H.E. (Region 10 Environmental Protection Agency). 5 July
     1978. Letter to Thuring, D.W. (Seattle District Corps of
     Engineers). 2 pages.

21. Koch, F. E. (Seattle District Corps of Engineers). 31 October
     1986. Letter to Soike, H.E. (Port of Grays Harbor). 2 pages.

22. Likes, J.G. (U.S. Fish and Wildlife Service-Olympia Area
     Office). 7 November 1975. Memo to Field Supervisor and
     Regional Director. (U.S. Fish and Wildlife Service). 2 pages.

23. Malek, J. (former ERS, Seattle District Corps of Engineers). 18
     December 1990. Personal conversation.

24. Martinson, R. K. (U.S. Fish and Wildlife Service). 28 May 1975.
     Letter to Eineigl, Col. R. (Seattle District Corps of
     Engineers). 3 pages.

25. --. 3 November
     1975. Letter to District Engineer (Seattle District Corps of
     Engineers). 2 pages.

26. Matheson, B. (Sierra Club). 19 March 1976. Letter to Eineigl,
     Col. R. J. (Seattle District Corps of Engineers). 1 page.

27. McNeely, M. (Seattle District Corps of Engineers). 26 May 1979.
     Phone conversation record with Likes, J. (U.S. Fish and Wildlife
     Service-Olympia Area Office). 1 page.

28. Moraski, Col. L. (Seattle District Corps of Engineers). 17 Jan
     1980. Letter to Blum, J.F. (U.S. Fish and Wildlife Service). 2
     pages.

29. Ortman, D.E. (Northwest office, Friends of the Earth). 4

November 1975. Letter to Permit Section. (Seattle District
Corps of Engineers). 2 pages.

30. ---. 18 March
1976. Letter to Eineigl, Col. R. J. (Seattle District Corps of
Engineers). 2 pages.

31. ---. 19 April
1979. Letter to Poteat, Col. J.A. (Seattle District Corps of
Engineers). 2 pages.

32. ---. For a
discussion of the Grays Harbor Estuary Management Planning
Process see: 1983. "The Public Be Damned: Grays Harbor
Revisited". Coastal Zone '83 Symposium Proceedings. Orville T.
Magoon (ed.). Vol II (A.S.C.E.). p. 1694ff.

33. ---. For a
discussion of the Washington State Coastal Zone Management
Program see: 1987. "Washington's CZMP - The First Shall Be
Last". Coastal Zone '87 Symposium Proceedings. Orville T. Magoon
(ed.). Vol III (A.S.C.E.).

34. Parenteau, P.A. (Northwestern School of Law, Lewis and Clark
College). 19 January 1976. Letter to Eineigl, Col. R. (Seattle
District Corps of Engineers). 2 pages.

35. Poteat, Col. J.A. (Seattle District Corps of Engineers). 30
August 1976. Statement of Findings, Port of Grays Harbor and
Kaiser Steel Corporation - 071-OYB-1- 002533. 3 pages.

36. ---. 30
August 1976. Department of Army Permit issued to Port of Grays
Harbor. 4 pages.

37. ---. 22
September 1976. Letter to Jones, D. (Northwest office, Friends
of the Earth). 2 pages.

38. ---. 7
August 1978. Environmental Assessment, Dredging of Upper
Chehalis River Navigation Channel and Temporary Barge Moorage in
Hoquiam River, Grays Harbor, and Chehalis River, Washington. 6
pages.

39. Reed, L. A. (Region 10 Environmental Protection Agency). 14
October 1975. Letter to Latta, R.H. (Seattle District Corps of
Engineers). 2 pages.

40. Reese, D, (U.S. Fish and Wildlife Service). 8 December 1975.
Letter to Ortman, D.E. (Northwest office, Friends of the Earth).
1 page.

41. Robinson, S.A. (U.S. Fish and Wildlife Service). 7 August 1987. Letter to Senner, S.E. (International Council for Bird Preservation). 2 pages.

42. Seattle District Corps of Engineers. 10 March 1976. Questions Regarding Project Description for Kaiser Permit EIS (INCL 1). 2 pages.

43. -------------------------------------. July 1976. Port of Grays Harbor/Kaiser Steel Corp. Final Environmental Impact Statement, Grays Harbor.

44. -------------------------------------. Undated. Circa 1976. Fact Sheet, PORT OF GRAYS HARBOR PERMIT APPLICATION - EIS PREPARATION. 3 pages.

45. Sellevold, R.P. (Seattle District Corps of Engineers). 20 June 1976. Memo to Office of Counsel. (Seattle District Corps of Engineers). 2 pages.

46. Senner, S.E. (International Council for Bird Preservation). 22 June 1987. Letter to Dunkle, F. (U.S. Fish and Wildlife Service). 2 pages.

47. Sherburne, -. (Seattle District Corps of Engineers). 7 April 1976. Memo for Record. 12 pages.

48. Soike, H.E. (Port of Grays Harbor). 12 March 1976. Letter to Dice, S.F. (Seattle District Corps of Engineers). 3 pages.

49. -------------------------------------. 25 April 1979. Letter to Poteat, Col. J. A. (Seattle District Corps of Engineers). 2 pages.

50. Stevens, J. (Port of Grays Harbor). 17 April 1987. Letter to Dunkle, F. (U.S. Fish and Wildlife Service). 1 page.

51. Stout, D.J. (U.S. Fish and Wildlife Service). 22 July 1986. Memorandum to Assistant Regional Director. (U.S. Fish and Wildlife Service). SUBJECT: Meeting with Port of Grays Harbor Regarding Kaiser Fill. 2 pages.

52. Terpstra, Lt. Col. J.J. Jr., (Seattle District Corps of Engineers). 8 November 1976. Letter to Ortman, D.C. (N.W. Office, Friends of the Earth). 2 pages.

53. Thompson, -. 23 August 1976. Note RE: Kaiser Permit meeting to Dice, S.F. (Seattle District Corps of Engineers). 1 page.

54. U.S. Fish and Wildlife Service. Undated. Circa 1975. Unsigned pages in U.S. F&WS files. U.S. F&WS numbers 000172-176.

55. U.S. Senate. Senate Bill 1979. Section 3(c)(2).

56. -----------. Committee on Environment and Public Works.
     Committee Report to accompany S. 1979. 9 pages.

57. -----------. Subcommittee on Environmental Protection. 25
     February 1988. Testimony on S. 1979. Sen. Hrg. 100-571.

58. Walsh, J. F. (Seattle District Corps of Engineers). 25 June
     1976. Memo to Ch. Engineering Div. (Seattle District Corps of
     Engineers). 2 pages.

59. Weber, F. (Seattle District Corps of Engineers). 13 August 1975.
     Memo to Chief of Opns. Division/Ch. Off. Counsel. 1 page.

Comparison of Two Marsh Management Plans
in Louisiana

L. Phil Pittman[1] and Ricardo W. Serpas[2]

Abstract

Marsh management plans are currently being
utilized in Louisiana's coastal wetlands to attempt to
reduce land loss and increase productivity for
vegetation, fish, and wildlife. Louisiana's three
million acres of coastal wetlands are currently
experiencing significant land loss and habitat change
from both natural and man-made causes. Many of these
acres are being managed extensively to reduce land loss
as well as for the production of fish or shellfish,
waterfowl, fur bearers, and alligators. These marsh
management plans are attempting to increase
productivity in the marshes and swamps by reducing
erosion and land loss, increasing vegetative
production, preventing habitat deterioration and/or
change, regulating plant and animal populations for
both size and composition, and mitigating for impacts
due to our coastal use permitting process.

The Louisiana Coastal Management Division began
a sampling and monitoring program on specified marsh
management plans in October, 1988 and continued through
September, 1990. Two of these plans, Vermilion Bay
Land in Vermilion Parish and West Creole Canal in
Cameron Parish, will be discussed and compared in this
paper. Sampling stations were selected both inside and
outside the management areas and included collection
data for the following parameters: fisheries, water

----
1. Section Manager, Louisiana Department of Natural
Resources/Coastal Management Division, P. O. Box 44487,
Baton Rouge, LA 70804
2. Field Investigator, Louisiana Department of Natural
Resources/Coastal Management Division, 1313 E. Tunnel
Blvd., Houma, LA 70363

chemistry, hydrology, and structural component
operation. Samples were taken monthly and the data
placed on our computer for statistical analysis.

The complexities associated with setting up and
managing an area for marsh management are well
documented. Our sampling and monitoring program was
designed to determine if these plans are fulfilling
their goals to create, restore, or enhance wetlands and
its associated productivity. This study may show how
fisheries standing crop is affected by the operation of
two marsh management plans in southwestern Louisiana.

Introduction

The Louisiana coastal zone encompasses all or
parts of nineteen parishes in the southern portion of
the state. Almost 5.3 million acres of marshes,
swamps, bottomland forests, barrier islands, pine
forests, chenier plains, and various types of
waterbodies exist within this zone (Turner and
Gosselink, 1975). This represents over 41% of the
nation's coastal wetlands and 25% of all wetlands in
the United States, thus making it one of the largest,
richest, and most complex estuarine areas in the
world. These wetlands support a very large commercial
and recreational fisheries, a large and diverse fur and
trapping industry, and they serve as a nursery ground
for various species of fish or shellfish (Pittman and
Piehler, 1989). In addition, they provide important
waterfowl habitat for countless ducks and geese, they
support and enhance an important tourist industry, and
the outermost barrier islands act as a buffer to reduce
the mainland impacts from storms and hurricanes (Howey
and Blackman, 1987).

Louisiana's wetlands are currently experiencing
severe land loss (over 32,000 acres/year), wetland
alteration, and water quality degradation throughout
the coastal zone. With the approval of the state
Coastal Resources Management Act of 1978, Louisiana
began implementing steps to alleviate these impacts.
Marsh management plans are one means of attempting to
ameliorate such adverse impacts.

Since 1980, over 120 marsh management plans
have been received by the Coastal Management Division
of DNR for permitting (Clark, deMond, Bourgeois, Barras
and Blackman, 1988). These plans involved the
management of over 460,000 acres of Louisiana's coastal
zone. They range in size from 0.3 acres to over 68,000
acres, with the mean being 4,133 acres per plan. All

plans must include the following components: management goals; the history of area habitat related problems; a habitat description; a design scheme for water control structures, including location and operation schedule; non-marsh management activities; and a monitoring schedule. Interagency field evaluations and meetings are held to review all large marsh management plans prior to issuance of the permit.

To be consistent with the Louisiana's Coastal Resources Program, as well as other state and federal agencies, those plans must be designed to protect, restore, and/or enhance coastal renewable resources. Management plan goals must include land loss or erosion prevention, increased wildlife production, increased wildlife and fisheries diversity, regulation of plant and animal population, increased viable habitats, and overall productivity enhancement.

Management Plan Profiles

A total of six (6) marsh management plans were originally sampled from October, 1988 to September, 1989. Of these six, two (2) were selected for an additional year's study. These two (West Creole Canal in Cameron Parish and Vermilion Bay Land Corporation in Vermilion Parish) were chosen due to their differences in management profiles (active versus passive) and their similarities in marsh types. A more indepth description of these two plans is given below.

The Coastal Management Division (CMD) of the Louisiana Department of Natural Resources issued Coastal Use Permit P830450 to the Cameron Parish Gravity Drainage District #4 for the West Creole Canal Marsh Management Plan in January of 1984. The primary objective of the plan was to prevent extensive shoreline erosion and interior marsh erosion by stabilizing water levels and reducing tidal gradient and salinity in the channelizing of brackish wetland southwest of Oak Grove, Louisiana. Oak Grove is located approximately 30 miles south-southwest of Lake Charles, Louisiana.

West Creole Canal Marsh Management Plan (MMP) encompasses approximately 1500 acres of brackish to saline marsh. Dominant vegetation in the marsh consists mainly of Oystergrass (Spartina alterniflora), Wiregrass (S. patens), and Silverling (Baccharis halimifolia). The sole water control structures regulating this marsh consists of five (5) 48-inch corrugated aluminum culverts with variable crest

inlets, four of which are further equipped with outside
flap gates to prevent inflow. Permitted operation of
the structure allows for the inlet crests to be set at
0.5 feet below mean marsh elevation, with the open
culvert to be altered by the management area manager
during adverse climatic conditions (i.e., extremes in
water levels and salinities). Although the plan does
not contain a schedule for artificially induced
drawdown natural climatic conditions cause the water
bottoms in the marshes to dry during some drought
periods.

        The Vermilion Bay Land Corporation MMP Permit
(P860129) was issued in December of 1986. This
management plan is located in Vermilion Parish
approximately 7 miles east of Intracoastal City,
Louisiana or approximately 20 miles south of Abbeville,
Louisiana. It was primarily intended to regulate and
exchange water levels in order to re-establish marsh
vegetation, enhance waterfowl production, increase
fur-bearer and alligator productivity, and accommodate
migrations of aquatic organisms. Approximately 1,000
acres are located within the management area. The
predominant vegetation in the plan includes Wiregrass,
Three-cornered Grass (Scirpus olneyi), Hogcane
(Spartina crynosuroides), and Bulltongue (Sagittaria
lancifolia).

        To accomplish the management area's goals,
existing levees were plugged, cuts were made in the
spoil banks on an existing canal, three (3) water
control structures were removed and one (1) was
plugged. Barrel culverts with stop lock structures on
the interior and flap gates on the exterior were
installed at two of the sites where existing structures
were removed. Permit conditions also included
establishing water level gauges at water control
structures and adhering to a water level schedule to
allow for the migration of estuarine organisms and the
arrival of migrating waterfowl.

Methods and Materials

        Several criteria were used to select sites for
sampling. Geomorphology, site proximity, hydrologic
patterns, accessibility, and gear type suitability were
taken into consideration in selecting our sites. Each
of these criteria is discussed below.

        Geomorphologically, two basic aquatic habitats
were available for sampling - shallow pond bottoms and
man-made channels with corresponding spoil banks.

Sites were selected within each habitat type both inside the management area and outside the area as a control. Concerning site proximity, outside stations were selected giving proximal sites. These outside stations were our "control areas" and were recognized as the origin of any estuarine migrant entering the management area. Proximity was intended to reduce the chance of sample bias due to distance. In regards to hydrology, comparable sites were chosen within similar hydrologic units. The shallow water sites were essentially channelized drainage features, funneling net water flow out of headwaters and toward the mouth of the estuary.

Accessibility of the sites played a major role in the selection process. Due to the limited man-power available the total number of sites selected was limited. Logistical considerations made some potential sites inefficacious. Finally, suitability to the gear types selected was taken into account. Occasionally, the substrate proved too soft for the trafficability necessary for effective seine sampling. Other shallow water pond sites were deleted from consideration due to insufficient depth for trawling.

Gear types utilized throughout the two-year study consisted of trawls, seines, and castnets. Trawls measuring six feet along the corkline were used to sample deep water habitats. The trawl was operated at the respective stations three replicating times for intervals of five minutes at a speed of approximately one meter per second. Replicates were slightly overlapping. Bag mesh size of the trawls utilized was one-quarter inch bar, one-half inch stretch.

Seines were utilized to sample shallow water habitats. Three replicate samples were taken at each site using a 25-foot bag seine. Each sample was designed to cover an area 50 feet long by 25 feet wide. The drags were terminated against marsh shoreline to reduce escape potential. Seine mesh size was one-quarter inch.

The ecotone created by the presence of the operating water control structures was subjectively sampled by the use of a monofilament castnet with three-eighths inch bar mesh. A five foot radius castnet was used in the West Creole Canal MMP and a seven foot radius net was utilized for the Vermilion Bay Land MMP. Three throws comprised one station sample,

and samples were taken both inside and outside the
structure. The castnet samples provided evidence of
fish and shellfish presence within a critical
transition area.

On capture, all specimens were placed on ice to
preserve the freshness and prevent dessication. Samples
were usually analyzed within 24 hours of capture.
Each sample was separated according to species, and
each species was measured by individual length (to a
maximum of 25 individuals). In cases where the total
number of individuals exceeded 25, a random subsample
of 25 individuals was selected for measurement and then
weighed. Total species biomass was then obtained by
weighing all individuals of a species collectively.
Total number was back-calculated from this data.

In order to reduce effort during the second
year, individual lengths were only taken for
commercially and recreationally important fishery
species. This reduced the sample processing time by 25
to 50 percent.

Six commercially and recreationally important
estuarine fisheries species were selected for profile
analysis - Atlantic Croaker (Micropogonias undulatus),
Gulf Menhaden (Brevoortia patronus), Sand Seatrout
(Cynoscion arenarius), Brown Shrimp (Penaeus aztecus),
Blue Crab (Callinectes sapidus), and White Shrimp
(Penaeus setiferus). Six additional species were also
selected to quantify the resident organisms that use
the managed and unmanaged sites, namely Grass Shrimp
(Palaemonetes sp.), Sailfin Molly (Poecilia latipinna),
Sheepshead Minnow (Cyprinidon variegatus), Freshwater
Goby (Gobionellus shufeldt), Gulf Killifish (Fundulus
grandis), and Rainwater Killifish (Lucania parva).

In addition to analyzing fish and shellfish
collected, the following ambient parameters were
measured at each station monthly: air temperature,
water temperature, salinity, turbidity, relative water
levels, tide stage and movement, and wind speed and
direction. Temperature, salinity, and turbidity were
logistically compared using statistics.

Ocular estimates of vegetative species
composition were taken quarterly at each station and
the number of wildlife species utilizing the habitat at
each station was recorded whenever an observation
occurred.

Statistical analysis of the raw data was performed using the MacIntosh "Filemaker" by "Statview". Friedman, Kruskal-Wallis and Mann-Whitney U-test was used to analyze non-parametric data, while t-test and Scheffe's F-test were used for parametric data.

Results

In regard to the physical parameters that were analyzed, water level, turbidity, and salinity differed noticeably between the managed area and the controls for each management site. Salinity readings at VBL site were consistently higher for unmanaged marsh than managed sites. Turbidity was significantly less for managed sites than unmanaged sites at Vermilion Bay Land site. Station 4 (unmanaged seine station) at West Creole was the only station that was more turbid than any other station (Figure 1) and had consistently higher salinities due to its close proximity to the Gulf of Mexico.

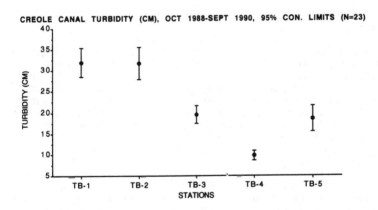

Figure 1.

Water levels between managed and unmanaged sites were significantly higher and longer in duration after a high water event (+30 cm above mean marsh level, MML) at Vermilion Bay Land (Figure 2). This was not observed at West Creole during the same innundation event.

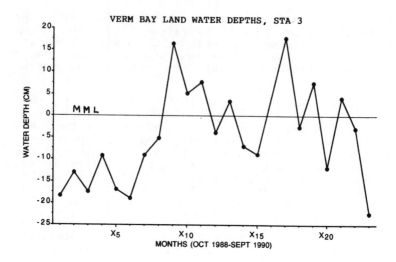

Figure 2.

Selected Species (Commercial & Recreational)

At Vermilion Bay Land, Atlantic Croaker populations were reduced by 78% for managed stations when compared to "available" population at the control (unmanaged) stations. Gulf Menhaden were 76% less for managed stations in comparison to the surrounding marshes (control). Presence of Sand Seatrout at managed stations was reduced by 97%. Brown Shrimp were 86% fewer, but considerably larger for managed stations. White Shrimp were reduced 98% and significantly larger for managed stations. Blue Crab increased 12% and were definitely larger for managed stations.

At West Creole Canal, Atlantic Croaker were reduced by 53% when comparing the available population (control) against the managed stations, particularly at the outside seine station. Brown Shrimp had considerable numbers of individuals at unmanaged stations to the point that they were reduced by 72% at managed stations. White Shrimp were reduced 73% and was most obvious at the trawl stations. Sand Seatrout were reduced 69% at managed stations. Blue Crab had increased by 48% at managed stations and were significantly larger at the managed trawl station. Gulf Menhaden increased 375% above the available "outside" population at the managed trawl and castnet stations (Table 1).

Table 1. Percent change in catch when comparing outside (unmanaged) stations against inside (managed) stations for important fishery species.

Species	West Creole Canal			Verm. Bay Land		
	Out	In	%	Out	In	%
Atl. Croaker	26,609	12,400	-53	4,077	882	-78
Blue Crab	673	995	+48	625	702	+12
Brown Shrimp	3,070	875	-72	956	137	-86
Gulf Menhaden	1,665	7,911	+375	4,543	1,109	-76
Sand Seatrout	828	255	-69	193	5	-97
White Shrimp	7,179	1,942	-73	4,981	108	-98

Residents

Grass Shrimp populations at Vermilion Bay Land were 190% greater for managed stations than unmanaged stations. Sailfin Molly increased 31,600% for managed stations, while Sheepshead Minnow increased 841% for managed stations. Freshwater Goby increased 375% for managed stations but were significantly smaller. Gulf Killifish were significantly larger for managed stations and increased 30% compared to unmanaged stations. Rainwater Killifish increased 42,250% for managed stations.

For West Creole Canal, Gulf Killifish were reduced 80% at managed seine stations, while Freshwater Goby were 74% fewer at managed trawl stations. Sheepshead Minnow were only 6% less at managed stations and Grass Shrimp were 17% greater at managed stations. Sailfin Molly increased 127% at managed stations and Rainwater Killifish increased 484% for managed stations (Table 2).

Table 2. Percent change in catch between outside and inside stations for resident species.

Species	West Creole Canal			Verm. Bay Land		
	Out	In	%	Out	In	%
Freshwater Goby	35	9	-74	54	257	+376
Grass Shrimp	8,058	9,409	+17	9,973	28,993	+191
Gulf Killifish	1,351	275	-80	173	225	+30
Rainwater Killifish	87	508	+484	2	847	+42,250
Sailfin Molly	549	1,246	+127	13	4,121	+31,600
Sheepshead Minnow	5,911	5,530	-6	120	1,130	+842

Managed and control stations were analyzed for community parameters, namely total biomass, total numbers, numbers of species, and diversity indices. Both study sites had significant differences in total

biomass for managed and unmanaged trawl stations. This
is due to numerous predators (i.e., 15-40 garfish at ca
0.5 to 1.5 kg) that were captured regularly during
winter months or low water levels at managed stations.
Total number and number of species was greater for the
managed trawl station at Vermilion Bay Land, while West
Creole Canal had the reverse with the unmanaged trawl
station having the greater number of individuals and
more species. For managed seine stations, VBL had a
greater number of species and higher indices of
diversity; at WCC, the number of species and the number
of individuals were significant. For WCC, castnet
stations had significant diversity difference between
stations. The managed castnet station for VBL had
significantly greater numbers, weight, number of
species, and diversity. All managed stations showed a
noticeable reduction of migrant fishery species that
could utilize the different study areas, and an
increase in resident species (Tables 1 and 2).

Discussion

        Louisiana's marsh management plans have a wide
variety of management goals that have been implemented
under various state and federal policies. These plans
have serious inconsistencies which are detrimental to
the resources that are being managed for and protected
by the regulatory agencies.

        Fishery species will have the greatest impact
exerted on them by marsh management to the point that a
large portion of the available population of any given
species will be unable to utilize a managed area in
order to complete its life cycle. These two site
studies have further confirmed the direct affect of
marsh management on fisheries that was previously
documented by Bradshaw (1985), Weaver and Holloway
(1974), and Herke, et. al. (1987a, 1987b).

        As a result of our study, the Vermilion Bay
Land marsh management plan should be considered as a
worse case scenario in regard to marsh management
practices. First, the site was impounded without full
implementation of what was permitted for the plan;
namely, only one water control structure was installed
instead of the required two structures which reduced
the amount of weir opening needed for the plan by
half. Secondly, duck hunters/leasees manipulated the
stop-log structure so that water levels were at or
exceeded mean marsh elevation, thus keeping the
vegetation inundated. These two things lead to
prolonged (three month) flooding of marsh plants after

an unusual tidal event (Figure 2). This type of inundation was found to be detrimental to the type of marsh plants found at both study sites (Mendelssohn and McKee, 1987; Mendelssohn and McKee, 1989). This deleterious impact was curtailed when a perimeter levee gave way in May of 1990, allowing full and unimpeded tidal exchange to take place between the managed area and the intracoastal waterway. Any further study of managed and unmanaged marsh at VBL was curtailed at this point.

The West Creole Canal management plan should be considered an adequate plan that is actively managed, i.e., lowering stop-logs during peak periods of ingress and egress of estuarine organisms. WCC was fully implemented without any discrepancies to permit conditions and fully operational for six years. After eight months of sampling, the manager for WCC informed us that commercial trawling was taking place in the main canal for the management plan (Coriel, 1989). This created considerable bias in our trawl sample data by removing those individuals large enough to capture in commercial gear and eliminated them from the managed population that was being sampled. The commercial effort was 14 times greater than our sampling effort, as they were trawling every night one week prior to and one week after each full moon during the spring and fall shrimp seasons.

The impact of marsh management on commercial and recreational fishery species is considerable when over ten percent of Louisiana's coastal zone is under management of this type. This could be doubled if only the type of wetlands conducive to estuarine fisheries production were considered. Intermediate to brackish marshes are where most marsh management plans are currently situated in the state.

For better utilization of coastal marshes by fishery species, there must be more active or, even better, aggressive management of water control structures on marsh management plans so that ingress and egress can be more proficient for those commercially and recreationally important fish species. Redesign of water control structures, namely weirs, has been proven to improve use of marsh nursery area for brown shrimp when a slotted weir was used (Rogers, et. al., 1987).

416   COASTAL WETLANDS

References

Bradshaw, W.H. 1985. Relative abundance of small
brown shrimp as influenced by semi-impoundment. M.S.
Thesis. Louisiana State University, Baton Rouge. 61
pp.

Clark, D., J. de Mond, P. Bourgeois, J. Barres, and B.
Blackman. A review of the Louisiana Coastal
Resources Marsh Management Program. W.G. Duffy and
D. Clark, eds. Marsh management in coastal Louisiana:
effects and issues - proceedings of a symposium. U.S.
Fish and Wildlife Service and Louisiana Department of
Natural Resources. U.S. Fish Wildl. Serv. Bio. Rep.
89(22) 378 pp.

Coriel, Paul. 1989. Personal communication.

Herke, W.H., E.E. Knudsen, Z.X. Chen, N.S. Green, P.A.
Knudsen, and B.D. Rogers. 1987a. Final report for
the Cameron-Creole watershed fisheries investigations.
Louisiana State University Agricultural Center, Baton
Rouge, Louisiana. 419 pp.

Herke, W.H., E.E. Knudsen, P.A. Knudsen, and B.D.
Rogers. 1987b. Effects of semi-impoundment on fish
and crustacean nursery use: evaluation of a
"solution." Pages 2562-2576 in O.T. Magoon, H.
Converse, D. Miner, L.T. Tobin, D. Clark, and G.
Domurat, eds. Coastal Zone '87: proceedings of the
fifth symposium on coastal and ocean management.
American Society of Civil Engineers, New York.

Howey, T., and B. Blackman. 1987. Use of a geographic
information system as a tool for making land use
management decisions for coastal wetlands in a state
regulatory program. Coastal Zone '87. 1:399-413.

Mendelssohn, I.A., and K.L. McKee. 1987. Experimental
field and greenhouse verification of the influence of
saltwater intrusion and submergence on marsh
deterioration: mechanisms of action. Pages 8/1-8/34
in R.E. Turner and D.R. Cahoon eds. Causes of wetland
loss in the coastal central Gulf of Mexico. II:
Technical narrative. Final report to the Minerals
Management Service, New Orleans, Louisiana. 400 pp.

Mendelssohn, I.A., and K.L. McKee. 1989. The use of
basic research in wetland management decisions. Pages
354-364 in W.G. Duffy and D. Clark, eds. Marsh
management in coastal Louisiana: effects and issues -
proceedings of a symposium. U.S. Fish and Wildlife

Service and Louisiana Department of Natural Resources.
U.S. Fish Wildl. Serv. Bio. Rep. 89(22) 378 pp.

Pittman, L. and C. Piehler. 1989. Sampling and
monitoring marsh management plans in Louisiana.
Coastal Zone '89. 1:351-368.

Rogers, B.D., W.H., Herke, and E.E. Knudsen. 1987.
Investigation of a weir-design alternative for
coastal fisheries benefits. Louisiana State
University Agricultural Center, Baton Rouge,
Louisiana. 98 pp.

Turner, R., and J. Gosselink. 1975. A note on
standing crops of *Spartina alterniflora* in Texas and
Florida. Contr. Mar. Sci. 19:113-118.

Weaver, J.E., and L.F. Holloway. 1974. Community
structure of fishes and macrocrustaceans in ponds of a
Louisiana tidal march [sic] influenced by weirs.
Contrib. Mar. Sci. 18:57-69.

Acknowledgements

        Chris Piehler and Charlie Mestayer for taking
and sorting samples, and generating the data. Cindi
Jones, Laura Roberts, and Darren Whatley for data
input, and Darryl Clark for data management and
integrity. Mark Swan for statistical analysis and his
profound computer expertise. Debi Serpas for typing
and proofing this manuscript.

Restoring Prospect Island Wetlands:
A Proposal for the
Sacramento-San Joaquin Delta of California

Fred Kindel[1]

Abstract

A proposal was prepared in 1988 by the Corps of Engineers to restore wetlands on 1,276 acres of Prospect Island. It is proposed to reduce operation and maintenance costs of a navigation project and restore fish and wildlife habitat and other benefits. Federal legislation is needed to provide land acquisition authority to the Corps for implementation. The proposal was developed as part of a joint Nationwide pilot study by the National Marine Fisheries Service and the Corps of Engineers.

Introduction

The Administrator of the National Oceanic and Atmospheric Administration and the Assistant Secretary of Army (Civil Works) agreed in 1985 that the National Marine Fisheries Service and the Corps of Engineers would conduct a three-year Nationwide pilot study. Its purpose was to determine if a program could be devised to modify Corps of Engineers projects to restore and improve marine, estuarine and anadromous fisheries habitats, and accomplish this within existing authorities, resources, and funding. Six pilot study sites were selected in California, Maryland, North Carolina and Texas. A report was issued jointly by the two agencies (National Marine Fisheries Service and Corps of Engineers, 1990). It concluded that fisheries habitat restoration and creation opportunities can be implemented at no net increase in Corps project costs. It was recommended that an expanded Nationwide program of this kind should be implemented.

----

[1]Chief, Environmental Resources Branch, Sacramento District, Corps of Engineers, 1325 J Street, Sacramento, CA 95814

Prospect Island was one of the six study sites. The
proposal to restore wetlands and fisheries habitat as part
of a Corps of Engineers project at no increase in cost is
described below.

Corps of Engineers Project

The Sacramento River Deep Water Ship Channel was
authorized by Congress in 1946 and essentially completed
in 1963. It provides a 30-foot deep channel for ocean-
going ships to reach the Port of Sacramento. Another
authorization was provided by Congress in 1985 for
deepening the channel to 35 feet and construction was
initiated in 1989. The channel extends a distance of
about 43 miles from the Port of Sacramento to existing
depths of 35 feet at New York Slough near the cities of
Antioch and Pittsburg.

The authorizations by Congress direct that the
projects be implemented as a partnership between the Corps
of Engineers and the non-Federal sponsoring agency. The
non-Federal agency sponsoring the project is the Port of
Sacramento. The Port was authorized to acquire all lands
needed for the channel, build and maintain a public port
and associated facilities, and share other costs of the
project. The Corps of Engineers was authorized to plan,
design, construct, and operate and maintain the completed
channel works.

The Corps of Engineers maintains the channel to
design depths with periodic dredging and performs other
necessary operation and maintenance work. One feature of
the channel project is a cut through fastland at Prospect
Island which divided the island into a large eastern
segment and a long narrow remnant along Prospect Slough to
the west. The eastern segment of Prospect Island occupies
a total of 1,580 acres. Some 1,276 acres are privately
owned and farmed; the remaining 304 acres are owned by the
Port of Sacramento and used for disposal of dredged
materials and fish and wildlife mitigation for the
deepening project.

As an obligation of the original project, the Corps
maintains the Prospect Island levee fronting the east side
of the deep water ship channel. The cost of this
maintenance has averaged $311,000 annually since 1964
(1988 prices).

Restoration of Wetlands and Fisheries Resources

The Sacramento District of the Corps of Engineers and
the Southwest Region of the National Marine Fisheries

Service conducted studies in 1986, 1987 and 1988. The
objective of the studies was to compare the continuing
costs of maintaining the Prospect Island levees with the
alternative costs of acquiring the farmland and converting
it to a wetland/fisheries habitat. A report was prepared
and furnished to higher headquarters of both agencies
(Sacramento District, Corps of Engineers, 1988). The
results are highlighted below.

Background. Prospect Island is one of many islands
reclaimed from the historic marshes of the Sacramento-San
Joaquin Delta and converted to farmlands surrounded by
levees and waterways over the past 150 years. The island
is elongated north to south and in somewhat of a diamond
shape with the narrowest point at the south end. The deep
water channel is on the west side of the island; Miner
Slough is on the east side; a levee separates the north
end from other lands further north. In the interior of
the island, an elevation of mean sea level occurs about
one-half way from north to south; elevation rises
northward to about one foot above mean sea level and
lowers southward to about one foot below mean sea level.

Costs (1988 prices). The cost of continuing levee
maintenance was estimated at $311,000 annually. The cost
of acquiring 1,276 acres at fair market value and creating
improvements for wetlands, fisheries and other needs was
estimated at about $3.4 million (capital costs) or
$301,000 on an annual basis (at the prescribed Federal
water projects interest rate of 8 5/8%). The Federal
maintenance costs would cease, and cost savings are thus
estimated to be at least $10,000 on an annual basis.
Refinements in the improvements necessary could result in
additional cost savings.

Restoration Improvements. The principal features
necessary to restore wetlands and fisheries habitat
include the following. The lands would be acquired at
fair market value. After improvement, levees would be
breached to allow free access of Delta tidal waters.
Improvements would include shallow excavations to create
diverse fish habitat and marsh; the material removed would
be placed along the Miner Slough levees and biotechnical
slope stabilization would be added to counteract future
wavewash erosion of the levees. The levees would be
planted with riparian vegetation. A power line would be
relocated. Based on other recent studies in the vicinity,
it is anticipated that volunteer wetland plants will
invade rapidly following inundation by tidal waters, thus
restoring a diverse wetlands habitat.

Administration of Completed Wetlands/Fisheries
Habitat. The California Department of Fish and Game

strongly supports the proposal and has indicated it will
accept the newly restored wetlands and fisheries habitats
and operate and maintain them.

Summary of Additional Cost Savings and Benefits of
the Wetlands/ Fisheries Habitat Proposal. In addition to
the $10,000 annual savings in maintenance costs, other
cost savings and benefits would also accrue.   (1)
Additional Cost Savings: There is a potential for savings
in costs of the new deepening project now under
construction. The Port of Sacramento owns 304 acres at
the southern end of Prospect Island on which dredged
materials and mitigation developments needed for the new
deepening are expected to be placed. This area could be
combined with the 1,276-acre area, and dredged materials
and mitigation improvements could be relocated throughout
the larger 1,580-acre area. Both types of improvements
are already included in the wetlands/fisheries habitat
proposal. If this were done, the 304 acres would be
removed from the cost side of the new project and counted
as a benefit to fish and wildlife resources. This
potential cost savings is estimated at $21,000 annually;
if combined with the estimated costs of Prospect Island
proposal, this would reduce its annual costs from $301,000
to $280,000.   (2) Cost Savings Benefits: There is a
potential for cost savings to the nearby Sacramento River
Bank Protection Project which is under construction and
provides bank protection for extensive Federal levees in
the Delta area.   This project requires substantial
mitigation costs generally consisting of replacement of
riparian habitat. The riparian habitat included in the
Prospect Island proposal along the Miner Slough levees
could be counted as a cost savings in the other project's
mitigation needs, thus resulting in an additional benefit
to the Prospect Island proposal. About 450 acres of
riparian vegetation could be provided and it is estimated
that benefits from such costs saved to the other project
could exceed $39,000 annually based on experienced costs
to date.   (3) Fisheries Benefits: We estimated that
fisheries benefits would accrue from the restored
fisheries resources and accompanying recreational fishing.
Based on a contingent value survey of fishermen conducted
by Corps of Engineers employees specifically for the
Prospect Island studies, we estimated that the proposed
wetlands/fisheries habitat improvements would increase
fishing values by $2.85 million annually.   (4) Wetlands
Waste Assimilation Benefits. The wetlands will assimilate
herbicides, pesticides, fertilizer residues in the soil,
and other contaminants imported to the site by currents
and tidal exchanges.   The National Marine Fisheries
Service estimated the value for waste assimilation to be
about $39,500 annually for the 1,580 acres. Total Annual

<u>Costs, Savings, and Benefits.</u> Accordingly, the minimum
cost savings anticipated is $10,000 annually to implement
the wetlands/fisheries habitat proposal. Additional cost
savings could reduce the annual costs of the proposal from
$301,000 to $280,000, and three types of benefits could
exceed $2.9 million annually, resulting in a benefit-cost
ratio greater than 10 to 1.

<u>Need for Additional Federal Legislative Authority.</u> At
this point, based on the cost savings and benefits which
could be realized, you are probably asking yourself, "Why
don't they get on with it?"   That seems a logical
question.   I have asked it myself.   But as most
bureaucrats find, there is always a bureaucratic problem
hiding around the corner in such good proposals as this
appears to be.

The problem in this case arises from the description
of the authorized projects which I mentioned a moment ago.
The Corps of Engineers was authorized to build the ship
channel project and maintain it.   The Port of Sacramento
was authorized to pay for and acquire all the lands for
the project.   Our higher headquarters offices have
reviewed the proposal in comparison with the project
authorizations.   They have concluded that the Corps of
Engineers is not currently authorized to pay for or
acquire lands to be able to implement the cost savings.
The 1988 report is on hold until a way is found to solve
the problem on lack of authority for the Corps to pay for
and acquire lands to implement it. In addition, capital
funds of about $3.4 million would be needed in order to
realize the annual cost savings of $10,000 or more and the
estimated benefits.

We in Sacramento District recommended to Corps
headquarters in mid-1990 that legislation be sought to
provide the authority needed for this land acquisition so
that the proposal can be implemented when funds are
available.   The National Marine Fisheries Service and the
California Department of Fish and Game support such added
authority and implementation of the proposal.

<u>Concluding Note</u>

It is not the purpose of this paper to repeat details
of previous reports. The purpose is rather to summarize
such   information   and   communicate   this   to   persons
interested in this type of wetland/fisheries habitat
restoration.   Those interested in obtaining the complete
1988 and 1990 reports cited should contact: Major General
Patrick J. Kelly, Director of Civil Works, U.S. Army Corps
of Engineers, Washington, D.C. 20314-1000.

References Cited Appendix

1. National Marine Fisheries Service, Silver Spring, MD and U.S. Army Corps of Engineers, Washington, D.C., "Pilot Study to Determine the Feasibility of Establishing a Nationwide Program of Fisheries Habitat Restoration and Creation", February 9, 1990.

2. Sacramento District, Corps of Engineers, Sacramento, CA "Prospect Island Fish Habitat Restoration Pilot Program of the Corps of Engineers and National Marine Fisheries Service", April 6, 1988.

THE MUZZI MARSH, CORTE MADERA, CALIFORNIA
LONG-TERM OBSERVATIONS OF A RESTORED MARSH
IN SAN FRANCISCO BAY

Phyllis M. Faber[1]

ABSTRACT

In 1976, the Golden Gate Bridge, Highway, and
Transportation District purchased a 200 acre diked
former marsh site in Corte Madera on San Francisco Bay
for spoils disposal and as a mitigation site for dredging a channel to a newly constructed ferry terminal.
One hundred and thirty acres were returned to tidal
activity. Long-term observations have been made on
physical changes (drainage channel formation, sedimentation in channels and on the marsh plain) and vegetation
distribution. Some of these observations are presented
in this paper.

INTRODUCTION

Observations have been made intermittently over a
ten-year period from 1979 to the present on a 130 acre
portion of a formerly 200 acre diked salt marsh, the
Muzzi property, Corte Madera, California (Faber 1979,
1980, 1982, 1984, 1985, 1988). The restoration of tidal
activity to this diked portion of a once extensive marsh
draining the Ross Valley, served in part, as mitigation
for dredging an access channel to a ferry terminal built
by the Golden Gate Bridge, Highway and Transportation
District (GGBHTD) on Corte Madera Creek, a quarter mile
to the north. The remaining of 70 acres was used as a
receiver site for dredge spoils. Early observations on
the restoration site focused on the process of natural
revegetation and the rate of colonization of both pickleweed (Salicornia virginica and S. rubra) and cordgrass
(Spartina foliosa).

In 1980, the GGBHTD carried out a site modification
project of creating channels and two embayments around
the periphery of the restoration site to enhance the

----
[1]Phyllis M. Faber, 212 Del Casa, Mill Valley, CA 94941

tidal flows to landward portions of the marsh. Following this work, additional studies were initiated to examine channel morphology, marsh plain topography, and its relationship to vegetation distribution.

I. BACKGROUND

A. Loss of Salt Marsh

Approximately ninety five percent of the tidal marshes of San Francisco Bay have been leveed or filled since the 1850s. Tidal salt marshes once extended upstream over three miles into the Ross Valley from San Francisco Bay along Corte Madera Creek in Marin County, California. Since the rapid urbanization of Marin County in the 1950s, most of the salt marshes were filled and developed. Figures 1 and 2 show progressive filling of the Ross Valley Marsh in aerial photos taken in 1952 and 1982. The 130 acres of the Muzzi restoration site at the bay edge (arrow, Figure 2) and the Corte Madera Ecological Reserve, a 66-acre remnant of intertidal marsh to the north, are less than one eighth of the historic Ross Valley marsh system, and are presently the only remaining parcels over 20 acres in size in the Ross Valley wetland system.

Figure 1 (left). 1952. Tidal salt marshes once extended 3 miles into the Ross Valley. Extensive filling and development of the marsh occurred in the 1950s and 1960s. Figure 2. 1982. By 1976, a 70 acre portion of the Muzzi Marsh filled with dredge spoils and dikes were breached in four places to return tidal access to the remaining 130 acres. An undiked remnant of ancient marsh, the Heerdt Marsh, is adjacent to the north.

This system has had a series of alterations ranging
from the installation of a railroad line in the 1870s, a
highway (now Highway 101) in the 1930s, extensive filling for housing in the 1950s and 60s, and finally diking
of a 200 acre bayward portion in the late 1950s. With
tidal access eliminated, salt marsh vegetation had
largely died and the easterly portion of the marsh plain
behind the dike had dried up and subsided to elevations
ranging from +1.2 to +2.9 feet NGVD. By comparison, the
adjacent Heerdt Marsh, undiked and abundantly vegetated,
has elevations of +2.5 to +4.0 feet NGVD.

B. A Typical Salt Marsh in San Francisco Bay

Three plants dominate natural tidal salt marshes in
San Francisco Bay: cordgrass (Spartina foliosa) colo-
nizes the margins of tidal mudflats at about mean sea
level (MSL) and dominates the lower salt marshes with
virtually pure stands; pickleweed (Salicornia virginica)
replaces cordgrass at about mean high water (MHW) and
forms dense bushy cover in almost pure stands over the
vast majority of tidal marshes; saltgrass (Distichlis
spicata) dominates plant life in the highest portions of
marshes at extreme high water (EHW).

The distribution of marsh plants species and their
vigor is affected by marsh plain elevations as well as
by the tidal regime; by the frequency and duration of
inundation. These, in turn, affect other parameters
such as salinity and pH, as well as the movement of
detrital products, nutrients, and seeds (Atwater 1979;
Chapman 1960). MacDonald (1974) and Zedler (1977) both
note the relationship between maximum continuous submersion and elevation on vegetation distribution in a tidal
marsh. The tidal regime of landward portions of a marsh
is influenced both by channel distance to the bay and by
marsh plain elevations.

At the Muzzi restoration site both pickleweed
(Salicornia virginica) and cordgrass (Spartina foliosa)
grow outside the dikes along the bay edge providing a
seed source for marsh plants. Inside the dikes, prior
to the 1976 breach, the marsh surface was largely
cracked mud with little vegetative cover. Isolated
pickleweed plants grew in areas of lower elevation,
maintained by ponded winter rains. A few cordgrass
plants grew near a tide gate that allowed a small amount
of bay water to leak into the site.

II. GEOMORPHIC EVOLUTION

The Muzzi restoration site is surrounded by dikes:

those installed by Muzzi in 1959 were strengthened by
the addition of spoils on three sides and breached in
four places on the fourth bayside. Other dikes were
added by GGBHTD in 1975 on the northwest side to create
a spoils disposal area. A mid-marsh "training dike" was
constructed on the restoration site to restrain spoils
and only partially removed following the spoiling activities. The dike was to prevent spoils from increasing
elevations of bayward portions of the site and reducing
the restoration potential for a cordgrass type habitat.
The remnant of this dike can be seen in Figure 3.

Figure 3. A 1990 aerial view of the Muzzi site showing
natural meanders and the remnant of a mid-marsh "training dike" that significantly affects drainage patterns.

A. Meander formation on the Marsh Plain

     When dikes along the bay were breached in 1976,
there was evidence of a remnant meander system predating
the 1959 dike on the marsh plain towards the bay in
areas not covered by the dredge spoils placed in 1976.
West of the training dike, however, spoils eliminated
any topographic features and left an essentially flat
plain. Unlike natural marshes, incoming tidal waters
flowed in sheet fashion, not through a system of meanders. By 1979, the meander system in bayward areas was
expanding into areas west of the training dike.

     These channels continue to extend upstream. Sur-
face mud was initially very soft but has become more
firm as vegetation has become established and plant
evapo-transpiration dehydrates mud in the root zone.

Where channels extend landward, water seeps from under-
ground at the channel end at exposed tides. Upstream
beyond the channel head, small holes develop in the mud
for varying distances, usually for a quarter to a third
of a meter, which enlarge and eventually cause the
surface to collapse and the channel to extend upstream.
(See Figure 5) Collins and Collins (1989) attribute
this phenomenon to the soft unconsolidated mud from
dredge spoils underlying the marsh surface made more
solid by means of plant evapo-transpiration. They
suggest that groundwater piping below the root zone
might serve as a dominant mechanism of tidal channel
headward erosion in young marshes developed on spoils.
The effect of this has not been studied but if the soft
underlying mud results in channel bank collapse and
channel width becomes greater relative to depth, the
result may adversely affect the quality of habitat for
species such as the clapper rail. Clapper rail are
secretive birds and prefer the cover and greater safety
of deep cut channels of a natural marsh.

B.   Marsh Plain Elevations - Subsidence Vs. Sedimentation

    Marsh plain elevations have been lowered by subsi-
dence and plant evapo-transpiration and increased by
sediment deposition. Subsidence and lowered elevations
occurred over the entire 200 acres after dikes excluded
tidal waters in the 1950s because surface moisture
evaporated. Subsidence also occurred when 1975 dredge
spoils consolidated and moisture evaporated. Mud to the
depth of plant roots has compacted where vegetation
grows through plant evapo-transpiration. On the other
hand, marsh plain elevations are increased in a decreasing gradient from channels by deposition of suspended
sediment carried onto the marsh plain by tidal waters.
Final elevations of landward portions of the marsh
following the 1976 dredge disposal operation were +3.5
to +3.9 feet NGVD. Elevations for a direct flow pathway
where suspended dredge spoils overflowed in the dewatering process into the restoration site from the dredge
spoil site exceeded +4.0 feet NGVD. In 1986 elevations
in these areas were reduced by subsidence to +2.8 to
+3.2 feet with the direct spoils overflow path reduced
to +3.7 feet NGVD.

    The marsh plain surface is a lower energy environ-
ment than channels bringing tidal waters, hence, suspended sediments and debris are deposited from tidal
waters as they overflow channels and are slowed over the
flat marsh plain surface. Sediment deposition on the
marsh plain surface was measured over a two-year period.
At selected sites across the marsh plain surface, flat

plastic plates were secured level with the marsh surface by a long spike. Sediment deposits were measured in September 1989 and 1990. Interior portions of the marsh with elevations exceeding 3.0 feet NGVD and over 45 meters distant from a major channel received little if any sediment in the two-year period while areas adjacent to a main channel received as much as 6 cm. Plants adjacent to channels grow taller and more luxuriantly and it is hypothesized that transported sediments bring additional nutrients while tidal flushing reduces soil salinities and increases soil aeration. Figure 4 shows figures for sediment deposition and maximum canopy height (measured in September) relative to channel distance.

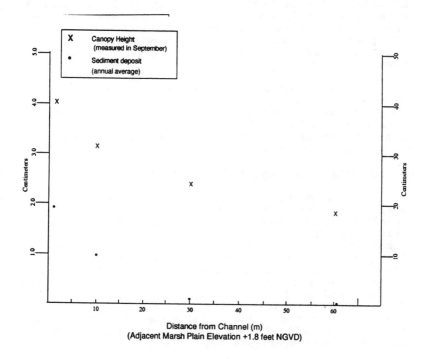

Figure 4.   Sediment Deposition vs Canopy Height Relative to Distance from Tidal Channel

III.   VEGETATIVE CHANGES FOLLOWING CHANNEL MODIFICATION

The modification project (1980) that created over-sized channels to increase tidal flows to landward parts

of the marsh, were designed to improve hydrologic efficiency and improve conditions in the marsh for growth of
vegetation.

A. Cordgrass Colonization of Excavated Channels

     Colonization of channels by cordgrass plays a sig-
nificant role in channel and marsh plain morphology.
Plant stems slow tidal flows so that suspended silt
loads are deposited around the base of the plants. This
increases elevations and expands cordgrass colonization
into and along channels.

     An inventory of cordgrass invading the modification
channels and embayments created in 1981 was made in
September of 1982, 1984, and 1986. Cordgrass stems were
grouped into 4 classes (clumps with 1-10, 10-500, over
500, over 1000 stems) and all cordgrass plants along
newly constructed channels were mapped. Results are
shown in Figure 5.

     Cordgrass vigorously colonized new channels within
the first year, 1982, and by the fifth year, 1986, individual clumps merged to form large stands along the
edges of channels. By 1986, six years after construction, the smaller of the excavated embayments had a
complete cover of cordgrass and the larger a 30% cover.

B. Vegetative Changes on the Marsh Plain

     To follow long-term vegetative changes on the
marsh plain, both from natural colonization and from
secondary effects resulting from construction of modification channels, an 730 meter line transect was established in 1986 across the marsh from the most landward
dike to the bay edge. Elevations were measured every 30
meters and cover percentages for each species were
determined by estimating cover within 10 meter segments
for the length of the 730 meter line (Faber et al.,
1988). Sediment plates were installed at 150 meter
intervals along the transect and across the marsh plain.
Soil salinity and pH measurements were made along the
transect in September 1986 (Figure 6).

     A visual estimate showed a dramatic increase in
pickleweed growth following channel construction, from
75% percent cover in 1981 before the site modification
to close to 100% in 1982 and from 23 centimeter average
canopy height in 1981 to 40 centimeters in 1982. By
1982 numerous small clumps of cordgrass emerged through
the pickleweed cover at elevations between +2.5 and +3.2
feet NGVD. Rainfall in 1981-1982 was 60% above normal

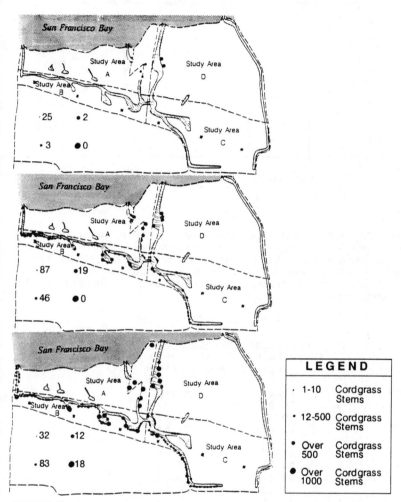

LEGEND

·	1-10	Cordgrass Stems
·	12-500	Cordgrass Stems
•	Over 500	Cordgrass Stems
●	Over 1000	Cordgrass Stems

FIGURE 5.

Inventory of cordgrass colonization of channels created in 1981: upper
figure 1982; middle figure 1984; lower figure 1986. Channel to the far
right was not inventoried.

Figure 6. Transect across Muzzi marsh showing pH and salinity values -1986

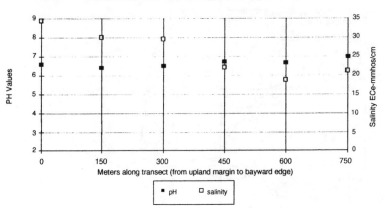

which may have increased seed germination for both species over drier years and no doubt contributed to the
immediate and dramatic vegetation response. While 1987-
1988 rainfall was 60% of normal, cordgrass cover was
maintained and reached an average canopy height of 33
contimeters.

Table 3.

California Rainfall Summary
(Department of Water Resources)
Corte Madera Station*

Year	Inches
79-80	42.93
80-81	21.13
81-82	61.64
82-83	66.55
83-84	36.00
84-85	36.63
85-86	53.02
86-87	22.31
87-88	27.87
88-89	28.79
89-90	17.65

* Long-term normal = 40 inches

Results from the 1986 transect survey show that
abundant stands of pickleweed (area coverage of 60 to
100%) grew at elevations ranging from +1.8 to +2.2 feet

NGVD. Cordgrass grew at these elevations but with coverage consistently less than 15% and with individual plant height averaging 50 centimeters. The average of 10 plants selected randomly within 1 meter of the transect line were measured to the tip of the inflorescence in September, beyond the end of the growing season. At elevations of +1.8 to +2.2 feet, cordgrass grew with cover ranging from 15% to 100% and individual plant height averaging 90 centimeters; while pickleweed cover values never exceeded 20% and annual pickleweed (Salicornia rubra) frequently replaced the perennial form. Closer to the bay in the last 300 meters of the transect (elevations ranging from +1.5 to +2.0 feet NGVD), only isolated patches of annual pickleweed or cordgrass grew.

The vegetation survey was repeated in September 1989 and 1990. Changes in species distribution from 1986 to 1990 are shown in Figure 7. Cordgrass cover decreases at elevations of +2.5 to +3.2 feet NGVD and pickleweed cover increases. At elevations below 2 foot, colonization by annual pickleweed and cordgrass increases.

An increase in submersion time appears to have the same effect on vegetation distribution as lowering the elevation. At the edge of San Francisco Bay, cordgrass typically grows within an elevational range from -0.5 to +2 feet NGVD. Following the channelization work, cordgrass colonized new areas of the marsh with elevations up to +3.0 feet NGVD; average pickleweed canopy height increased from 23 centimeters, before, to 58 centimeters after tidal increase. Figure 8 shows a schematic relationship between cordgrass growth, time of submergence and elevation.

D. Plant Competition

Plant species distribution is affected by plant competition as well as by environmental stress. Purer (1942) suggested that the upper distributional limit of cordgrass is determined by competition with pickleweed. In a computer modeling system Brenchley-Jackson, Foin, & Zedler (unpub.) examined interannual variability of marsh vegetation in the Tijuana Marsh in San Diego County and found that environmental stress alone did not account for variations in biomass productivity of cordgrass and pickleweed, but that with the addition of plant competition to the model, biomass production of cordgrass was consistent with what was seen in the field. When varying the amount of rainfall in a series of simulation experiments, two years of above average rainfall resulted in pickleweed disappearance from

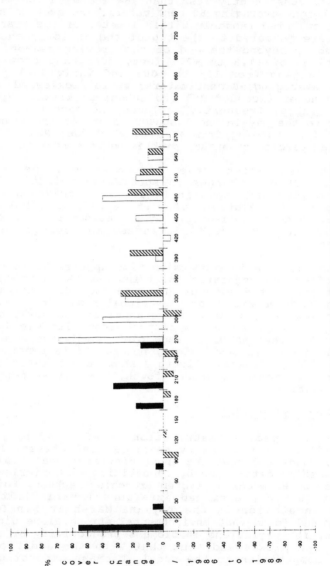

Figure 7. Change in % Cover in Species Distribution between 1986 and 1989
Determined per 30 meter segments of Transect

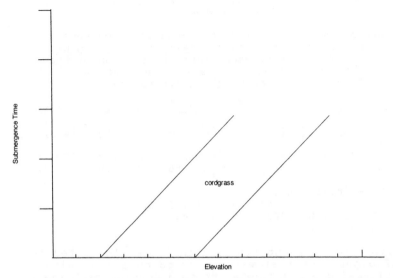

Figure 8. Schematic relationship between cordgrass cover, time of submergence and marsh plain elevation

the lower marsh and the biomass of cordgrass increasing over the entire marsh in response to lower salinity and better soil moisture. Two years of below average rain resulted in cordgrass disappearance from the upper and middle marsh, where tidal flushing moderated salinity and provided moist soils, and pickleweed biomass increasing over the entire range in response to drier soils. Alternating years of above and below average rainfall resulted in a reduction of biomass of cordgrass but not a disappearance of the species. Pickleweed also occurred over the entire range but with reduced biomass.

These simulation experiments may be useful in interpreting distribution patterns on the Muzzi Marsh. In 1982, a year of above average rainfall, and following the 1981 channel modification program, cordgrass colonized several acres of landward portions of the marsh that had only supported pickleweed in the past. Beginning in 1986, cover and frequency of occurrence were determined for both species for a 233 meter portion of the upper marsh where elevations range from +2.5 to 3.2 feet NGVD. Data was collected following one year of above average rainfall (1986) and four years of below average rainfall (1986-1990).

Table 4.

Percent Cover and Frequency (quadrant occurrence)
for Spartina foliosa and Salicornia virginica
on a 233 meter transect landward portion of the
Muzzi Marsh with elevations ranging from
+ 2.5 to 3.0 NGVD for the years 1986-1989

	1986[*]		1987[*]		1988[*]		1989[*]		1990[*]	
	%C	Fre.	%C	Fre.	%C	Fre.	%C	Fre.	%C	Fre.
Sali.	83	100	NA	NA	94	100	95	100	100	97
Spart.	6	56	NA	NA	4	56	3	50	2.5	26

[*] Rainfall in inches:
53.02 (1985-1986); 22.31 (1986-1987);
27.78 (1987-1988); 28.79 (1988-1989);
17.65 (1989-1990)

Shifts in plant distribution were also noted in a
1982 experiment that was conducted on a plot with 90%
pickleweed cover but adjacent to a control plot where a
clump of cordgrass 1 meter in diameter was growing naturally at the landward margin of the marsh (approx. +3
feet NGVD). To observe the growth of transplants under
San Francisco Bay conditions, an experimental transplanting of 24 clumps of cordgrass (Spartina foliosa)
(approximately 20 centimeters in diameter) from marshes
south of San Francisco Bay and 6 of (Spartina densiflo-
ra) from Humboldt Bay was made. Of the 30, 28 transplants survived and flowered in 1983 and again in 1984.
By 1985, however, all transplanted clumps had died along
with the adjacent control clone. Sedimentation in this
area is less than 0.1 centimeter per year, and rainfall
was nearly normal for the three year period; thus,
factors other than rainfall and elevation must account
for the disappearance of these stands. Changes in the
tidal prism reaching this area must be considered since
oversized channels constructed in 1981 increased the
duration of tidal flows to all parts of the marsh;
however, plant competition is most likely to be a major
factor with pickleweed winning out at these higher
elevations. Zedler (pers. comm.) comments on the con-
founding aspects of salinity and species competition in
accounting for plant distribution.

V. SUMMARY

Channels develop in areas of dredge spoil deposits
in a different manner than in a natural marsh. Sediment
deposition from channel overflows onto the marsh plain

is greater close to channels; canopy height is also greater close to channels. Cordgrass colonized newly constructed channels within the first year and formed dense stands by the fifth year. Distribution of vegetation appears to be affected by rainfall, and tidal access. The role of plant competition in vegetation distribution needs more study.

ACKNOWLEDGMENTS

Funding for this project was received from the Golden Gate Bridge, Highway and Transportation District, the Main Audubon Society, the San Francisco Foundation, and the Marin Community Foundation.

REFERENCES

Atwater, B.F., S.G. Conrad, J.N. Dowden, C.W. Hedel, R.L. MacDonald, and W. Savage. 1979. History, landforms, and vegetation of the estuary's tidal marshes. Pp 347-386 in T.J. Conomos, ed. San Francisco Bay: the Urbanized Estuary. Pacific Div. Am Assoc. Adv. Sci., San Francisco, CA

Chapman, V.J. 1960. Salt marshes and salt deserts of the world. Interscience. 392pp.

Collins, J.N., L.M. Collins. 1989. Site Visit to Muzzi Marsh; correspondence dated February 8, 1989

Faber, P. 1979, 1980, 1982, 1984, 1985, 1988. Status Reports on the Muzzi Marsh Restoration Project prepared for the Golden Gate Bridge, Highway and Transportation District.

Faber, P. 1980. Marsh Restoration as a Mitigation. Pp.167-178 in Proceedings of the Coastal Society, Arlington, VA 22201

Faber, P. 1983. Restoration with Natural Vegetation: A Case Study in San Francisco Bay. Pp. 729-734 in Proceedings of the Third Symposium on Coastal and Ocean Management. June, 1983.

Faber, P., A. Shepherd, P. Williams. 1988. Monitoring a Tidal Restoration Site in San Francisco Bay - The Muzzi Marsh. Pp. 351-355 in Kusler (ed.) Proceedings of the National Wetland Symposium: Urban Wetlands. Association of State Wetland Managers, Byrne, N.Y.

MacDonald, K.B., M.G. Barbour. 1974. Beach and salt marsh vegetation of the North American Pacific Coast. In Ecology of Halophytes. R.J. Reimold & W.H. Queen, (eds.) John Wiley & Sons, NY.

Purer, E.A. 1942. Plant ecology of the coastal salt marshlands of San Diego County, California. Ecol. Monogr. 12:82-111.

Zedler, J.B. 1977. Salt marsh community structure in the Tijuana Estuary, California. Estuarine and Coastal Marsh Science 5:39-53.

Zedler, J.B. 1986. Catastrophic flooding and distributional patterns of Pacific cordgrass (Spartina foliosa Trin.) Bull. So. Calif. Acad. Sci. 85(2):74-86.Zedler, J.B. January 1990. Personal Communication

Zedler, J.B. in press. Interannual variability in the Tijuana Estuary lower salt marsh.

Seagrass Decline: Problems and Solutions

Frederick T. Short[1], Galen E. Jones and David M. Burdick

Abstract

Human pollution has contributed to seagrass declines in estuaries around the world. Populations of eelgrass, *Zostera marina*, on the northeastern coast of the U.S. are threatened by coastal pollution and the wasting disease. Both factors impact eelgrass health and each has destroyed eelgrass populations in some locations. Evidence suggests that these two stresses interact, causing eelgrass to die-off rapidly. However, the effects of pollution stress appear to have the greatest potential for reducing population size and distribution below naturally recoverable levels.

Of the major pollution related effects, reduced available light resulting from eutrophication is by far the greatest threat to seagrass populations. Reduced water clarity limits plant production and distribution. Therefore, water clarity must be rectified before eelgrass can recolonize or be restored to an estuarine area. The reconstruction of eelgrass habitat destroyed by disease or by marine related human activities like dredging and boat storage is not feasible as long as water clarity remains poor. In cases where water quality is improved, mitigation of environmental damage through the establishment or restoration of eelgrass habitat can be very successful and should be encouraged.

Introduction

Seagrass beds are an important component of estuaries, providing primary production in a unique physical structure that is essential habitat for many important estuarine species including fish, shellfish, and waterfowl (Phillips 1984, Thayer et al. 1984). Unfortunately, the seagrass habitat is disappearing. The problem of seagrass decline is worldwide in scope

[1]Jackson Estuarine Laboratory, University of New Hampshire, Durham, NH 03824.

and is rapidly increasing in intensity (Thayer et al.
1975). The major causes of seagrass decline: pollution,
disease, and increased human activity within the coastal
zone, were described at the previous Coastal Zone
meeting in South Carolina (Short et al., 1989). Of
these, pollution poses the most widespread and greatest
long-term threat to coastal seagrass populations. The
decline of seagrass from pollution-related events is
evident in tropical and temperate environments in both
the northern and southern hemispheres.

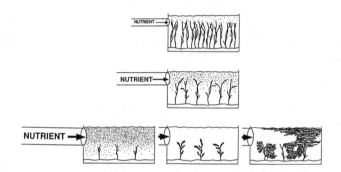

Figure 1. Diagram showing the effects of eutrophication on the
seagrass community. Increased nutrient loading negatively impacts
seagrasses and promotes a shift to phytoplankton, epiphyte, or
macroalgal dominance.

        Eutrophication and suspended sediments are the two
pollution problems which have the greatest impact on
seagrass decline in our coastal zone. Of these,
eutrophication has the greatest long-term impact on
seagrass health and survival, though both restrict
seagrass production and distribution by reducing light
reaching these rooted plants. Under low to moderate
nutrient loading conditions, seagrass can effectively
filter large amounts of nitrogen and phosphorus from the
water column (Short and Short 1984) However, the
excessive loading of nitrogen and phosphorus into the
coastal environment eliminates seagrass beds and
prevents their reestablishment. The direct impact of
eutrophication is the stimulation of various plant forms
which results in reduction of light (Twilley et al.
1985, Orth and van Montfrans 1983, Short et al. 1989).
The plants that compete with seagrasses for light are:
1) free-floating diatoms and dinoflagellates,
collectively termed phytoplankton, 2) macroalgae, and 3)
diatoms and algae that are attached to seagrass blades
and are therefore called epiphytes. Excessive nutrient
loading can result in a shift to any of the three algal

forms (Fig. 1). When these other plant forms thrive,
light reaching the eelgrass is greatly reduced, thereby
reducing the overall eelgrass production and plant
density, and ultimately reducing the distribution of
seagrass populations. Areas having identifiable
seagrass declines resulting from pollution include
Chesapeake Bay (Orth and Moore 1983, Kemp et al. 1983),
Indian River, Florida (Kenworthy et al. 1989) Cockleburn
Sound, Australia (Cambridge and McComb 1984) San
Francisco Bay (Wyllie-Echeverria et al. 1989), Waquoit
Bay, Massachusetts (Short and Jones 1990), and
Aburatsubo Bay, Japan (Short pers. obs.).

Disease also affects eelgrass, *Zostera marina*
over a wide geographic area. Catastrophic declines in
eelgrass caused by a disease known as the wasting
disease occurred in the 1930s on both coasts of the
North Atlantic and have recurred on the east coast of
the U.S. during the 1980s (Short et al. 1988). The
identification of the microorganism responsible for the
eelgrass wasting disease, and refinement of techniques
for its culture (Muehlstein et al. 1988) have led to the
recent discovery that this marine parasite is worldwide
in distribution and has infected eelgrass plants in
locations where no actual decline has yet been observed
(Short et al. 1987, 1989).

Other human activities within coastal waters that
negatively impact seagrasses include boating, dock
construction, deployment of moorings, fishing, and
various forms of recreational sport activity.
Mechanical disturbance of seagrass beds is a worldwide
problem typified by localized impacts directly
associated with construction or areas of activity.

The Problem

Nutrient Loading
    The ultimate impact of eutrophication on the
seagrass community is the loss or degradation of the
seagrasses themselves and a shift in the community of
primary producers away from seagrass dominance. Under
conditions of elevated nutrient loading, phytoplankton
may bloom (reproduce rapidly) so that their abundance in
the water column is so great that the seagrass is
effectively shaded. Our experiments with eelgrass have
shown that reduction in light decreases growth, promotes
a reduction in plant density (Fig. 2), and ultimately
can eliminate an eelgrass population altogether.

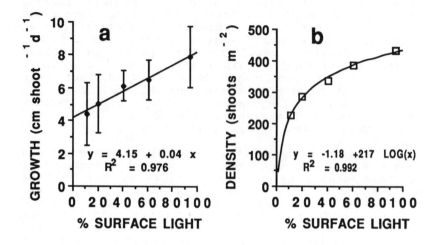

Figure 2. Eelgrass shoot growth (a) and density (b) vs. percent light for plants grown under shaded conditions in the 1988 mesocosm (1.5 m^2 tanks) experiment at UNH. Regression analysis of these data demonstrates the influence of reduced light.

A second alternative, a shift in the community toward domination by green or red macroalgae, is typically seen following rapid growth of either small plants initiated as epiphytes attached to seagrass blades, or free-floating algal mats (Fig. 1). Growing rapidly when stimulated by excess nutrient loading, seaweeds become thick mats that overcome and weigh down the eelgrass blades, causing shading stress (Fig. 3). In addition, algae take up nutrients more rapidly than eelgrass leaves under eutrophic conditions and outcompete the eelgrass (Harlin and Thorne-Miller 1981). The excessive production increases the organic flux to the bottom, where decomposing algal material creates anoxic conditions. Such an absence of oxygen negatively affects the entire plant and animal community normally sustained by seagrass.

A third possible condition brought on by excess nutrient loading is the growth of epiphytic diatoms and small algal forms on eelgrass blades which limit both light and nutrients reaching the eelgrass leaves (Fig. 3). Dense epiphytic growth also becomes a detriment to the long-term success of the eelgrass population primarily through light reduction (Sand-Jensen 1977).

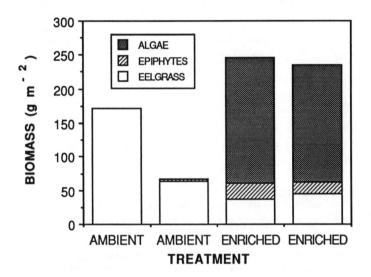

Figure 3. Shifts in the plant community of replicate experimental mesocosms in 1990 following eight weeks of nitrogen and phosphorus enrichment that was six-fold greater than ambient.

With respect to excess nutrient loading, the factors determining the type of primary producer that proliferates as seagrass declines in dominance are complex and as yet not clearly understood. Evidence from our experimental mesocosms suggests that the type of nutrient entering the system, the availability of various algal propagules to initiate growth, the presence of herbivorous snails and amphipods, as well as the presence of carnivorous fish which regulate amphipod populations, all play a role in determining the plant community resulting from eutrophication (Table 1). Whatever the pathway, the ultimate result of community changes in response to eutrophication is the decline of seagrass and the highly productive community it sustains.

Table 1. Outcomes of mesocosm enrichment experiment, 1989, based on descriptive observations. Eelgrass remained dominant in the three replicate control tanks, but was replaced by other plant forms in the three enriched (six-fold ambient N and P) tanks.

PLANT DOMINANCE	CONTROL	NUTRIENTS	NUTRIENTS & AMPHIPODS	NUTRIENTS, AMPHIPODS & FISH
Eelgrass	X X X			
Plankton		X		
Macroalgae			X	
Epiphytes				X

Suspended Sediments

The effects of suspended sediments and sediment loading into coastal waters negatively impacts seagrass beds in coastal estuarine areas through light limitation (Short et al 1989). Sediments wash into estuaries from upland development by overland runoff and through drains and runoff ditches. In many areas, suspended sediments in the water column drastically reduce the depth to which seagrasses can grow and in some areas may eliminate seagrass populations altogether. Seagrass beds in Hobe Sound, Florida have declined rapidly, leading to an investigation that focused on changes in the estuarine environment. The resuspension of sediments, resulting in part from increased boating activity in the intracoastal waterway, was found to limit light penetration into these waters, significantly reducing the depth to which the seagrasses could survive (Kenworthy et al. 1989).

Wasting Disease

Within the past decade, the eelgrass wasting disease has led to several major declines of eelgrass populations in New England (Short et al. 1989). The causal organism is a slime-mold-like protozoan of the genus *Labyrinthula* (Muehlstein et al. 1988) which infects and kills eelgrass. The spread of the wasting disease followed in outdoor mesocosms showed that decreases in seawater salinity below 10 ppt improved eelgrass survival (Muehlstein et al. 1988). As available light is decreased, eelgrass becomes more susceptible to infection by wasting disease which in turn can kill plants and eliminate whole populations (Fig. 4). Thus, as decreased water quality eliminates eelgrass from upper portions of estuaries, the only populations likely to survive wasting disease outbreaks are being destroyed (Short and Jones. 1990).

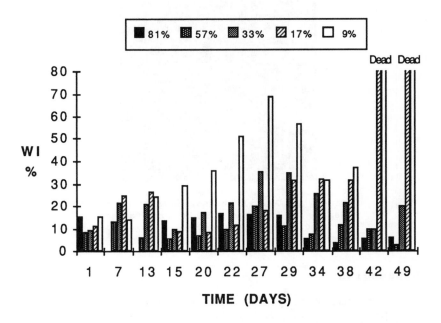

Figure 4. Wasting Index (WI) as percent surface area for eelgrass shoots at five light levels (81, 57, 33, 17, 9%) in a mesocosm shading experiment, starting October 11, 1988. WI is a quantitative measure of the extent of wasting disease infection. Shoots at the lowest light levels were dead after 42 days.

Mechanical Disruption

The success of seagrasses is also limited by mechanical processes in which various kinds of human activity eliminate seagrass through direct disruption of leaf and root material. Boating causes mechanical damage where propellers encounter seagrass beds at low tide. Propellers and their propwash dig up and erode sediments, uprooting seagrass from the bottom and leaving linear bare patches (Zieman 1976). Dock facilities for boats also have detrimental effects on seagrass by concentrating boating activity and shading

beds that were able to survive or recolonize following
dock construction. Mooring chains swinging around their
mooring blocks create circular bare patches within
seagrass meadows (Fig. 5). Other activities like fin
and shellfishing using trawls and nets also disrupt
eelgrass beds.

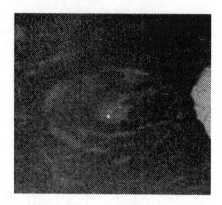

Figure 5. Aerial photograph showing the
damage to eelgrass beds from a boat
mooring in Waquoit Bay ,MA.

Discussion

     A number of problems face the long-term survival
and health of seagrass populations within our coastal
zone. Sufficient evidence exists to demonstrate that
loss of seagrass habitat will mean losses in the
productivity and health of our estuaries (Short and
Short 1984, Short et al. 1989). The task is to identify
and rank the severity of these problems and investigate
ways of dealing with and reducing these types of
environmental impacts to the level where long-term
success of seagrass communities is assured.

     The major long-term threat to seagrass populations
both in the U.S. and around the world is derived from
coastal pollution. Recently compiled data (NOAA/EPA
1989) lead us to suggest that the majority of estuaries
within the continental United States are eutrophied.
High levels of nutrient discharge into coastal waters
deriving from both point and non-point sources are
fundamentally changing the ecological character and
structure of our estuarine environments (NOAA's
Eutrophication Workshop at Narragansett, RI, January

1991). In estuaries dominated by seagrass, the shift is dramatic and results in declining health of the seagrass community which includes many commercially important estuarine species as well as the ecological character of the estuary. The extreme case that leads to elimination of seagrass from the estuary is accompanied by declines in estuarine health and function.

The eutrophication of our coastal zone is brought on by the large scale discharge of nitrogen and phosphorus into coastal waters, predominately from sewage and human wastes, and to a lesser degree from applications of fertilizers from agriculture and domestic use. Accompanying the increased rate of development within our coastal zone, the disposal of sewage using wastewater treatment facilities has increased dramatically. As these facilities are upgraded from primary to secondary treatment, the discharge of organic particles will be reduced. These organic particles contribute to the biological oxygen demand within estuarine systems and can lead to anoxic or hypoxic conditions. However, few if any of the major treatment facilities have been able to effectively remove the inorganic nutrients from sewage, so these facilities continue to discharge nutrients at increasing rates.

The solution to the problem of eutrophication within estuaries is clear: a reduction in nutrient loading is required. The difficulty arises from the economic and social challenge of actually eliminating or redirecting the discharge of human wastes away from the shallow coastal environment. Curbing nutrient discharge from point sources such as wastewater treatment facilities entails large, capital-intensive projects. Estuaries have the ability to recover from eutrophication if excess nutrient loading is removed. The rapid recovery of an estuary in Connecticut demonstrates the potential for recovery of coastal ecosystems after loading stops (French et al. 1989). In 1987, Mumford Cove was receiving the sewage discharge from the Groton, CT treatment plant and the formerly healthy eelgrass community had been overgrown by extensive mats of nuisance macroalgae. In 1988, after the wastewater treatment outfall was removed from Mumford Cove, the nuisance seaweed was gone and eelgrass began to recover (French et al. 1989).

Non-point source nutrient discharge into estuarine areas requires a decentralized approach to address specific problems as they arise. Low cost solutions to most of these small scale problems already exist. For example, in the case of fertilizer input through run-

off, physical structures coupled with vegetation can be
designed to retain floodwaters and intercept excess
nitrogen and phosphorus. In some areas, regulations are
needed to regulate fertilizer use in order to minimize
contamination of estuaries.

Another input route of non-point source nutrients
to coastal waters is through groundwater discharge. In
densely populated, sandy coastal areas, where houses
have septic systems, the groundwater naturally
discharging into coastal lagoons and estuaries carries
with it high concentrations of nitrate and other
nutrients. The nitrogen carried by the groundwater
fluxes into estuarine waters and contributes to rapid
algal growth within these ecosystems. The source of
nitrate to groundwater is leachate from septic systems.
The input of nitrate into groundwater does not require
the failure of septic systems but more directly results
from an inadequately designed septic system. Standard
septic systems installed in areas of course sand soils
do not adequately slow down water percolation. Septic
leachate flows through the sand so fast that nitrates
are not removed by the usual denitrification process and
they flow directly into the groundwater. As a result,
these septic systems pollute groundwater *even though
they meet state regulations.*

The discharge of nitrogen-laden groundwater is a
major concern along much of southern New England where
excessive development has occurred along areas of sandy
coastline. Eutrophication from this type of nitrogen
loading has been documented in Waquoit Bay,
Massachusetts (Foreman, et al. 1990), in Great South
Bay, Long Island (Capone and Bautista 1985), and in the
coastal ponds of Rhode Island (Lee and Olsen 1985, Nixon
and Pilson 1983).

In order to reduce the nutrient discharge into
coastal areas, lessen the degree of eutrophication, and
return the health of our coastal marine environments,
regulations must be established and enforced that will
contain and eliminate the problems of nutrient loading.
The days of free disposal of human waste have passed.
As is the case with solid waste, we must all face the
responsibility and costs of properly disposing of our
human effluent in a way that is not going to lead to the
long-term degradation of the environment.

The next question to consider is whether we can
make seagrass habitats work for us in the process of
rehabilitating our coastal environment. Seagrass beds
are enormous absorbers of inorganic nutrients (Short and

Short 1984, Short et al. 1989). It is only when excessive amounts of nutrients have been added to the water that these systems fail to remove nitrogen and phosphorus from the water sufficiently fast to outcompete nuisance algal species. If we can reduce the nutrient load to estuarine systems that have lost their seagrass habitats, then current technology makes it feasible to reestablish seagrass beds in these areas.

Reconstruction of a seagrass meadow that will again effectively filter nitrogen and phosphorus from incoming water begins the process of rehabilitation for these coastal environments. Methods for the transplantation and restoration of seagrass beds have been developed over the last ten years (Fonseca et al. 1982). Projects that have attempted to reestablish submerged aquatic vegetative communities have become increasingly successful (Fonseca et al. 1988; Short 1988; Short and Carlson in prep.). Recent efforts in seagrass transplanting and a more detailed knowledge of the environmental conditions required to sustain healthy seagrass beds have increased the success and efficiency seagrass habitat restoration.

Judging the suitability of a site for seagrass restoration is critical. Not all shallow estuarine areas are suitable for seagrass transplantation. Careful assessment of environmental conditions at a specific site is crucial to ensure repayment of the investment necessary for restoration. Although nutrient sources and flux into an estuary may be decreased to an appropriate level, other conditions such as sediment instability or high levels of suspended solids may inhabit a successful restoration effort. After conditions in an estuary are determined to be suitable to sustain seagrass, a bed may be transplanted into that location. A complete discussion of such conditions is lacking, and knowledge is based largely on field experience and anecdotal information from the literature. Seagrasses can, within one year transplantation, effectively function to accelerate the cleanup process and more quickly return the estuarine environment to a healthy productive ecosystem.

The long-term survival of seagrass meadows and the reestablishment of healthy estuarine environments can only be achieved by undertaking a series of steps to reduce the nutrients entering these systems. Efforts must now be started to develop new methods for disposal, or preferably recycling, of human waste resources. When recycling of these resources becomes financially desirable, the process of cleaning up our coastal waters will proceed rapidly, since the reduction of the nutrient load will turn the tide in favor of seagrass recolonization and improved estuarine function.

Reestablishment of seagrass beds can provide a cost-effective, self-sustaining mechanism to improve water quality, but nutrient loading must be reduced for this to be possible. The goal of estuarine management is to ensure that an estuary is a healthy productive ecosystem. In many estuaries, seagrass beds used to be more widely distributed than they are today, but their ability to support and sustain productive food webs in the face of current eutrophication is doubtful. The choice is ours: to reclaim, restore and preserve estuarine environments for productive use of their resources, or to suffer the loss of seagrass habitats. We can regulate the losses of seagrass beds from mechanical damage and boating activities, but addressing the problem of eutrophication means changing our methods for disposal of human waste. If estuaries become less eutrophied and seagrass beds are restored, the reestablishment of this highly productive estuarine habitat can be realized.

Acknowledgments

Thanks to Catherine Short for editing and constructive criticism and to James Kaldy and Jaimie Wolf for valuable contributions. The research reported was supported by NOAA's National Estuarine Research Reserve Program and by NOAA's Coastal Ocean Program. Jackson Estuarine Laboratory C.N. 234.

Literature

Cambridge, M.L. and A.J. McComb. 1984. The loss of seagrasses in Cockleburn Sound, Western Australia. I. The time course and magnitude of seagrass decline in relation to industrial development. Aquatic Botany 20: 229-243.

Capone, D.G. and M.F. Bautista. 1985. A groundwater source of nitrate in nearshore marine sediments. Nature 313:214-216.

Fonseca, M.S., W.J. Kenworthy and G.W. Thayer. 1982. A cost-evaluation technique for restoration of seagrass and other plant communities. Environment 9:237-241.

Fonseca, M.S., W.J. Kenworthy, and G.W. Thayer. 1988. Restoration and management of seagrass systems: a review. In: D.D. Hook, et al. (eds.) The Ecology and Management of Wetlands. Vol 2: Management, Use and Value of Wetlands. Timber Press, Portland, OR. pp. 353-368.

Foreman, K.H., J.E. Costa, C. D'Avanzo and J. Kremer. 1990. Preliminary measurements of phytoplankton biomass and production in subestuaries of Waquoit Bay receiving different nutrient loadings. (Abstract) New England Estuarine Research Society, Newport, RI.

French, D., M. Harlin, E. Gundlach, S. Pratt, H. Rines, K. Jayko, C. Turner, and S. Puckett. 1989. Mumford Cove water quality: 1988 monitoring study and assessment of historical trends. A.S.A., Inc. Final Report, 126 p.

Harlin, M.M. and B. Thorne-Miller. 1981. Nutrient enrichment of seagrass beds in a Rhode Island coastal lagoon. Mar. Bio. 65:221-229.

Kemp, W.M., W.R. Boynton, R.R. Twilley, J.C. Stevenson and J.C. Means. 1983. The decline of submerged vascular plants in Upper Chesapeake Bay: Summary of results concerning possible causes. Mar. Soc. Tech. J. 17:78-89.

Kenworthy, W.J., M.S. Fonseca, D.E. McIvor and G.W. Thayer. 1989. The Submarine Light Regime and Ecological Status of Seagrasses in Hobe Sound, Florida. Annual Report. N.M.F.S., NOAA, Southeast Fisheries Center, Beaufort Laboratory, Beaufort, NC.

Lee, V. and S. Olsen. 1985. Eutrophication and management initiatives for the control of nutrient inputs to Rhode Island coastal lagoons. Estuaries 8:191-202.

Muehlstein, L.K., D. Porter and F.T. Short. 1988. Labyrinthula sp., a marine slime mold producing the symptoms of wasting disease in eelgrass, Zostera marina. Mar. Biol. 99:465-472.

Nixon S.W. and M.E. Pilson. 1983. Nitrogen in estuarine and coastal marine ecosystems. In E.J. Carpenter and D.G. Capone (eds.) Nitrogen in the Marine Environment. Academic Press. pp 565-648.

NOAA/EPA. 1989. Strategic Assessment of Near Coastal Waters. Susceptibility of East Coast Estuaries to Nutrient Discharges: Passamaquoddy Bay to Chesapeake Bay. Summary Report. NOAA Rockville, MD

Orth, R.J. and K.A. Moore. 1983. Chesapeake Bay: An unprecedented decline in submerged aquatic vegetation. Science 22:51-52.

Orth, R. J. and J. van Montfrans. 1983. Epiphyte-seagrass relationships with an emphasis on the role of micrograzing: A review. Aquat. Bot. 18:43-69.

Phillips, R.C. 1984. The Ecology of Eelgrass Meadows in the Pacific Northwest: A Community Profile. U.S. Fish Wildl. Serv. FWS/OBS-84/24. 85 pp.

Sand-Jensen, K. 1977. Effect of epiphytes on eelgrass photosynthesis. Aquat. Bot. 3:55-63.

Short, F.T. 1988. Eelgrass-scallop research in the Niantic River. Final Report to the Waterford-East Lyme, CT Shellfish Commission. November 15, 1988. 12 pp.

Short, F.T., B.W. Ibelings, and C. den Hartog. 1988. Comparison of a current eelgrass disease to the wasting disease of the 1930's. Aquat. Bot. 30:295-304.

Short, F. T. and G. E. Jones. 1990. Eelgrass Decline in National Estuarine Research Reserves along the east coast of the United States. NERR/MEMD/NOAA Final Report, in preparation.

Short, F.T., L.K. Muehlstein and D. Porter. 1987. Eelgrass wasting disease: cause and recurrence of a marine epidemic. Biol Bul. 173:557-562.

Short, F.T. and C.A. Short. 1984. The seagrass filter: purification of estuarine and coastal waters. In V. S. Kennedy (ed.) The Estuary as a Filter. Academic Press. pp. 395-413.

Short, F.T., J. Wolf, and G.E. Jones. 1989. Sustaining eelgrass to manage a healthy estuary. Proc. Sixth. Symp. Coast. Ocean Manag. 3689-3706.

Thayer, G.W., W.J. Kenworthy, and M.S. Fonseca. 1984. The ecology of eelgrass meadows of the Atlantic coast: a community profile. U.S. Fish Wildl. Serv. FWS/OBS-84/24. 85 pp.

Thayer, G.W, D.A. Wolff, and R.B. Williams. 1975. The impact of man on seagrass. American Scientist, Vol. 63:288-296.

Twilley, R.R., W.M. Kemp, K.W. Staver, J.C. Stevenson and W.R. Boynton. 1985. Nutrient enrichment of estuarine submerged vascular plant communities. I. Algal growth and effects on production of plants and associated communities. Mar. Ecol. Prog. Series. 23:179-191.

Wyllie-Echeverria, S., R.C. Phillips and T. Wyllie-Echeverria. 1989. Populational structure of eelgrass (*Zostera marina* L.) in San Francisco Bay, California. (Abstract), Tenth Biennial International Estuarine Research Conference, Baltimore, MD.

Zieman, J.C. 1976. The ecological effects of physical damage from motorboats on turtle grass beds in southern Florida. Aquat. Bot. 2:127-139.

Exploitation of Ecological Growth Models
For Management of Seagrass Resources

Randall S. Alberte and Richard C. Zimmerman[1]

Seagrasses form the basis of critical, yet extremely fragile ecosystems in shallow coastal embayments and estuaries throughout the world. Although highly productive, these systems are particularly vulnerable to reduced light availability caused by eutrophication, chronic upstream erosion and periodic dredging of coastal environments. Light is recognized generally as the most important environmental factor regulating the depth distribution, density and productivity of these submerged aquatic macrophytes (Backman and Barilotti, 1976; Orth and Moore, 1983; 1988; Dennison and Alberte, 1982; 1985; 1986).

The success of any seagrass management program is strictly dependent upon the presence of a physical environment that will ensure initial establishment and support long-term growth. Unfortunately, our ability to define environmental growth requirements for seagrasses with any degree of confidence is poor. As a result, seagrass management practices that have centered on revegetating disturbed sites and creating new grassbeds in unvegetated areas, have had successes in these generally degraded environments because habitat suitability could not be determined in advance. The fundamental problem is that environmental requirements for successful growth and recruitment have not been defined adequately for most seagrass species, so revegetation programs have been forced to use a trial-and-error approach. Attention to growth requirements has focused on eelgrass Zostera marina L., and a few simple growth models have been constructed. However useful as research tools, the relatively simple

----

1. Department of Molecular Genetics & Cell Biology
   The University of Chicago
   Chicago, IL 60637

models have not been useful management tools because they
are unable, at this time, to address the complexity of
issues associated with growth responses to temporal
variations in environmental conditions, genetic
differentiation of geographically isolated populations
and environmental stresses that affect processes other
than whole-plant carbon balance. The next-generation
model we are developing is based on a better
understanding of the complex response of seagrasses to
environmental variation on several time scales and the
subsequent metabolic factors that regulate plant growth
and survival. Our goal is a quantitative understanding of
the physiological mechanisms by which light availability
regulates the growth, productivity and survival of
eelgrass in nature.

Even without the aid of predictive models, however,
it must be recognized that environmental quality of most
degraded environments needs significant improvement
before submerged aquatic vegetation resources can be
returned to historical levels. In San Francisco Bay, for
example, coefficients of light attenuation often exceed
3.0 m^{-1} (Secchi depths ≤ 0.75 m), making the euphotic
zone (depth of the 1% light level) less than 2 m deep.
With mixing depths frequently in excess of 5 m, rates of
primary productivity are relatively low in the Bay
(Alpine and Cloern, 1988). Rapid light attenuation by
the water column also limits the depth distribution of
important aquatic macrophytes, such as Zostera marina, to
extremely shallow fringes and shoals (Zimmerman et al.,
1991). Most existing water quality standards only serve
to maintain the status quo of light transparency will not
permit environmentally-sound management of eelgrass
resources or expansion of primary production in estuarine
environments such as San Francisco Bay that are heavily
affected by a history of anthropogenic turbidity. Thus,
in addition to guiding revegetation efforts, reliable
models of light requirements for SAV could also be used
to manage human activities that lead to eutrophication
and sediment loading into coastal waterways and estuary
habitats.

Simple model calculations of carbon-balance based on
assessment of daily metabolic activity indicate that $\underline{Z}$.
marina requires somewhere between 3 and 10 h of
irradiance-saturated photosynthesis each day (termed
H_{sat}; Dennison and Alberte, 1982) to meet the demands of
respiration and growth (Dennison and Alberte, 1985; 1986;
Marsh et al. 1986; Zimmerman et al. 1989). The
uncertainty in this estimate reflects the combined
effects of environmental influences (temperature and
light) and endogenous seasonal rhythms on metabolic
activity, the distribution of biomass between roots and

shoots, and genetically-determined adaptive characters.
Thus, it is difficult to define a single "critical" value
of H_{sat} that predicts or ensures long-term growth and
survival of <u>Zostera</u> <u>marina</u> in all environments. Although
these relatively simple models can be useful management
tools in specific habitats after extensive field-
calibration, they are not general enough to be applied
directly to management-oriented problems in other regions
without extensive testing. Furthermore, the data base of
field observations from most habitats required to test
even these simple models is currently inadequate,
particularly with respect to temporal variations in light
availability and the physiological response of different
eelgrass populations.

A better understanding of the short term (daily)
scales of variation in the physical environment certainly
will improve estimates of eelgrass light requirements.
Within San Francisco Bay, <u>Zostera</u> <u>marina</u> is limited to
different depths at 5 different sites (Fig. 1, from
Zimmerman et al., 1991). Rather than falling

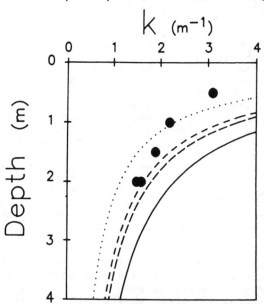

Figure 1. Depth limits for <u>Zostera</u> <u>marina</u>
plotted against mean attenuation coefficient
for 5 sites in San Francisco bay. Lines of
constant H_{sat} are illustrated as follows:
(-- -- --: 2 h); (- - -: 6h); (· · ·: 10h);
(————: depth of the euphotic zone)

along a single "critical" value of light availability,
however, depth limits of eelgrass at the different sites
cross lines of mean constant H_{sat} as the mean attenuation
coefficient ($\bar{k}$) increases. Thus, minimum daily light
requirements for eelgrass survival appear to increase
with mean value of the attenuation coefficient, perhaps
because the variance in light attenuation also increases
with the mean value. As a result, transient periods of
extreme turbidity may be critical in determining the
depth limits of eelgrass in some locations. These
transient events occur on time scales that are generally
undersampled by traditional weekly or monthly monitoring
programs (Zimmerman and Kremer, 1984). Undersampling can
result in a serious underestimation of the turbidity
regimes and a significant overestimate of H_{sat}.
Consequently, detailed information regarding relevant
time scales of variation in critical environmental
parameters, such as light availability, are absolutely
essential for reliable application of any model
calculations to the management of submerged aquatic
macrophytes.

Recent experiments have expanded our appreciation of
the critical role that roots play in the question of
eelgrass survival. Eelgrass roots exist within the
anoxic sediment layer and are dependent on the transport
of photosynthetically-produced oxygen from the shoots to
maintain aerobic respiration during the light period
(Smith et al., 1984). Roots must be able to tolerate
prolonged periods of anoxia each night and even through
days of extremely low light availability. While anoxic,
most metabolic processes (ATP synthesis, protein
synthesis, carbohydrate transport and growth) are
inhibited or greatly reduced (Smith, 1989). We have also
found that extending the anoxic period to 18 h or more
(simulating $\bar{k} \geq 3.0$ or Secchi depths ≥ 0.75 m) may
disrupt the transport of carbohydrate to root tissue
(Zimmerman and Alberte, in prep.). Prolonged exposure
(up to 30 d) to short daily periods of photosynthesis (<6
h) leads to carbohydrate depletion in roots even though
significant carbon reserves remain in the shoots (Fig.
2, from Zimmerman and Alberte, in prep.). Although Z.
marina appears to posses some adaptive capacity to
increase rates of carbohydrate transport to roots under
shortened daily light periods, this adaptation is not
sufficient to prevent carbohydrate depletion of the roots
under short H_{sat} periods (<6 h). Thus, these data
indicate that light availability may regulate the depth
distribution and productivity of eelgrass by controlling
rates of carbon transport to roots independent of whole-
plant carbon balance.

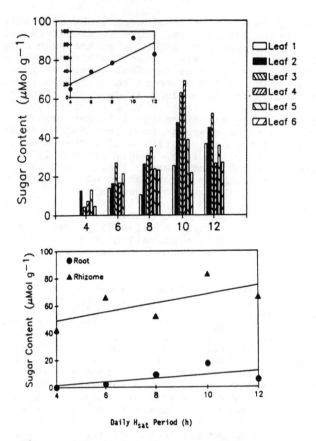

Figure 2. Sugar content of shoots (top)
distributed among individual leaves as well
as the average sugar content of integrated
shoots (inset). Sugar content of rhizomes
and roots (bottom).

As a result of these studies, our modeling efforts
have gone beyond simple issues of shoot and/or whole
plant carbon balance to more detailed examinations of
carbon partitioning between shoots and roots, and how
these dynamics are affected by the length of the daily
light (or aerobic) period. The model, as currently
conceived, simulates the transport of carbohydrate from
leaves to roots as regulated by the daily light period
through the activities of the enzymes sucrose phosphate
synthase (SPS) and sucrose synthase (SS) (Fig. 3). These
enzymes are reliable indicators of the rate of source-to-
sink sucrose transport in a variety of higher plants,
including eelgrass (Huber et al., 1985; Lowell et al.,

1989; Zimmerman and Alberte in prep.). Rates of
photosynthesis, growth, sucrose synthesis, transport and
catabolism are driven by light availability and coupled
to each other by a series of partial differential
equations based conceptually on the cell quota model
proposed by Droop (1973). Laboratory experiments are
currently under way that will provide physiological data
to parameterize the model. We have just initiated a
field program to collect the necessary data on _in situ_
light availability, growth and carbon partitioning with
both subtidal and intertidal populations of eelgrass that
can be used to test the model.

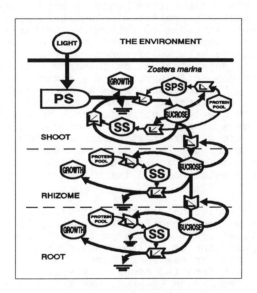

Figure 3. Energy circuit diagram of the
physiological dynamics regulating sucrose
partitioning in _Zostera marina_. Symbols are
after Odum (1983). Arrows represent flow of
energy and/or resources from one storage
compartment to another. Circles represent
resources under control outside the
formulated model, tank symbols represent
accumulation of resources into specific pools
that are affected by interactions with other
terms. The nature of the interaction is
indicated by the graph symbols inside the
block arrows. Positive slopes indicate that
the interaction enhances the flow from one
compartment to another, while negative slopes
indicate the opposite.

Genetic variability among eelgrass populations can also have significant impacts on growth and physiology, as isolated populations may have evolved specific adaptations to unique features of their individual habitats. This is known as ecotypic differentiation. Zostera marina, a true marine "weed", has a cosmopolitan distribution in temperate oceans of the northern hemisphere. Thus, this species is expected to show genetic diversity at the population level that may have important regional implications for the restoration and management of this resource. Based on genomic DNA fingerprinting approaches, we have found that distinct populations of eelgrass from Monterey Bay, Tomales Bay and Elkhorn Slough represent distinct genotypes (Fain et al. in review). In order to determine whether such differences are expressed as performance differences, we are examining the effects of genetically-based differences in physiological performance characteristics of 8 eelgrass populations from the Pacific coast growing in experimental common gardens in Elkhorn Slough near Monterey Bay, California. These experiments will allow us to quantify the level of performance variation that is genetically determined and provide insights into the roles of genetic differentiation and ecotypic variation in controlling the growth of different populations. In addition, such information may be critical for assessment of losses in populations and for mitigation/restoration activities.

Although individual criteria of water clarity can be developed to protect specific seagrass populations in specific habitats, the ability to transfer these criteria to other populations or environments cannot be considered universal. Thus, in the short term it will be necessary to continue with detailed studies of specific populations and habitats threatened by environmental change while simultaneously working toward the long-term goal of developing a general model (or set of criteria) based on a full mechanistic understanding of the role light availability plays in regulating the productivity and distribution of seagrasses.

ACKNOWLEDGEMENTS

We thank Robert Smith, Jason Smith, Steven Fain, Susan Britting, Catherine Dea, Anthony DeTomaso, John Reguzzoni, Sandy Wyllie-Echeverria, Michael Josselyn, Mark Silberstein and the Elkhorn Slough Foundation for assistance with various aspects of this research. This work has been supported in part by NOAA, National Sea Grant College Program, Department of Commerce, under grant number NA85AA-D-SG140 project number R/CZ-83 through the California Sea Grant College Program, by NOAA, Sanctuaries and Reserve Division, Department of Commerce, under grant numbers NA88AA-D-CZ027 and NA90AA-H-CZ191, and by the National Science Foundation, under grant number OCE-8603369. The U.S. Government is authorized to reproduce and distribute for governmental purposes.

REFERENCES

Alpine, A.E. and Cloern, J.E. 1988. Phytoplankton growth rates in a light-limited environment, San Francisco Bay. Mar.Ecol. Prog. Ser., 44:167-173.

Backman, T.W. and Barilotti, D.C. 1976. Irradiance reduction: effects on standing crops of the eelgrass Zostera marina in a coastal lagoon. Mar. Biol., 34:33-40.

Dennison, W.C. and Alberte, R.S. 1982. Photosynthetic response of Zostera marina L. (eelgrass) to in situ manipulations of light intensity. Oecologia, 55:137-144.

Dennison, W.C. and Alberte, R.S. 1985. Role of daily light period in the depth distribution of Zostera marina (eelgrass). Mar. Ecol. Prog. Ser., 25:51-61.

Dennison, W.C. and Alberte, R.S. 1986. Photoadaptation and growth of Zostera marina L. (eelgrass) transplants along a depth gradient. J. Exp. Mar. Biol. Ecol., 98:265-282.

Droop, M.R. 1973. Some thoughts on nutrient limitation in algae. J. Phycol. 9:264-272.

Fain, S.R., DeTomaso, A. and Alberte, R.S. In review. Characterization of disjunct populations of Zostera marina L. (eelgrass) from California. II. Genetic structure resolved by restriction fragment length polymorphisms. Mar. Biol. In Review.

Huber, S.C., Kerr, P.S., Kalt-Torres, W. 1985. Regulation of sucrose formation and movement. pp. 199-215. In. R.L. Heath and J. Priess (eds), Regulation of carbon partitioning in photosynthetic tissues. Am. Soc. Plant Phys., Rockville, Md.

Lowell, C.A., Tomlinson, P.T. and Koch, K.E. 1989. Sucrose-metabolizing enzymes in transport tissue and adjacent sink structures in developing citrus fruit. Plant Physiol. 90:1394-1402.

Marsh, J.A., Dennison, W.C. and Alberte, R.S. 1986.
Effects of temperature on photosynthesis and
respiration in eelgrass (Zostera marina L.). J. Exp.
Mar. Biol. Ecol., 101:257-267.

Orth, R.J. and Moore, K.A. 1983. Chesapeake Bay: an
unprecedented decline in submerged aquatic
vegetation. Science, 222:51-53.

Orth, R.J. and Moore, K.A. 1988. Distribution of Zostera
marina L. and Ruppia maritima L. sensu lato along
depth gradients in the lower Chesapeake Bay, U.S.A.
Aquat. Bot.,
32:291-305.

Smith, R.D. 1989. Anaerobic metabolism in the roots of
Zostera marina L. Ph.D. Dissertation, The University
of Chicago, Chicago, 241 pp.

Smith, R.D. Dennison, W.C. and Alberte, R.S. 1984. The
role of seagrass photosynthesis in aerobic root
processes. Plant Physiol. 74:1055-1058.

Zimmerman, R.C. and Kremer, J.N. 1984. Episodic nutrient
supply to a kelp forest ecosystem in Southern
California. J. Mar. Res. 42:591-604.

Zimmerman, R.C., Smith, R.D. and Alberte, R.S. 1989.
Thermal acclimation and whole plant carbon balance
in Zostera marina L. (eelgrass). J. Exp. Mar. Biol.
Ecol., 130:93-109.

Zimmerman, R.C., Reguzzoni, J.L., Wyllie-Echeverria, S.,
Josselyn, M. and Alberte, R.S. 1991. Assessment of
environmental suitability for growth of Zostera
marina L. in San Francisco Bay. Aquat. Bot. In
Press.

Zimmerman, R.C. and Alberte, R.S. In prep. Effects of
daily light period on sucrose metabolism and
distribution in Zostera marina L. (eelgrass). Plant
Physiol.

Successful Tidal Wetland Mitigation
In Norfolk, VA

Carvel Blair *

Abstract

In the 1940's the U.S. Navy dredged a portion of
Willoughby Bay in Norfolk, VA and deposited the sediment
on the Bay's eastern side. In the ensuing years about 2.8
hectares of the spoil area sank below mean high tide,
were periodically inundated by salt water, and grew a
cover of marsh grasses. In 1982 the Navy was granted permission to place additional spoil on the marsh provided
that a compensation wetland was established adjacent to
the spoil area. After depositing the dredged material the
Navy employed a contractor to create a wetland of the
same area about 1000 m. north of the former marsh. The
land was graded to an elevation below mean high water
(MHW), planted with saltmarsh cordgrass (Spartina
alterniflora) seedlings, and fertilized.

In 1988 the city of Norfolk obtained funds to
evaluate the ecological success of the new wetland by
comparing it to similar nearby marshes. Investigators
from Old Dominion University (ODU) and Virginia Institute
of Marine Science (VIMS) made comparative surveys of the
vegetation, invertebrates, fish, birds, and bottom sediments. All concluded that the new marsh had become an
effective component of the estuarine system of lower
Chesapeake Bay.

Without giving blanket approval to the concept of
compensation wetlands, the Virginia Marine Resources
Commission (VMRC) in 1989 established guidelines for
evaluating such proposals. The wetland we surveyed met
several of the pertinent criteria. It is judged to have

----

*Associate Professor Emeritus, Old Dominion University,
and former Chairman, Norfolk Wetlands Board; P.O.Box 95,
Cobham, VA 22929.

attained the goals for which it was created. The choice
of its site, careful design and execution of the grading
and planting, and close attention subsequent to its
establishment all played important roles in this success.

Background

            Willoughby (Little) Bay, a shallow body of water
about 2 km^2 in area, lies east of Hampton Roads.
Willoughby Spit separates it from Chesapeake (Big) Bay on
the north. Land areas of the Norfolk Naval Station bound
it to the east, south, and southwest (Fig. 1). In the
1940's, at some date prior to March 1948, the Navy conducted a major dredging operation in Willoughby Bay.
Spoil deposited in its eastern end converted about one-
sixth of the Bay's original subaqueous area to a low-
lying upland. This section, for reasons that have not
survived, came to be known as Monkey Bottom. As the
dredged material compacted and dried, an area of about
7.6 ha sank below MHW level. Daily inundation of the new
wetland allowed colonization by S. Alterniflora. The new
surrounding uplands grew a dense stand of common reed
(Phragmites australis). (Priest et al, 1982). By 1980 the
Navy needed to dispose of more spoil from aircraft carrier berths adjacent to the Norfolk Naval Base. In 1972,
however, the Commonwealth of Virginia had enacted the
Wetlands Act requiring State permission to fill any tidal
wetland.   Federal law likewise required authorization
from the Corps of Engineers. As a result, the Naval
Facilities Engineering Command in 1981 submitted requests
to the Corps and to the City of Norfolk Wetlands Board
(to whom State approval authority had been delegated).

Compensation Marsh

            This unusually large request was on the agenda of
the initial meeting of the Norfolk Wetlands Board (of
which I was chairman). The Board granted permission to
fill the existing Monkey Bottom marsh on condition that a
marsh of equal area be created adjacent to the new fill.
The Corps granted permission with the same proviso. At
that time there were no State standards for compensation
marshes, so the two agencies developed a common specifi-
cation for the new wetland. Its main points were (1) that
the new area should be at least equal to the old, and (2)
that the Navy repeat the plantings if the initial plants
did not survive.

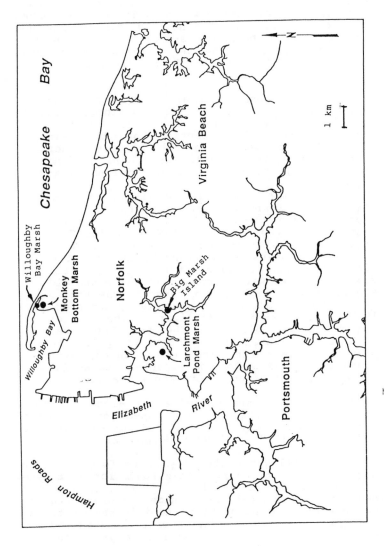

Figure 1. Location of Monkey Bottom and control marshes. (From Feigenbaum and Swift 1989)

The Navy contracted with Great Lakes Dredge and
Docking Corp., who subcontracted to Higgerson-Buchanan
Inc., for the creation of the wetland. The cost was about
$325,000. (Naval Facilities Engineering Command, personal
communication). Work began in summer 1984 at the same
time as construction of the berm around the disposal
area. Priest describes the marsh construction (Priest,
1989:

>   The compensation area was graded to
> elevations at or below MHW. Drainage was
> accomplished by sloping the area to four
> lateral ditches which emptied into the main
> ditch that connected to the culvert under
> [interstate highway] I-64. During the grading
> and planting, the area was isolated from tidal
> inundation.
>   The marsh was planted with saltmarsh
> cordgrass, <u>Spartina alterniflora</u>, during
> September and October 1984, using a tree
> planter. The two sections closest to the City
> of Norfolk's Visitors Center were planted on
> two foot centers with transplants from the
> site. The two sections furthest from the
> Visitors Center were planted with six month old
> seedlings also on two foot centers. The entire
> area was broadcast with an especially prepared
> 19-5-12 slow release fertilizer at the rate of
> one ounce per planting. The fertilizer was
> mechanically raked into the soil prior to
> planting. The entire area was planted, even
> areas above and below the expected successful
> elevation, so that good coverage was ensured.

Material from the upland spoil along the south
edge of the wetland subsequently slumped into the marsh
and most of it was removed by dragline. The configuration
that developed by 1989 is shown in Figure 2.

For several years the City of Norfolk, ODU, and
VIMS unsuccessfully tried to obtain funds for a study of
the new wetland as it developed. At last the City
obtained a modest Coastal Zone Management Grant from the
National Oceanic and Atmospheric Administration through
the Virginia Council on the Environment. In 1988 and 1989
scientists from the two universities compared selected
aspects of the new Monkey Bottom marsh with corresponding
aspects of control marshes. Because only a cursory survey
of the destroyed marsh existed (Priest et al. 1982)and
because the destroyed marsh was itself man-made, natural

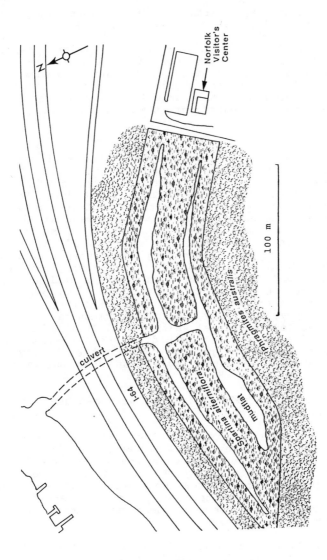

Figure 2. Monkey Bottom compensation wetland.
(From Priest 1989)

COASTAL WETLANDS

marsh areas were selected for control. The characteristics chosen for survey were sediments, plants, invertebrates, fish, and birds. Funding allowed only limited field work and simple comparative studies.

Swift investigated benthic sediments. Feigenbaum studied fishes and invertebrates, both mobile and benthic (Feigenbaum and Swift 1989), and I surveyed birds (Blair 1989). Priest conducted vegetative studies (Priest 1989). Methods, findings, and conclusions, set forth at length in the cited references, are summarized below.

Methods

Control Marshes. For the reasons stated above, Monkey Bottom marsh was compared primarily not to the destroyed wetland but to one or more of three nearby natural marshes (Fig. 1). There was some scant data (Priest et al. 1982) on the destroyed wetland, but much more information was needed. Vegetation was surveyed at Willoughby Bay marsh, a small pocket 500 m north of Monkey Bottom compensation marsh. These two wetlands lie at opposite ends of the culvert connecting the latter to Willoughby Bay. All field characteristics (except bird populations) were surveyed at Larchmont Pond marsh, a wetland fringing Weir Pond in the city's Larchmont section. On the Lafayette River about 7.5 km south of Monkey Bottom, its physiography, hydrology, and plant communitites are typical of Norfolk's natural saltmarshes. The bird survey was conducted at Big Marsh Island, farther upstream in the Lafayette. Long and narrow, it resembles Monkey Bottom marsh in area and shape more than does Larchmont Pond marsh.

Temperature and salinity. These were measured in August-September 1988.

Sediments. Three reference elevations were selected and levelled in at Monkey Bottom and at Larchmont Pond marsh: mean high water springs, mean low water (MLW) springs, and half tide level (HTL). Box cores were taken at each level in each marsh. Cores were X-rayed and analyzed, at surface and 20-cm depth, for grain size. Another set of cores was frozen for subsequent measurement of carbon content by CSN analyzer .

Plants. Priest established transect lines in Monkey Bottom, Willoughby Bay, and Larchmond Pond marshes, then selected random sampling plots. At each an 0.25 m^2

quadrat was surveyed for cover, density, and peak standing crop. Significant elevations were measured.

Fish and mobile invertebrates. Monkey Bottom and Larchmont Pond marshes are each fed through single culverts. During the August-September survey, mobile fauna were trapped in stop nets set on the ebb tide across the lower ends of the culverts. Fish and invertebrates were identified, counted, and released. Since <u>Fundulus</u> species did not leave on the ebb in great numbers, stop net data were supplemented with information from baited minnow traps.

Benthic invertebrates. The investigators took sets of cores at Larchmont Pond and Monkey Bottom marshes in August-September and again in December 1988. Replicate cores, 10 cm in diameter and at least 20 cm deep, were taken at three sites along each of the three selected elevations. Sediment was seived through an 0.5 mm screen and remaining material preserved in 15% Formalin/rose bengal solution. Organisms, stained pink by rose bengal, were removed and identified by genus or family.

Birds. I made four short bird counts at Monkey Bottom and at Big Marsh Island, one in each marsh in June 1988 and one in each in January 1989. Both marshes are long and narrow in shape. The route for each count was a circuit around the wetland, by foot at Monkey Botom, by canoe at Big Marsh Island. I counted all individuals seen in, above, or within about 200 m of each marsh.

Results

Temperature and salinity, were similar at all sites, with salinities in the 21-23 ppt range.

Sediments. Core X-rays at Monkey Bottom showed a layer of newly accumulated sediment varying from 2-11 cm deep above compacted older material. New sediment had relatively high sand content and low water content. Older, deeper deposits were finer grained, also with low water content. All cores at Larchmont Pond had high sand content at all depths. The samples taken at low tide elevation contained 53-64% water; those at half- and high-tide elevation were better drained with 20-30% water. Depth-averaged total organic carbon (Table 2, Feigenbaum and Swift 1989) was distributed as follows:

	MONKEY BOTTOM	LARCHMONT POND
MLW	1.9 %	10.4%
Half-tide	1.3	2.9
MHW	1.8	2.0
Average	1.7	5.1

Plants. Fig. 3 summarizes Priest's measurements of cover, density, and standing crop. An analysis of variance showed no significant differences at the 0.05 level among these parameters at the three sites (Monkey Bottom, Willoughby Bay, and Larchmont Pond.) Spartina alterniflora was the only species found in the quadrats at Monkey Bottom and Willoughby Bay, and dominated those at Larchmont Pond. At Monkey Bottom the stand of Spartina alterniflora, on the average, ranged from 37 cm to 82 cm above MLW, approximately from half tide to MHW. It was bordered above by P. australis.

Fish and mobile invertebrates. The trap net at Monkey Bottom caught 1279 individuals comprising 13 species; at Larchmont Pond, 229 individuals comprising 8 species. Unfortunatley catches per unit time were not reported. At Monkey Bottom, mullet (Mugil cephalus and M. curema)made up 55% of the catch; Callenictes sapidus (blue crab) was next most numerous at 20%. No other species accounted for more than 10%. At Larchmont Pond Brevoortia tyrannus (menhaden), 45%, Fundulus (mummichog and killifish), 31%, and Menidia menidia (silverside), 15%, dominated. Mullet were seen jumping in the pond and over the net, but none was caught. The principal catch of the minnow traps was F. heteroclitus; 0.7 were taken per minute at Monkey Bottom, 2.5 at Larchmont Pond.

Benthos. Feigenbaum and Swift summarize these results as follows:

At Monkey Bottom, the low and medium tidal samples were characterized by three species of clams (hard, soft, and razor) and relative abundance of polychaete worms. High tide samples had very few organisms...At Larchmont Pond, samples were characterized by small nematode worms, particularly in the late fall samples. Fiddler crabs were extremely abundant, but too mobile to be caught in any numbers in the core samples. Few Littorina were observed in the [Larchmont] marsh...If nematodes are

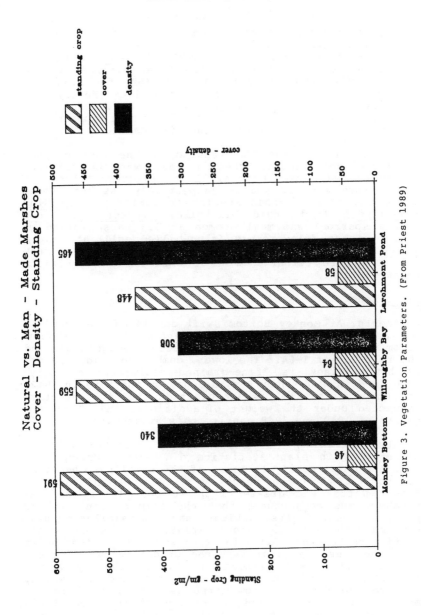

Figure 3. Vegetation Parameters. (From Priest 1989)

excluded, Monkey Bottom had far more organisms than Larchmont Pond.

Benthic samples taken by Priest et al. in December 1981 were from the tidal ditches in the Willoughby Disposal Area (before the area was filled in and the mitigation [Monkey Bottom] marsh created...The number of infaunal organisms in the new marsh in winter appear more or less the same as the 1981 marsh.

Birds. Table 1 shows the results of the four counts. Population densities in breeding season were 30 individuals per ha at Monkey Bottom, 21 at Big Marsh Island. In winter they were 12 and 15 respectively. At both locations the most numerous species in summer was Agelaius phoenicus (red-winged blackbird). Passer domesticus (house sparrow) was most common at Big Marsh Island in winter; Larus delawarensis (ring-billed gull) at Monkey Bottom. Except for the domestic goose and duck, all species seen are either common or abundant in Virginia's coastal plain.

Discussion

Sediments. At Monkey Bottom Swift concludes that sand in the new, upper layer has been carried by rain runoff "from the steep slopes of the adjacent dredge spoil and the road bed of I-64." The lower, compacted, older layer, he hypothesises, is "fine-grained sediment, initially deposited in Willoughby Bay, dredged by the Navy in 1942, and dumped into its present position and, subsequently, compacted under the weight of additional fill." Survey charts of 1940 and 1948 confirm this hypothesis.

Plants. Although not statistically significant, some differences in plant statistics did exist. Priest explains them primarily in terms of the sampling scheme. He points out that the range of elevations at man-made Monkey Bottom was greater than at the flatter natural marshes, and conjectures that the former "can tend to provide a more diverse habitat than some similar natural systems with less [elevation] variability." Comparing his statistics to those found in other wetlands, Priest notes a "tremendous" range of reported productivity, with his values falling well inside the range. Density values likewise are within the ranges reported elsewhere. Because the Monkey Bottom marsh was established in the midst of an extensive and vigorous Phragmites bed, there was concern that this undesirable plant would take over from the newly planted Spartina. Priest attributes the

failure of Phragmites to invade to the conjecture that it
will not grow below MHW in mesohaline areas. As a guide
to planting Spartina he points out that here its
"effective upper and lower limits" were MHW and HTL.

Fish and invertebrates. Feigenbaum does not comment extensively on his results. He points out that Monkey Bottom marsh is three times larger than Larchmont Pond marsh
and drains more completely at low tide, then summarizes
as follows:

> The infaunal invertebrates of the two
> marshes appear to represent different
> communities, presumably due to differences in
> location and sediment characteristics. The
> mobile fauna were quite similar, with the
> exception of fiddler crab populations. The
> absence of a large number of fiddler crabs at
> Monkey Bottom may possibly be explained by
> lower level of sediment organics which also
> affects sediment grain size and water
> characteristics. These can be expected to
> improve as the marsh ages.

> Fishes and mobile invertebrates at Monkey
> Bottom were abundant during the summer sampling
> and compared favorably with those at Larchmont
> Pond marsh. The number of larger organisms
> (nematodes excluded) in the sediment was higher
> than at the Larchmont Pond marsh and, for
> December, was similar to the numbers in pre-
> marsh samples of the Willoughby Bay Disposal
> Area (Priest et al 1982.)

Birds. Of the 42 species observed, 16 were recorded at
both Monkey Bottom and Big Marsh Island. Suburban yards,
trees, and lawns produced several species at the latter
that would not be expected in the relatively barren terrain surrounding the former. Most of the other species
seen at only one site during these brief counts are known
to occur regularly at the other. Longer, more detailed
field observations would probably confirm that species
using both wetlands, and their diversity, are similar.

Virginia's policy.

After a study by Barnard et al.(1987) VMRC in
October 1989 adopted a "Wetlands Mitigation-Compensation
Policy." The operative parts are (1) a set of criteria
which must be met if a proposal for compensation is to be

accepted and (2) a list of 16 "Supplemental Guidelines" for establishing such a wetland. Some of these are relevant to our study results.

Guideline 3 requires that each compensation marsh be established "by experienced professionals knowledgeable of the general and site-specific requirements for wetland establishemnt and long-term survival." This guideline was followed at Monkey Bottom.

Guideline 5 requires a compensation marsh of at least equal area with "similar plant structure to that being lost." Plant structure at Monkey Bottom turned out as desired. Lack of prior agreement on tidal datum caused some confusion in measuring wetland areas and several earth-slides temporarily reduced the compensation area, but the Navy removed the slides and restored the areas to their pre-slide dimensions. The resulting marsh has approximately the same area as the filling marsh.

Guideline 6 states that the "compensation should be accomplished prior to, or concurrently with, the construction of the proposed project." At Monkey Bottom filling and marsh construction began simultaneously.

Guideline 14 concerns monitoring of the new wetland, to be funded in some cases by the applicant. As discussed earlier, money for our studies was meager and hard to obtain. Had this guideline been in existence and followed in 1982, better surveys would have resulted.

Guideline 16 requires perpetual conservation easements for the new marsh. None exists for Monkey Bottom. The City of Norfolk and the Navy are planning to construct a marsh walk over the area, with public access and educational material. It is possible that an agreement to retain Monkey Bottom as a wetland in perpetuity can be included in the negotiations.

Conclusions

1.   "The Monkey Bottom plant community appears to be a viable and productive component of the estuarine system." (Priest 1989)

2.   "The design criteria of planting Spartina alterniflora between MTL [mid-tide level] and MHW in mesohaline areas to successfully compete with Phragmites australis appears to be confirmed." (Priest 1989)

3.   "The Monkey Bottom marsh appears to be healthy
and functioning in several beneficial ways...as a nursery
area...[and]...forage area...The marsh is accumulating
organic materials in the sediments some of which will
eventually wash into nearby waterways enhancing their
food chains." (Feigenbaum and Swift 1989)

4.   Although the man-made wetland at Monkey Bottom is
similar in most respects to nearby natural wetlands, it
does not reproduce every geological, physical, and bio-
ogical aspect of the latter.

5.   For a period of several years following de-
truction of the former wetland, there was a net loss of
wetland acreage adjoining Willoughby Bay. It is unlikely
that any of the individual flora or fauna of the old
marsh survived and made their way to the new marsh.

7.   In this particular case, given the geological
setting, the geographical location, and the skillful
construction of the Monkey Bottom wetland, the decision
to compensate for a destroyed wetland by creation of a
replacement was environmentally useful and acceptable.

Appendix I.--References

Barnard, T.A., Priest, W.I., and Watkinson, T.K., "Report of
    the VMRC/VIMS Subcommittee on Wetlands Mitigation-
    Compensation Policy," Virginia Institute of Marine
    Science, Gloucester Point, VA, Aug., 1987.
Blair, C.H., "Comparative Bird Counts in Natural and Man-made
    Wetlands in Norfolk, VA," The Raven(Journal of the Vir-
    ginia Society of Ornithology), Vol. 59, July,1989, pp.3-9.
Feigenbaum, D.L., and Swift, D.J.P., "Wetlands Mitigation
    Project: A Comparison of a Mitigation Marsh with an
    Established Marsh in Norfolk, VA; An Evaluation of the
    Invertebrates, Fish, and Sediments," Applied Marine Re-
    search Lab, College of Sciences, Old Dominion University,
    Norfolk, VA, May, 1989.
Priest, W.I., Terman, C.R., and Ihle, W., "A Natural
    Resources Survey and Habitat Evaluation of the Willoughby
    Disposal Area," Final report prepared for Naval Facilities
    Engineering Command, Norfolk, VA, 1982.
Priest, W.I., "Wetlands Mitigation Evaluation Vegetation
    Studies: Monkey Bottom Disposal Area," Virginia Institute
    of Marine Science, Gloucester Point, VA, July, 1989.

SPECIES	SCIENTIFIC NAME	LAFAYETTE RIVER JUNE	LAFAYETTE RIVER JANUARY	MONKEY BOTTOM JUNE	MONKEY BOTTOM JANUARY
Brown Pelican	Pelecanus occidentalis			18	
Canada Goose	Branta canadensis	8			
Domestic Goose	Anser cygnoides (?)	1			
Mallard	Anas platyrhnchos	5	3	1	
Domestic Duck	?	1			
Bufflehead	Bucephala albeola .		4		
Hooded Merganser	Lophodytes cucullatus				2
American Kestrel	Falco sparverius				1
Great Egret	Casmerodius albus	4		1	
Great Blue Heron	Ardea herodias		2		
Green-backed Heron	Butorides striatus	2		1	
Yel-crwn Night Heron	Nycticorax violaceus	1		3	
Clapper Rail	Rallus longirostris	1		1	
Kildeer	Charadrius vociferus			1	
Common Snipe *	Gallinago gallinago				1
Great Blackbacked Gull	Larus marinus				1
Herring Gull	Larus argentatus	1		4	4
Ring-billed Gull	Larus delawarensis		6		7
Laughing Gull	Larus atricilla	3		2	
Least Tern	Sterna antillarum	1		1	
Forsters Tern	Sterna forsteri	2			
Royal Tern	Sterna maxima			3	
Black Skimmer	Rynchops niger			1	
Rock Dove	Columba livia				5
Mourning Dove	Zenaida macroura	2	6	4	5
Belted Kingfisher	Ceryle alcyon		1		
Northern Flicker	Colaptes auratus		1		
Barn Swallow	Hirundo rustica			1	
Purple Martin	Progne subis			3	
Blue Jay	Cyanocitta cristata	1			
American Crow	Corvus brachyrhyncos	7	14	1	
Tufted Titmouse	Parus bicolor	1			
Marsh Wren	Cistothorus palustris	3			
Northern Mockingbird	Mimus polyglottos	2			
American Robin	Turdus migratorius	3		1	
European Starling	Sternus vulgaris	1		3	
Red-winged Blackbird	Agelaius phoeniceus	27	2	22	1
Common Grackle	Quiscalus quiscula	3		4	
Northern Cardinal	Cardinalis cardinalis	2			
House Finch	Carpodacus mexicanus			3	
Song Sparrow	Melospiza melodia	1		5	6
House Sparrow	Passer domesticus		20		
Total	Species 42 Individuals	83	59	84	33

* probably this species

Table 1 Bird Species Observed at Each Site During
Summer and Winter Surveys

ESTABLISHING EELGRASS BEDS IN CALIFORNIA

Michael D. Curtis[1]

Abstract

Eelgrass (Zostera marina) beds are environmentally sensitive habitats in southern California. Many species of marine life utilize this important resource for nursery grounds, protection, and living space. However, anthropogenic disturbances of coastal bays and wetlands have resulted in a substantial reduction in this habitat. Therefore, resource agencies have placed particular emphasis on restoration or establishment of eelgrass beds as mitigation for habitat loss due to development. During the past several years, biologists from MBC Applied Environmental Sciences have restored or established eelgrass beds in bays and wetlands of southern California as mitigation for habitat loss. This paper summarizes the extent of this restoration effort, problems associated with the effort, and findings of subsequent monitoring studies.

Introduction

Eelgrass (Zostera marina) is one of several marine flowering plants that occurs in temperate coastal areas in the northern hemisphere. During studies spanning the last 50 years, eelgrass had been recognized as an important habitat and nursery area for a variety of marine life. Because more than an estimated 50% of eelgrass habitat and associated fauna has been lost over the last century due to dredging and filling of bays and estuarine areas in California, resource agencies have taken the position that there will be no further net loss of eelgrass. Any eelgrass that they permit to be removed for the purpose of development or dredging must be replaced on at least a one-to-one basis. Future requirements will probably factor in the amount of time the eelgrass is lost

---

[1]   MBC Applied Environmental Sciences,
      947 Newhall Street, Costa Mesa, CA 92627.

to the environment, which will require even greater
amounts of eelgrass to be transplanted than removed (R.
Hoffman, National Marine Fisheries Service, Terminal
Island, CA, pers. comm, 25 January 1991). All transplant
projects are also required to be monitored periodically
to ensure that eelgrass is established and survives in
densities comparable to that in the bed removed. Under
limited circumstances mitigation of other marine-oriented
encroachments by transplanting eelgrass into estuaries and
harbors may be permitted, if a suitable area can be found
or prepared. However, other mitigation measures may not
be credited towards or substituted for eelgrass
transplants.

     Areas presently without eelgrass are probably light-
limited or depth-limited, or the physical/chemical
properties of the sediments are inappropriate for the
successful establishment of new eelgrass beds.
Accordingly, there is little support among the resource
agencies for planting eelgrass in totally barren areas;
it is assumed that eelgrass would be growing at the
location or nearby if one or more of these factors were
not operating (R. Hoffman, per. com.). Finding an
appropriate area is difficult and when found, physical,
chemical, and biological parameters need to be carefully
examined to determine why eelgrass is not present.
Sediment characteristics, water circulation patterns,
turbidity, and depth considerations must also be addressed
prior to a transplant.

     At present, the most readily available sites are
areas where eelgrass is restricted by depth. Resource
agencies have been willing to relinquish, on a site-
specific basis, certain deeper areas of harbors or
estuaries for transplant purposes. When an appropriate
site is found and agreed upon by all agencies, the area
must be prepared to accept the newly transplanted
eelgrass. The most effective means of preparing a site is
to place available marine, dredged sediments onto an area
near existing viable eelgrass beds. These sediments are
then either contoured artificially or allowed to settle
until they resemble the natural habitat.

     Ideally, sediment grain size characteristics and
chemical loads are determined prior to fill operations.
The area should be monitored periodically to determine
settling patterns and possibly, if time permits, a small-
scale transplant should be undertaken. If conditions
remain favorable after six months to one year, the full-
scale transplant may then be performed with a reasonable
chance for success.

Locating Transplant Areas. New transplant areas should be located in areas free of boat traffic, and where circulation patterns indicate an appreciable tidal water exchange will take place. Boat traffic in shallow areas of developed bays and harbors is probably an inherent risk as many boaters are unaware of what navigation buoys signify. Swaths through the eelgrass beds from boat propellers attest to boaters' failure to heed the warning buoys. Sediment characteristics which support eelgrass range from coarse sands to loose clays (Schubel 1973). Coarse sands are usually indicative of a high energy environment and are not stable environments; the best soils appear to be sand-silt.

Although eelgrass may be successfully planted during any time of the year in southern California, the rhizomes are relatively dormant except in late-spring and summer (Boone and Hoeppel 1976). The planting units need to be firmly anchored as currents have a tendency to dislodge the turions or bury them as unstable sediments settle or are lost to tidal flux. The preferred season to perform a transplant is late winter or early spring, as they come out of dormancy. Growth is rapid during this period and the developing eelgrass plants spread rhizomes and vegetatively produce more turions (shoots), thereby anchoring and stabilizing the sediments. In addition the turion blades form a barrier, which reduces current velocities, causing sedimentation and accretion which further increases the potential for beds to expand naturally (Fonseca et al. 1982).

Transplanting Techniques

Grid and Reference Lines. Prior to conducting any transplant operations a grid delineating the exact area to be transplanted should be deployed. Poor visibility inherent in underwater operations in harbors and estuaries make it necessary to place reference lines as an aid to navigation. The transplant areas are usually very silty and sediments are easily disturbed by divers. Even very experienced divers can become disoriented when working in these conditions and must have reference lines to ensure proper spacing of the transplant units. The reference lines should encompass the entire site and should subdivide the site into manageable units for the transplant operations. The size of the transplant should determine the proper spacing. One project of approximately 13,000 ft^2 (1208 m^2) had reference lines every 2 m, whereas another project area totaling 406,000 ft^2 (37,700 m^2) had reference lines every 15 m. Of course the transplanted rows of eelgrass were better defined in the area with 2 m reference lines.

Collection of Eelgrass. Harvesting techniques may be
as simple as coring or spading eelgrass with sediment
attached during low tides (Boone and Hoeppel 1976). Plugs
of eelgrass with sediment and rhizome mass intact may then
be transported to the transplant site. This method was
relatively successful in low current areas, however, the
cores are very heavy, space consuming, and awkward to
transport. The need to wait for low tides and the
relatively short collection period usually allows little
flexibility in a schedule for any, but the smallest
transplant operations. A transplant program larger than
one acre (0.4 hectares) with this technique would be very
time-consuming.

A more useful technique which is relatively
independent of tides, was to use scuba-equipped divers
to collect eelgrass in large mesh bags. The diver scooped
sediment with rhizomes and blades intact into the bag.
With practice, they were able to filter most of the
sediment out of the mesh bags and deliver relatively
sediment-free eelgrass to the next phase of the transplant
operation. With this technique enough eelgrass could be
collected by two divers in six hours to supply a crew of
10 eelgrass bundlers for 10 hours.

Transplant Units. Several transplant units have been
used in the past with varying results. The sediment cores
(sediment with rhizomes intact) were placed into
biodegradable pots approximately 15 cm in diameter. These
were then transplanted into a hole prepared underwater by
divers. Although this method was relatively successful
in some projects (Phillips 1974), on the single project
MBC used this method, biologists were only marginally
successful. Apparently the rhizomes were unable to
penetrate the sides of the pot.

The most promising method tried was the sediment-
free bundling method (Fonseca et al. 1982). Eelgrass was
harvested by divers and the individual rhizomes were
washed off sediment carefully leaving the root-rhizome
system intact. These were then grouped together in bundles
of 15 turions (shoots). Metal coat hangers were cut into
20 cm lengths and the top 5 cm was bent to form an L-
shape for anchoring purposes. This was placed on the
bundle and a strip of paper (file cards cut into strips)
was wrapped around the bundle and coat hanger. A metal-
paper twist tie was then wrapped around this to finish the
individual bundle. This method proved to be effective and
large areas were transplanted more quickly. However, the
anchor system was not particularly successful in high
current areas: although faster than coring and placing
eelgrass in pots, it was still time consuming to fabricate
and plant eelgrass bundles. In high current areas, almost

10% of the transplanted units were lost (higher than expected). There was also concern of possible injuries to bathers or waders, as well as the heavy metal loads that would be imparted to the sediment.

A slight modification of the Fonseca et al. (1982) methodology resulted in stronger anchoring, eliminated hazardous metal points in the sediment, and left no trace within six months. Each bundle was tied, snug but not tight just above the rhizome mass, using cotton string approximately 25 cm long. The other end of the string was tied to the middle of a common popsicle stick (Figure 1). A supply of these was prepared with one end tied to the popsicle stick prior to the transplant. In the actual transplant, the free end was tied around the naked bundle of approximately 15 turions. Fonseca et al. (1982) reported bundling rates of 100 per hour, however, our experience indicated that 30 per hour was a more realistic expectation.

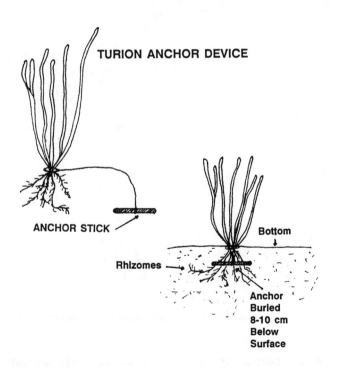

TURION ANCHOR DEVICE

ANCHOR STICK

Bottom

Rhizomes

Anchor
Buried
8-10 cm
Below
Surface

Figure 1. Diagrammatic representation of eelgrass turion anchor device.

Diver-deployed Eelgrass Bundles. Seventy-five turion
bundles made using the modified sediment-free, turion-
anchor method, were placed on a plastic Vilas-board
designed for ease in carrying and fast deployment
underwater (Figure 2). Transplanting could be accomplished
in either of two ways: 1) one handed — cup the bundle in
one hand, 2) with the popsicle stick parallel to the
sediment make a hole with the back of the hand, 3) place
the turion bundle inside, and 4) use the same hand to
backfill the hole to just above the rhizome mass. The
other method worked well in low visibility situations: 1)
deploy all bundles by sticking the popsicle stick upright
into the sediment, 2) then retracing (your) path and bury
the rhizome mass with the popsicle stick across the
rhizome mass and parallel to the sediments to provide the
anchor power. An experienced diver could plant
approximately 150 bundles an hour using either method.

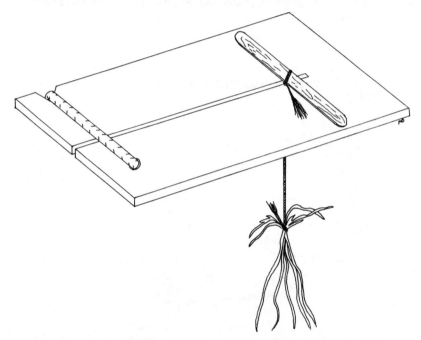

**Figure 2. Vilas-board used to transport eelgrass turion
bundles.**

To reduce turbidity, divers adjusted their buoyancy
to float just above the bottom and planting bundles as
they went. Using a measuring stick, either 2 or 3 ft long
(0.6 or 0.9 m) planting was done on 2-ft (0.6 m) centers

(one unit per 4 ft^2 = 0.37 m^2) in high current areas or on 3 ft (0.9 m) centers (one per 9 ft^2 = 0.84 m^2) in low current areas. These distances were used by Fonseca el al. (1982), but were modified to fit the requirements of the specific project. It is worth remembering that more than twice as many bundles are needed to fill an area on 2-ft centers as on 3-ft centers.

Monitoring Transplant Site. A survey three months after planting will ensure that the transplant is viable. If the transplant was performed during the late-winter or early-spring, the individual turion bundles should have grown to cover 1 ft^2 (0.09 m) and the initial turion bundle should still be recognizable. Success at this point is measured by counting the number of turion units visible, or subsampling an area if very large. If losses are more than 10% or some gaps are large, it may be necessary to do a remedial transplant to fill in these gaps. However, it should be noted that although transplant shock may result in the loss of all turion blades, the plant may still have viable rhizomes which could produce a healthy eelgrass bed in time. One year after planting, it should be impossible to determine the location of planting units. Random density measurements (turions/m^2) combined with the area of any gaps should suffice to measure the success of the transplant. If growth is not depth-limited, coverage should exceed the boundaries by 10% by the end of the second year, and density should equal that of the donor bed or natural beds nearby.

Project Experience. Biologists from MBC Applied Environmental Sciences have been involved in several eelgrass mitigation transplants and have surveyed existing eelgrass conditions for a variety of projects from both government and private agencies in the past several years.

In 1983 an eelgrass transplant was performed at Peters Landing in Huntington Harbour, Orange County, following the accidental destruction of a marsh area. The sediment core in biodegradable pots method was used. The results of this transplant and subsequent surveys were mixed. Few of the eelgrass rhizomes penetrated the walls of the pot and a bed did not develop. In 1985, sediment-free eelgrass bundles were added to the area using the Fonseca et al. (1982) method. This transplant was successful and a bed approaching 500 ft^2 (46.5 m^2) developed.

In 1985, several eelgrass beds totaling 6000 ft^2 (560 m^2) were transplanted in Huntington Harbor, Alamitos Bay, Cabrillo Beach, Newport Harbor backbay to mitigate the loss of eelgrass habitat due to construction of docks in Huntington Harbour. In Huntington Harbor the eelgrass was placed along a sandy-silty beach adjacent to an existing bed. It flourished for six months, then dwindled to a very small area and has remained static for the past five years.

In Alamitos Bay the eelgrass was transplant near existing small beds along a sandy beach which graded to silt very near shore. This bed developed well and resembled the native beds within one year; it was still healthy when a survey was conducted in 1988. An eelgrass transplant was attempted in the backbay of Newport Harbor, however, due to turbidity, boat traffic, or unknown factors it did not survive the first year. A transplant at two shallow areas Cabrillo Beach in San Pedro Harbor was not expected to survive due to high turbidity and low current flow. No trace of eelgrass was found within the perimeter of the transplant site six months after the transplant. However, at one year, a perfunctory survey to ascertain the fate of the beds revealed a healthy bed. In late 1990, another survey of the area confirmed the continued presence of several small beds, although they had not spread much beyond the original boundaries (R.Hoffman NMFS per comm. 19 September 1990)

Construction of a new marina and docks in San Diego Harbor caused the loss of 3000 ft^2 (280 m^2) of eelgrass habitat and least tern (Sterna albifrons browni) foraging habitat. In 1987, as mitigation for these losses, a 1500-ft^2 bed of eelgrass was transplanted (experimentally) in several small plots to a site prepared by contouring sediments to resemble a natural bed. Based on the results of this effort, the transplant area was readjusted to accommodate two findings: 1) that part of the area was light-limited due to the presence of a surge wall and 2) that chemical constituents of the sediments in another portion were unacceptably high.

More than 13,000 ft^2 (1236 m^2) of eelgrass was then transplanted successfully from existing beds in San Diego Harbor. During the three-month post-transplant survey a heavy diatom mat had settled around the eelgrass transplants, shadowing the eelgrass in some areas, and causing losses estimated at 10%. However, by six months after the transplant, the diatom mat was gone and transplants were developing well with some growth outside of planting boundaries.

Although the bed was extremely lush in most areas, approximately 10% of the transplant was unsuccessful and remedial transplants using eelgrass from dense portions of the "new" bed was undertaken. One year after the transplant, the eelgrass bed had grown from a 10% loss at six months to a bed 12% larger than when planted. This transplant and all subsequent transplants were made using the modified sediment free turion anchor method very successfully.

Dredging operations in Sunset-Huntington Harbour in 1988 made it necessary for the County of Orange to transplant 13,000 ft^2 (1208 m^2) of eelgrass. Dredge spoils from the harbor with high proportions of shell and in some areas clay, were placed adjacent to an existing bed and left to settle for six months prior to the transplant. Eelgrass developed well, but losses were high near the crest of the spoils due to poor sediment conditions and erosion. During the six-month survey a few intact and living turion bundles were found growing in the water column, where they were suspended above the sediment by the cotton string and popsicle anchor system. One year after the survey, additional sediment had been deposited and the bed had expanded beyond the original boundaries. Boat traffic was high and propellers paths through the eelgrass were common. Losses represented by barren areas totaled approximately 10%, but the area appeared to be relatively stable.

In late 1989, MBC contracted with the U.S. Navy to replace 6.5 acres (2.6 hectares) of eelgrass which had been displaced by a variety of construction and dredging projects. Eelgrass was to be transplanted to sediments that had previously been placed in deeper water offshore of an existing eelgrass bed in south San Diego Bay. This is probably the largest eelgrass transplant ever attempted in southern California, and the logistics of transplanting to this large an area were significant. It took four divers several days just to define the transplant area, grid it into one-acre (0.4 hectare) plots, and subdivide each acre into 16 approximately 52 ft (16 m) squares. Fortunately the site is in a relatively calm part of the bay which was free of boat traffic; as potential donor beds were nearby, it was also an ideal location for a transplant. The transplant appears to have been very successful; more than 98% of the turion bundles survived for at least three months.

Conclusions

Eelgrass transplants are a viable option for offsite mitigation of bay and estuarine development, and for in-kind mitigation of eelgrass to degraded estuarine areas. The techniques developed by a variety of researchers have been shown to be amenable to different locales, different current regimes, and different planting densities. The major concern often logistical and therefore cost, which depend on many considerations especially size and location of the site. However, the cost per area decreases rapidly as the amount of area to be transplanted increases. Most transplants should be successful if previous lessons are learned and applied.

References

Adams. S. M. 1976. The ecology of eelgrass, Zostera marina (L.), fish communities. II. Functional analysis. J. exp. mar. Biol. Ecol. 22:293-311.

Boone, C. G., and R. E. Hoeppel. 1976. Feasibility of transplantation, revegetation, and restoration of eelgrass in San Diego Bay, California. Prepared for U.S. Army Eng. Dist., Los Angeles, CA. Cont. DACW09-75-B-0026. U.S. Army Eng. Waterways Exper. Stn., Vicksburg, MS 39180. Misc. Pap. Y-76-2. 42 p. plus appendices.

Fonseca, M. S., J. S. Fisher, J. C. Zieman, and G. W. Thayer. 1982. Influence of the seagrass, Zostera marina L., on current flow. Estuar. Coast. Shelf Sci. 15:351-364.

Fonseca, M.S., W. J. Kenworthy, J. Homziak, and G. W. Thayer. 1979. Transplanting of eelgrass and shoalgrass as a potential means of economically mitigating a recent loss of habitat. Pages 279-325 in D. P. Cole (ed.), Proc. 6th Ann. Conf. on Wetlands restoration and creation. May 1979, Hillsborough Comm. College. Environ. Stud. Ctr., and Tampa Port Authority, Tampa, FL. 357 p.

Fonseca, M.S., W. J. Kenworthy, and R. C. Phillips. 1982. A cost-evaluation technique for restoration of seagrass and other plant communities. Environ. Conservation 9(3):237-241.

Fonseca, M.S., W. J. Kenworthy, and G. W. Thayer. 1982. A low-cost planting technique for eelgrass (Zostera marina L.). U.S. Army, Corps of Eng., Coastal Eng. Res. Ctr., Ft. Belvoir, VA 22060. Coastal Eng. Tech. Aid 82-6. 15 p.

Goforth, H. W., and T. J. Peeling. 1975. Eelgrass (Zostera marina L.) beds along the western shore of North Island Naval Air Station, California. A study of the impact of pier construction and possible compensating actions. Mar. Environ. Mgmt. Off., Chem. Environ. Sci. Div., Naval Undersea Ctr., San Diego, CA. 21 p.

Kenworthy, W. J., M. S. Fonseca, J. Homziak, and G. W. Thayer. 1980. Development of a transplanted seagrass (Zostera marina L.) meadow in Back Sound, Carteret

County, North Carolina. Pages 175-193 in Proc. 7th Ann. Conf. Restoration and Creation of Wetlands. Contrib. 80-55B.

Kikuchi, T. 1961. An ecological study on animal community of Zostera belt, in Tomioka Bay, Amakusa, Kyushu. Records of Oceanographic Works in Japan. Spec. No. 5:211-219.

Kita, T., and E. Harada. 1962. Studies on the epiphytic communities. 1. Abundance and distribution of microalgae and small animals on the Zostera blades. Publ. Seto Mar. Biol. Lab., 10(2):245-257.

MBC Applied Environmental Sciences. 1985. Eelgrass transplant project, Phase I - Second reconnaissance study. September 1985. Prepared for Mola Development Corporation, Huntington Beach, CA. 5 p.

MBC Applied Environmental Sciences. 1986. Infauna and epibiota associated with transplants of eelgrass (Zostera marina) in southern California, March 1986. Prepared for Maguire Thomas Partners, Huntington Partnership, National Marine Fisheries Service and U. S. Fish and Wildlife Service. 48 pp.

MBC Applied Environmental Sciences. 1986. Eelgrass (Zostera marina) habitat mapping and reconnaissance survey, Sunset/Anaheim Bays and Huntington Harbor, Orange County, CA, May 1986. Prepared for Noble Coastal and Harbor Engineering, Ltd., Irvine, CA. 10 pp.

MBC Applied Environmental Sciences. 1987. Eelgrass (Zostera marina) habitat mapping survey, City of Coronado, San Diego Bay, California. Prepared for Engineering Science, San Diego, CA 92121. 14 p. plus appendices.

MBC Applied Environmental Sciences. 1987. Mola development corporation eelgrass transplant mitigation project: One year monitoring survey and evaluation of transplant success. March 1987. Prepared for Mola Development Corporation, Huntington Beach, CA 92646. 13 p.

MBC Applied Environmental Sciences. 1988. Eelgrass bed reconnaissance surveys, Mariner's Point fuel dock, Sunset Bay, Orange County. Letter report prepared for Shagren and Associates, 6 February 1988.

MBC Applied Environmental Sciences. 1988. Pre-dredge eelgrass survey project, Sunset Bay, Orange County.

488 COASTAL WETLANDS

February 1988. Prepared for Western Pacific Dredging Co., Division of Riedel Int'l., Portland, OR. 9 p. plus appendices.

MBC Applied Environmental Sciences. 1988. Sunroad Marina eelgrass transplant mitigation project: Habitat suitability study results and progress report for experimental eelgrass transplant. December 1988. Prepared for Sunroad Enterprises, San Diego, CA. 19 p.

MBC Applied Environmental Sciences. 1989. (A) Initial eelgrass bed habitat survey, 26 October 1989 (10 Nov 1989). (B) Four-month eelgrass bed survey, 22 February 1990 (27 Apr 1990). (C) Eight- month eelgrass bed survey, 27 June 1990 (3 Jul 1990). Prepared for Portofino Cove Yacht Association, Anaheim, CA.

MBC Applied Environmental Sciences. 1990. Sunroad eelgrass one-year post-transplant survey. February 1990. Prepared for Sunroad Enterprises, San Diego, CA. 3 p.

MBC Applied Environmental Sciences. 1990. Sunset-Huntington Harbour eelgrass transplant one-year survey final report. April 1990 (Rev. May 1990). Prepared for County of Orange, Environmental Management Agency. 4 p.

MBC Applied Environmental Sciences. 1990. Sunset-Huntington Harbour eelgrass transplant one-year survey final report. April 1990 (Rev. May 1990). Prepared for County of Orange, Environmental Management Agency. 4 p.

MBC Applied Environmental Sciences. 1990. Naval Amphibious Base eelgrass transplant, Coronado Island, California: Final report. Prepared for Naval Facilities Engineering Command, NAB, San Diego, CA., Contract N68711-89-C-5008. 10 p. plus appendix.

Nagle, J. S. 1968. Distribution of the epibiota of macroepibenthic plants. Cont. Mar. Sci. Univ. Texas 13:105-144.

Phillips, R. C. 1978. Seagrasses and the coastal marine environment. Oceanus 21(3):30-40.

Schubel, J. R. 1973. Some comments on seagrasses and sedimentary processes. International Seagrass Workshop, Leiden, Netherlands.

Stauffer, R. C. 1937. Changes in the invertebrate community of a lagoon after disappearance of the eel grass. Ecology 18(3):472-431.

Thayer, G. W., and R. C. Phillips. Importance of eelgrass beds in Puget Sound. Mar. Fish. Rev. 39(11):18-22.

Thorhaug, A., B. Miller, B. Jupp, and F. Booker. 1985. Effects of a variety of impacts on seagrass restoration in Jamaica. Mar. Pollut. Bull. 16(9):355-360.

Virnstein, R. W., P. S. Mikkelsen, K. D. Cairns, and M. A. Capone. 1983. Seagrass beds versus sand bottoms: The trophic importance of their associated benthic invertebrates. Florida Sci. 46(3/4):363-381.

Sea Management Patterns. Taxonomical Frameworks
Adalberto Vallega [1]

Abstract
Bearing in mind the progress in setting up
comprehensive views of coastal and ocean management,
taxonomical criteria will be clustered. The reasoning
will be developed considering:

i. the methodologies through which uses, relations
    between uses, and relations between uses and
    environment as well, are taxonomized;
ii. the objectives of management.

On this basis, the "differentiation-coordination
referent" is suggested in order to evaluate the
management patterns as a whole.

The sea structure

The sea structure can be imagined as consisting of two
modules: (i) sea uses, (ii) the natural context
constituted by the physical elements in which sea uses
interact. As a consequence, the knowledge of sea
structures is focused on four sets: (i) the set of
uses, (ii) the set of relationships between uses, (iii)
the set of physical and biological elements giving
shape to the natural context, (iv) the set of
relationships between sea uses, on the one hand, and
the natural context, on the other. In order to deal
with the first stage, sea uses are to be clustered. As
is well known, literature has usually built two
level-based classifications consisting of categories,
in their turn subdivided into kinds of uses. Anyway, in
order to provide more specific tools for sea management
a sea use framework can be set up grounding it on three

----
[1] Adalberto Vallega, Istituto di Scienze Geografiche,
Università di Genova, Lungoparco Gropallo, 16122
Genova, Italia.

levels:
i.   <u>upper level</u>, categories of uses;
ii. <u>intermediate level</u>, sub-categories of uses;
iii. <u>lower level</u>, kinds of uses.

In its total extent this framework includes
- 18 categories of uses;
- 67 sub-categories of uses;
- 263 kinds of uses.

The categories are
1. seaports
2. shipping, carriers
3. shipping, routes
4. shipping, navigation aids
5. sea pipelines
6. cables
7. air transportation
8. biological resources
9. hydrocarbons
10. metalliferous resources
11. renewable energy
12. defence
13. recreation
15. waste disposal
16. research
17. archeology
18. environmental protection and preservation.

Here there is no room to show the whole framework. As
an example, it can be taken into account that the
category   <u>5. sea pipelines</u> can be divided into 5
sub-categories:
5.1 slurry pipelines
5.2 liquid bulk pipelines
5.3 gas pipelines
5.4 water pipelines
5.5 waste pipelines.

Again as an example the sub-category <u>5.1 slurry</u>
<u>pipelines</u> can be divided into 6 kinds of uses:
5.1.1 fine coal slurry pipelines
5.1.2 coarse coal slurry pipelines
5.1.3 limestone slurry pipelines
5.1.4 phosphate slurry pipelines
5.2.5 ore slurry pipelines
5.2.6 copper slurry pipelines

Two methodological remarks are appropriate.

First, this framework, as well as any other framework
sketched by the same method, has only a tentative role,
i.e. it is a reference point to be borne in mind in

order to cluster the sea uses of a specific marine
area. Between the build-up of the sea use general
framework and its application in the investigation and
management of specific marine areas, a methodological
feed-back relation exists: starting from the general
framework, a framework relating to specific coastal or
ocean spaces can be sketched. In its turn, the areal
framework contributes to the adjustment and
implementation of the general framework, and so on
until: a circular methodological path comes to light.

Secondly, the sea use framework is necessarily referred
to a specific temporal phase. When it is concerned with
the short or medium term, the expected evolution of the
sea structure is to be investigated through
prospect-oriented analysis constructing scenarios on
man-sea interaction.

Thirdly, this framework relates to sea management as a
whole. In order to face coastal and ocean management
distinctly, specific frameworks could be formulated.
The coastal use framework includes all categories of
the general framework but only 64 sub-categories and
250 kind of uses. The ocean use framework embraces only
13 categories of uses:
2. shipping, carriers
3. shipping, routes
4. shipping navigation aids
5. pipelines
6. cables
8. biological resources
9. hydrocarbons
10. metalliferous resources
11. renewable energy sources
12. defence
13. recreation
15. waste disposal
16 research

that are divided into 28 sub-categories, which, in
their turn, embrace 198 kinds of uses.

The coastal use framework

Moving from the sea use framework to the framework
specifically relating to coastal management, two issues
are worth considering: (i) the location of coastal uses
and (ii) the natural context they involve.

As far as the former variable, i.e. the location of
coastal uses, is concerned, three kinds of extent are
relevant for coastal management: geomorphological,
environmental and legal.

i.   Geomorphological   extent.   Land,   shoreline,
      continental shelf,   slope and rise can be regarded
      as its main components.   Coastal uses can be
      clustered according to the geomorphological belts
      in which they are located,   or anyway the belts in
      which they are involved.   As an example of the
      latter case,   the pipelines for waste disposal can
      be recalled: they can bring about environmental
      implications in sites,   such as the slope or rise,
      which extend well beyond the place where they are
      located.

ii.   The extent of the biological context.   In this
      context the ecosystem is the reference point for
      taxonomic procedures.   Coastal uses are clustered
      according to whether they involve land,   the
      land-sea interface, sea surface, the water column,
      seabed and subsoil, since each of these components
      could host a specific ecosystem.

iii. Legal extent.   In this view internal waters,   the
      territorial sea,   the continental shelf,   the
      Exclusive Economic Zone and Exclusive Fishery Zone
      are considered as the variables according to which
      coastal areas are taxonomized.

As far as the latter variable,   i.e. the natural
context,   is   concerned,   it   is   methodologically
appropriate and practically relevant to take into
account whether the coastal use involves the sea and/or
the land.   In this context four sets of uses can be
distinguished:

i.   uses   located   onshore   and   not   generating
      implications   on   the   coastal   sea,   such   as
      industrial settlements not related to maritime
      transportation and not producing riverine or
      marine discharges;

ii.   uses located onshore and generating   implications
      on the coastal sea, such as seaports, recreational
      uses, sand and gravel dredging;

iii. uses   located   offshore   and   not   generating
      implications on the shoreline and/or land, such as
      navigation aids and research installations;

iv.   uses located offshore and generating implications
      on land, such as offshore oil unloading platforms;

The sea use-use relationship model

According to current methodological approaches, models
of sea use frameworks lead to the setting up of a
square matrix (Xi, Xj) in which the relationships
between uses are represented. This matrix can be based
on various approach levels, i.e. it can be built
according to the categories, or sub-categories, or
kinds of uses. As a result, it allows us to identify
more or less large ranges of sea uses. The more sea
uses are clustered in a specified manner the deeper the
analysis of their reciprocal relationships can be. The
sea use framework, which has just been suggested,
enables us to set up a quite detailed range of
relationships between sea uses through a large matrix
that can be called sea use-use relationships model.
There is no room to display the complete model and,
inter alia, it does not seem necessary. As an example,
the part of the model showing the relationships between
"defence" (category no. 12 of the sea use framework)
and "research" (category no. 16) is represented in
Figure 1--it includes only the sub-categories
(intermediate level of analysis).

It is no use stressing that this sea use-use
relationship model is a characteristic product of a
structuralist approach aiming at identifying sea
structures and investigating the relationships between
the elements of the structure. The objective is to give
methodological tools to sea management and, by its
nature, it undergoes two constraints: (i) in itself the
model is not a tool for diachronic analysis so, when
the evolution of sea use framework is to be
investigated, additional or alternative methodologies
are to be applied; (ii) only bilateral relationships
use-to-use can be represented, so complicated networks
of relationships, which simultaneously

RESEARCH FIELDS X$_j$ / DEFENCE X$_i$	water mass	seabed and subsoil	ecosystem	external environment interaction	special areas and particularly sensitive areas	sea management
exercise areas	B2	A	B2	A	B2	A
nuclear test areas	B2	B2	B2	B2	B2	A
minefields	B1	B2	B1	A	B2	A
explosive weapons	B2	B2	B2	A	B2	A

Figure 1. Relationships between defence and research

A: no existence of relationships
B: existence of relationships
B1: neutral relationships
B2: conflicting relationships
B3: reciprocally beneficial relationships
B4: relationships beneficial to use x$_i$
B5: relationships beneficial to use x$_j$

involve a set of uses cannot be investigated.

The use of the model implies that use-use relationships
are taxonomized. To this end the sea structure can be
described according to this frame:

1. Non existence of use-use relationships

2. Existence of use-use relationships
   2.1 neutral relationships
   2.2 conflicting relationships
   2.3 beneficial relationships.

It is self-evident that, as far as sea management is
concerned, conflicting relationships have the highest
relevance. Three sets of them are worth considering.

i.   Sea resource. In this case the conflict comes to
     light because two or more sea uses demand for the
     same kind of resources: e.g., conflicts between
     telephone cables and pipelines derive from the
     circumstance that both are placed on the seabed,
     which acts as a resource;

ii. Marine environment. In this case conflicting
     relationships are due to the alteration that a
     kind of use produces in the marine environment
     preventing another kinds of uses from developing:
     e.g. industrial effluents can alter the properties
     of the water column damaging fisheries and
     bringing about their collapse.

iii. Marine landscape. Conflicts involving the marine
     landscape depend on the aesthetic values that
     people want to maintain or create for developing a
     given use: e.g., the loading and unloading of
     minerals through offshore terminals are regarded
     as not appropriate to the development of
     recreational facilities.

Coastal use-use relationship model

Moving from the use-use relationships lato sensu to the
relationships existing between coastal uses,
conflicting relationships are worth of methodological
consideration. In this respect potential and actual
incompatibilities between uses occur when:

i.   two or more uses are located in the same place
     disturbing each other: in this case one can speak
     of locational incompatibility; for example,
     conflicts between mercantile navigation and naval
     exercise areas;

ii. to some extent two or more uses imply conflicting management patterns, bringing about <u>organizational incompatibility</u>; for example, conflicts between maritime transportation in oil and gas offshore exploitation platforms (crude oil and product carriers, supply vessels, etc.) on the one hand, and yacht racing and cruising on the other;

iii. a use brings forth environmental impacts that another use cannot tolerate so <u>environmental incompatibility</u> takes place; for example, the warm water discharges from coastal thermoelectric plants alter the coastal ecosystem preventing the establishment of marine parks;

iv. one use gives shape to a picture that the other use cannot tolerate (<u>aesthetic incompatibility</u>); e.g. this incompatibility occurs between coastal manufacturing plants and recreational facilities.

At this point it is worth considering the evolution of the whole entirety of coastal uses. In order to facilitate reasoning on this issue the <u>differentiation-coordination referent</u> can be taken into account. As can be seen from Figure 2, the referent consists of a three-axis diagram: differentiation, co-ordination, time.

The <u>differentiation axis</u> (Y) represents the level of complexity that the set of uses acquires because of the growth of the exploitation of sea resources and environment. It is self-evident that differentiation depends on three variables: (i) the number of kinds of coastal uses; (ii) the number of uses pertaining to each kind; (iii) the technological level on which uses are managed. Anyway, to make the model simple the differentiation axis represents only the number of kinds of coastal uses. The values increase from the origin of the diagram.

The <u>co-ordination axis</u> (X) shows the degree to which the coastal uses are involved by relationships bringing about cohesive energy. This implies that conflicting relationships are regarded as reducing the co-ordination level and neutral and beneficial relations as acting in the opposite direction. As a first approach, co-ordination could be expressed by the relation

$$(n + b): c$$

where $\underline{n}$ is the number of neutral relationships; $\underline{b}$ is the number of beneficial and $\underline{c}$ the number of

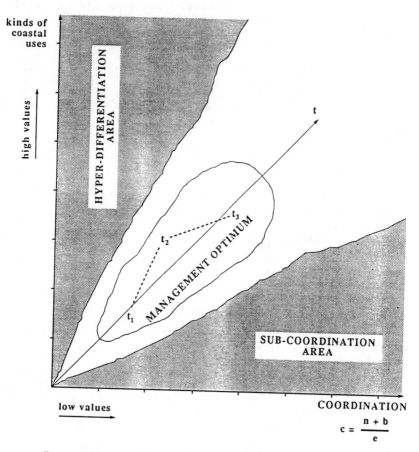

Figure 2. Differentiation-coordination referent

conflicting ones. The value of this ratio decreases moving from the origin of the diagram.

The time axis represents the evolution of differentiation and co-ordination.

At a given time, which is characterized by a given level of technological advance, organizational patterns, etc.:

i. the number of kinds of coastal uses cannot exceed a given threshold because beyond that the differentiation of the set of uses becomes too high to be managed;

ii. the ratio (n+b):c cannot slow down below a given threshold because below that the co-ordination between coastal uses becomes too weak to enable us to manage the set of uses.

The set of values of differentiation ranged above the differentiation threshold forms the hyper-differentiation area. The set of values of the ratio (n+b):c below the co-ordination threshold forms the sub-coordination area. Between these two critical areas the differentiation-coordination optimum is located. It is a management space within which the number of kinds of coastal uses and the relationships between them--and factors influencing them as well--change without transferring the set of coastal uses into critical points.

The sea use-environment relationship model

The environment can be thought of as consisting of (i) the ecosystem, (ii) the natural context. The latter can be assumed as the reference basis for evaluating if and where environmental impacts take place. Bearing in mind the physical and biological complexity of the marine environment and distinguishing the coastal area from the ocean one this scheme comes to life.

marine environment	
coastal area	ocean area

emerged land surface

periodically emerged land surface

sea surface	sea surface
water column:	water column:
upper layers	upper layers
intermediate layers	interm. layers
lower layers	lower layers
seabed	seabed
subsoil	subsoil

As a consequence, each kind of use can be evaluated according to this pattern in order to decide whether it is related to ocean and/or the coastal ecosystem and which components of the natural context are involved.

Relationships between uses and the natural context acquire unusual levels of complexity in coastal management. In this respect a matrix involving the mentioned components of the coastal natural context (lines) and coastal uses (columns) can be imagined. In this relationships are represented according to their directions, i.e.:

i.   inputs from the natural context to coastal uses;
ii. inputs from coastal uses to the natural context;
iii. inputs moving in both directions.

Inputs moving from coastal uses to the natural context are the most important for management. According to the implications they are able to generate, the following taxonomic framework can be taken into account:

- U_b, inputs beneficial to the coastal environment;

- U_n, inputs neutral for the coastal natural context, because they do not cause damage nor benefits to the coastal ecosystem and/or the geomorphological assessment;

- U_r, inputs harmful to the coastal natural context calling for protection and preservation;

- U_z, inputs hazardous to the coastal natural context calling for contingency plans or similar tools.

Transferring reasoning to the diachronic dimension the development of relationships is evaluated in order to reveal whether it leads only to simple adjustments of morphogeneses of the coastal natural context, particularly to the coastal ecosystem. Secondly, the evaluation of sea uses and the ecosystems is considered according to the interactions in which they are involved. In this respect coastal management can be regarded as consisting of two areas.

i. The adjustment area. In this area the set of neutral and beneficial coastal uses brings about inputs larger than those generated by the set of harmful and hazardous coastal uses. The latter categories of uses not only are less extended but are kept under control also thanks to the role played by beneficial uses. The amount of global biological and chemical oxygen demand--which can be regarded as a parameter of the pollution of the coastal marine area--grows so slowly that it does not alter the ecosystem. As a result, the ecosystem only undergoes adjustments which, in the absence of extra-uses factors, are not expected to produce risks.

ii. The collapse area. The set of harmful and hazardous coastal uses brings about inputs larger than those generated by the set of neutral and beneficial uses. In addition, the latter is not able to reduce the effects produced by harmful and hazardous uses As a consequence, the global demand for biological and chemical oxygen grows to the point of generating a collapse in the ecosystem. In this case the morphogenesis of coastal management takes place.

Deductions

Taxonomizing coastal and ocean uses, as well as their relationships with the environment, is the normal function of sea management. Making efforts in this direction research acquires more and more the consciousness that this concern is the product of a deeper one: the need to define which are the coastal and ocean structures and to describe their evolution. In this context it is self-evident that: (i) when attention is focused on sea uses and their relationships with the environment, methodologies are structuralistic; (ii) when the combined evolution of uses and the environment is the focus of knowledge, methodologies are grounded on the general system theory and the subsequent theory of complexity; (iii) between the two levels there is remarkable theoretical gap.

Acting on the system-based level implies that:

i.　　contrary to current approaches, the ecosystem is
　　　regarded as a not trivial machine, i.e. as a
　　　machine capable of giving a range of responses to
　　　the same input;

ii.　　the self-regulation and self-poietic properties of
　　　the ecosystem are considered to their whole
　　　extent;

iii.　sea uses and the ecosystem are regarded as a
　　　bi-modular system;

iv.　adjustments and morphogeneses are the two main
　　　reference categories in order to understand the
　　　evolution of the ecosystem, particularly when its
　　　implications in sea uses are the focus of
　　　analysis;

v.　　as a consequence, management patterns are grounded
　　　in a not deterministic view.

This concern has taken shape in scientific thought and
research on man-environmental relationships per se but
it acquires particular relevance when relationships
between human pressure and activities, on the one hand,
and the marine environment, on the other, are
concerned. The prospect of achieving results depends on
the interactions that can be set up between scientific
thought and management principles and methodologies.

COASTAL WETLANDS MANAGEMENT IN VENEZUELA

Mirady Sebastiani[1], Alicia Villamizar[2] and María de Lourdes Olivo[3]

Abstract.
Venezuela has different types of coastal wetlands: tidal salt marshes, tidal freshwater marshes and mangrove wetlands. Among the wetlands, overwash mangrove islands, fringe mangrove and salt tidal marshes are explicitly protected by law and land use is restricted. For the other types of wetlands, the available tools for land use planning are weak. In this context, the purpose of this paper is to:
- point out different land use planning strategies used in Venezuela, particularly, to protect coastal wetlands,
- present common situations that a working team can face in the process of given a permission to a project on a coastal wetland and
- suggest ideas to improve the planning process on these ecotones in relation to the legal aspects and the existing information.

Introduction

Venezuela is a Latin American country located north of South America. It has approximately 3726 km of coast line, mostly on the Caribbean Sea (Figure 1). Between the coast line and inland, there are different types of wetlands. According to Mistch and Gosselink (1987), these wetlands can be of different types: tidal salt marshes; tidal freshwater marshes and mangrove wetlands.

The purpose of this paper is to:
- point out different land use planning strategies used in Venezuela, particularly, to protect coastal wetlands and
-suggest ideas to improve the planning process on these ecotones.

Venezuela efforts on coastal wetland management.

Sorensen and Brandani (1987) point out that there have been coastal management efforts in Latin America based on different strategies:
• Sectorial planning to manage various coastal resources or activities.
• Designation of protected areas.
• Requirement for Environmental Impact Statement, for some projects.
• Shoreline restriction.
• Special area or regional plans.
• Compilation of a coastal atlas or a data bank.
• Nationwide or statewide land-use planning and regulation.

In Venezuela, all strategies are being used as tools to control intervention on coastal area, as a single or as multiple criteria. The data bank strategy is the most recent of all, it has been applied since 1989 specially to the Island of Margarita (Figure 1). These strategies are required, directly or indirectly, by law. Table I shows the relevant regulations existing in Venezuela to enforce these strategies, particularly, on coastal wetlands.

---

1 Geographer M.A, M.Sc.; Head of, 2 and 3 Biologists, Graduate Students. Instituto de Recursos Naturales Renovables. Universidad Simón Bolívar. Apartado Postal 89000. Caracas-Venezuela.

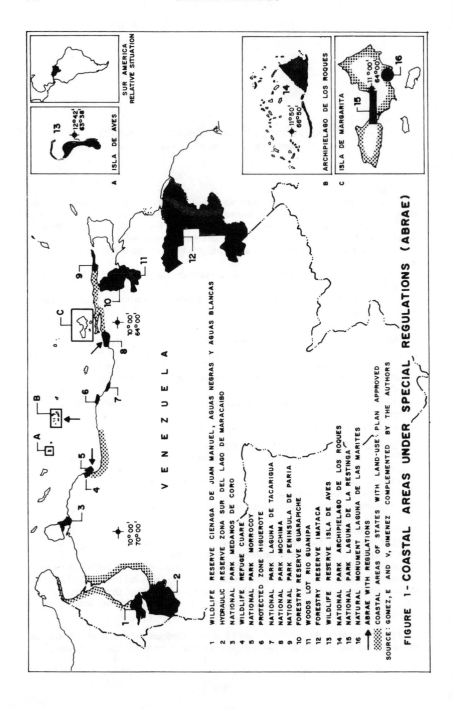

1   WILDLIFE RESERVE CIENAGA DE JUAN MANUEL, AGUAS NEGRAS Y AGUAS BLANCAS
2   HYDRAULIC RESERVE ZONA SUR DEL LAGO DE MARACAIBO
3   NATIONAL PARK MEDANOS DE CORO
4   WILDLIFE REFUGE CUARE
5   NATIONAL PARK MORROCOY
6   PROTECTED ZONE HIGUEROTE
7   NATIONAL PARK LAGUNA DE TACARIGUA
8   NATIONAL PARK MOCHIMA
9   NATIONAL PARK PENINSULA DE PARIA
10 FORESTRY RESERVE GUARAPCHE
11 WOODS LOT RIO GUANIPA
12 FORESTRY RESERVE IMATACA
13 WILDLIFE RESERVE ISLA DE AVES
14 NATIONAL PARK ARCHIPIELAGO DE LOS ROQUES
15 NATIONAL PARK LAGUNA DE LA RESTINGA
16 NATURAL MONUMENT LAGUNA DE LAS MARITES

▓▓▓ ABRAE WITH REGULATIONS

➤   COASTAL AREAS OF STATES WITH LAND-USE PLAN APPROVED

SOURCE: GOMEZ, E AND V, GIMENEZ COMPLEMENTED BY THE AUTHORS

FIGURE 1 - COASTAL AREAS UNDER SPECIAL REGULATIONS (ABRAE)

Legislation	Year	Key provisions
Ley Forestal de Suelos y Aguas (Law of Forestry, Soils and Waters)	1966	Creation of national parks, natural monuments, protected zones regional virgen reserves and reserves for forestry. It establishes a protection area of 50m form each side of a navigable river and 25m for non navigable river.
Decreto 110 (Decree 110)	1974	Prohibition to cut mangroves, dredged or fill mangrove areas and discharge of sewage waters on mangroves or in their vecinity.
Ley Orgánica del Ambiente (Environmental Act)	1976	Land use planning.Creation, protection, conservation of national parks, reserves for forestry, natural monuments, protected zones, natural wildlife refugies and sanctuaries.
		Activities that could cause environmental damage only can be authorized if there are guarantees, procedures and norms to correct or diminish the damage.
Ley Orgánica de la Administración Central (Central Administration Act)	1976	Creation of the Ministry of Environment and Natural Resources (Ministerio del Ambiente y de los Recursos Naturales Renovables). It is encharged of developing and ejecuting programs of conservation,defense and regulation of land-use in rural areas and of National Parks.

Table 1. Relevant Venezuelan regulations related to planning on coastal wetlans

Legislation	Year	Key provisions
Reforma Parcial del Reglamento de la Ley Forestal de Suelos y Aguas. (Parcial Reform of the Regulation of the Forestry Law of Soils and Waters)	1977	Protection area of minimun 50m from margins of lakes or natural lagoons to inland.
Ley Orgánica de Ordenación del Territorio (Organic Law of Territorial Ordening)	1983	There are areas that must have special regulation named Areas Bajo Régimen de Administración Especial (ABRAE). These are: National Parks, Protected Areas, Reserves for Forestry, Areas for Security and Defense, Reserves and Sanctuary for Wild Life, Natural Monuments, Turistic Zones, Areas under Internacional Treaty. There are also considerer ABRAE for explotation and intensive land use coastal wetlands: salt and fresh water marshes. ABRAES must be establish by a Presidencial Decree, afterward a a land use plan should be developed and a regulation have to support the plan in order for it to be applied.
Decreto 623 (Decree 623)	1989	A 80km wide " Protected Coastal Zone" is established for which a land use plan must be developed by the Ministry of Environment and Natural Reneawable Resources.

Table 1. Relevant Venezuelan regulations related to planning on coastal wetlans

Figure 1 shows the sections of the coast that have been declared ABRAE (area under special regulation). The scale of the map does not allow to show the protected zone of 80m wide from the shore to inland.

The working team in charge of land use permission procedures on coastal areas can face basically three situations in relation to the proposed project. The project can be on:

• Case 1: a coastal zone that is an ABRAE (area with special regulations) and it has already the plan required and the necessary regulation to make effective the land use restrictions.

• Case 2: an ABRAE that have not yet the plan and regulation to sustain restrictions.

• Case 3: an area with no specific restrictions yet defined.

These cases are going to be explored in relation to wetlands.

Before these cases are considered, it is important to know the following information:

The Land Use Act (Ley Orgánica para la Ordenación del Territorio, 1983) establishes that to have an area under special regulations its limits must be approved by a Presidential Decree. Afterwards, a land use plan should be proposed and must be made into a regulation. Without this conversion, the proposed restrictions in the plan are not valid.

If an area can be classified simultaneously under several types of ABRAE. The land use guidelines applied in such situation would be the most restrictive ones.

Case 1: A wetland protected by an ABRAE that has a plan made into a regulation.

In this case there are specific spatial boundaries and guidelines for land use to support the decision of the working team about the project. Figure 1 shows ABRAE that are under this situation. Wetlands in these areas are fully protected because they are national parks and their purpose is precisely to preserve the environment as natural as possible. Therefore, the land use restrictions are very rigorous (Table 2).

Case 2: A wetland protected by an ABRAE that has not yet the required plan.

• If the project is in an area delimited as a National Park and or a Natural Monument, there is Decree 276 (1989) that fully establishes the restrictions of land use and other activities in order to preserve the environment and hence wetlands. This Decree is complementary of the Land Use Planning Act.

• If the area selected is on a different ABRAE from the ones mentioned above, the working team has to look for supporting material to establish restrictions based mainly on: the existing regulations (Table 1) and the general guidelines for ABRAE issued by MARNR (1985). It is important to mention that this document is a guide not a regulation. Table 2 is a synthesis of the key provisions of the document.

One could think that if the area selected is on a mangrove wetland, the situation is easily solved because there is Decree 110 (Table 1). But this is partially true. The Decree refers to mangrove wetlands in general, nevertheless it is related to the Law for Fishing, section 25. This particular section prohibits the destruction of aquatic vegetation and natural formations where fish can hatch . This last remark has allowed legal interpretations of Decree 110 resulting in the protection of just an specific mangrove species: red mangrove (*Rhizophora mangle*) (MARNR, 1979). If the mangrove wetland in consideration is only formed by red mangrove, or this species is predominant, there is no problem. Thus, Decree 110 allows the protection, without question, of overwash mangrove islands and fringe mangrove wetlands because usually red mangrove predominates in these types of mangrove wetlands.

In the case of a basin mangrove wetland, which are in depressions behind fringe mangrove wetlands and in drainage depressions where water is stagnant and black mangrove (*Avicennia germinans*) dominates, the protection is poor. This is because of the fact that fish do not hatch in these type of mangrove wetland and in lands occupied by mangrove species as *Laguncularia racemosa* y *Conocarpus erectus*. The same statement can be used for inland dwarf mangrove wetlands.

In relation to a riparian mangrove wetland, the Decree can be complemented with the law of forestry, soils and waters (Ley Forestal de Suelos y Aguas, Table 1) and protect the vegetation up to 50m from both margins of a navigable river and up to 25m if the river is not navigable.

Forestry reserve:	Forestry sustainable exploitation,hydroelectricity, reforestation and plantations, use of water resources, exploitation of plants, animals and wild life forcommercial purposes, passive recreation,industrial activities with restrictions,and restrictive sport hunting.
Woods lot:	Agriculture, reforestation and plantation, use of water resources, exploitation of plant and animals and wild life for commercial purposes, passive recreation, industrial use without restrictions, forestry, human occupation with restrictions.
National Park:	Educational, cultural and recreational activities. Use of water resources with severe restrictions.
Natural Monument:	Use of water resources with severe limitations and recreational activities.
Protected zone:	Forestry sustainable exploitation, hydroelectricity, reforestation and plantations, use of water resources, exploitation of plants and animals and wild life for commercial purposes, passive recreation, industrial activities with restrictions, reforestation for agriculture and urban activity with severe restrictions, human occupation and restrictive sport hunting.
Hydraulic reserves:	Forestry sustainable exploitation, hydroelectricity,reforestation and plantations, mining with severe restrictions, exploitation of plants, animals and wild life for commercial purposes, passive recreation, industrial activities with restrictions, industrial use with restrictions, wood cutting for agriculture and urban activities with severe restrictions, limited human occupation and limited sport hunting.
Wild life reserve:	Extensive passive recreation and wild life utilization.
Wild life refuge:	Research and extensive passive recreation.
Zone for agriculture:	Agricultural activities.

Table 2. Guidelines for compatibility of land use and activities in coastal areas under special regulations (ABRAE) (Gomez and Giménez).

Up to now, the team has been able to isolate areas that can be limited from intervention based only on the protection of trees, with a mangle species centric approach due to the interpretation of the law. For the team the possibility of isolating mangrove areas is not sufficient, because the external forces that maintain this system has not yet been protected. To guard a mangrove wetland as a whole is necessary to maintain, at least: runoff from adjacent areas, quantity and quality of it; sedimentation rates and pattern and water circulation.

In the Decree 110, water quality in mangrove wetlands is considered when the prohibition to discharge sewage waters into mangroves or in their vicinity is established. But, it does not require the maintenance of the other factor mentioned above.

Runoff characteristics are partially protected by the Law of Forestry, Soils and Waters when it establish a buffer area for navigable and non navigable waters (Table 1). In this Law there is not a provision asking for maintenance of the quantity of the runoff.

In relation to sedimentation rates in mangrove wetland, the Regulation for Waters ("Reglamento Parcial No 4 de la Ley Orgánica del Ambiente sobre Clasificación de Aguas",1978) point out that: in waters to be used for recreational and fishing purposes, dissolved solids can not be changed in value of 33% from the natural concentration. This statement partially helps planning in mangrove areas because natural turbidity is already high. A change in value of 33% may cause environmental stress in this system. As one can see, the maintenance of the vital relationship between mangrove wetlands and other ecosystems is not clearly stated in the present regulation and is difficult for the team to state the condition on adjacent areas to protect the mangrove wetland.

The team could face another problem in relation to mangrove wetlands.

There are ABRAE that are for forestry sustainable exploitation. Mangrove is an important economical resource.

If the purpose of the project is to exploit a mangrove area, Decree 110 interferes with the activity because there is a prohibition to cut mangroves. If the argument to issue the permission for a sustainable exploitation is that red mangrove is the species to protect and not the other species, the mangrove forest could be in danger because a mangrove ecosystem is fragile and vital habitat no matter the species that are present. Therefore, provisions to allow a rational sustainable exploitation for mangrove in areas with feasibility to do so are missing in the present law.

If the case study is on a tidal salt marsh the following argument can be used: a salt marsh can be defined as a "natural or semi-natural halophytic grassland and dwarf brushwood on the alluvial sediments bordering saline water bodies whose water level fluctuates either tidally or non-tidally" (Beeftink, 1977 in Mistch and Gosselink,1987) . This marshes have been found to be highly productive and to support the spawning and feeding habits of many marine organisms (Mitsch and Gosselink). Because of this function of a tidal salt marsh, section 25 of the Law for Fishing can be applied to protect the area. Also, the decision can be reinforced by the Law of Forestry, Soils and Waters by establishing a 50m contour inland, around the waterbody.

If the case study is on a freshwater marsh the situation is difficult specifically in relation to these ABRAE: forestry reserve, woods lots and zones for agriculture . Mitsch and Gosselink (1987) point out that: fresh water marshes are inland from saline parts. They are unique ecosystems that combine many features of salt marshes and freshwater marshes. Because the reduction of the salt stress, plant diversity is high and more birds use these marshes than any other marsh type. Nutrient cycles are open. They have the capability to retain, purify and gradually release flood water (Clark,1977).

There is not an specific regulation for tidal fresh wetlands as there is for the others to regulate the actions to be taken. The law for Forestry, Soils and Waters and its regulation state that: a protection zone should be created at a 50m or more from the shore line of a natural lagoon and the activities should be authorized by MARNR based on a technical study. Actually, this protection zone can be complemented with the one declared in Decree 623 (1989) to 80m wide form the high water mark. Therefore, the restriction to manage the wetland with a sustainable exploitation criteria depends on the knowledge that the team has of characteristics and importance of this ecosystem. This will allow them to judge whether or not the result of the study give a valid alternative. Therefore, the permission is at discretion of the team.

• Case 3: an area with no specific land use restrictions yet defined.

The policy advanced by MARNR on this case is "supply oriented. Accordingly, the amount of activity and its location on the coast will be decided largely on the basis of environmental characteristics and conditions of each section of the coast " (Amir, 1983).

An input to be used to solve the case is: the existing regulation and available State and local plans (planes de ordenación del territorio). Figure 1 shows the States that have coastal areas that already have a plan. The available plans are on a working scale of 1:250.000 and 1:100.000. For the team, these scales are awkward to look for environmental problems that could arise with the project in coastal areas. Also because of the spatial scales of representation, the land use suggested in the plans are not necessarily correct . Therefore, information at scales from 1:5.000 to 1:10.000 is badly needed. In this context the team would have a difficult time to issue a permission on wetlands areas and it will be based on their discretion.

In relation to mangrove areas, the arguments and problems mentioned above are valid for this case too.

In relation to tidal fresh marshes the situation is really difficult. To create the protected zone established by law, the limits have to be approved by a Decree. This process will take time and the team can not stop the process to wait for the law. The only way to impose restriction is to apply:

- section 21 of the Environmental Act and ask for an EIA, and

- section 76 of the Land use planning Act that requires to take in consideration the ecological limitations impose by wetlands.

Again, these tools are considered if the team is fully aware of what is a tidal fresh marsh and which is their importance.

The worst situation to evaluate is when the terrain is both in a mangrove wetland and in a tidal fresh marsh; the tidal fresh marsh would be the section to negotiate because of its lack of a specific regulation.

The situation is also difficult if the marsh is more that 80m wide from the highest water mark, because part of it is protected and another is not. This shows the need for an integral approach to manage coastal wetlands.

Ideas to improve the process of coastal wetland management

To improve the management of coastal areas two aspects seem to be of importance: to enhance the legal aspects and to fulfil the missing detailed information.

In relation to the legal aspect, it is important to:

- pass an Act to legally support the land use guidelines issued by MARNR in 1985 for ABRAE. This will allow to have a tool to enforce the application of proposed restrictions even though the land use plan and regulation are not yet being established.

- Revise Decree 110 for mangrove in order to protect equally all species of mangrove and the vital interactions with other ecosystem that maintains the system and, to establish provisions for sustainable exploitation of mangrove.

- Pass regulations that effectively and specifically protect tidal fresh water and its vital interactions with other adjacent systems.

In relation to the existing information, there must be a compromise to give to the working team a comprehensive and flexible working tool. This is specially needed for the coastal areas that do not have guidelines for land use clearly established.

To issue a permission, the chain of steps required to do so starts with the land use conformation in the pertinent administrative zone of MARNR. At this the moment it is crucial for the planner to have information available. In what follows the work done by Sebastiani (1989), for a protected zone with no plan or regulation set yet, is suggested as a complementary strategy for land use conformation.

The planning team requires a handbook, specifically for every administration zone, to be aware of, at least (Figure 2):

. Landscape units. The identification of the units should be available on maps at scales less than 1:10.000. This should have a complementary precise information, preferable in matrix format, of the relevant physical biotic characteristics and actual land use.

. Critical relationships between the landscape units to maintain the coastal system as a whole. Landscape units do not function isolated. Therefore the planning team should be aware of the input and outputs required for the system to work. This should be presented in a graphical form. This type of presentation is very useful to explain to the promoter of the project the arguments being used by the planner.

. Sensitivity of the landscape units. This could be based on the importance and condition of the unit. Sensitivity based on the importance could be addressed based on:

-the role of the unit in the coastal system as a whole. This should be based on technical information;

-the facts established by the existing regulation. This information should be organized in a table where an aspect of the regulation is related to the pertinent landscape unit.

-critical parameters to maintain the unit functioning. This information should be organized also as mentioned above.

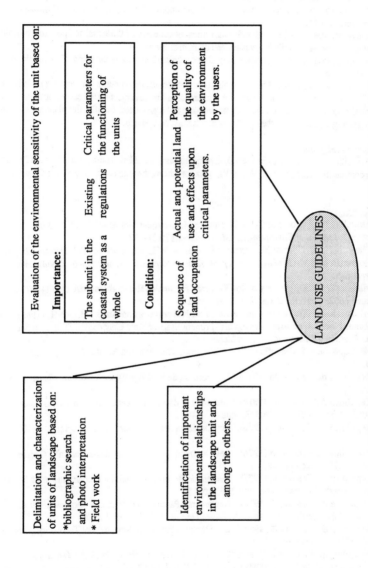

FIGURE 2. IDEAS TO PROPOSE LAND USE GUIDELINES (Sebastiani , 1989)

. Sensitivity because of condition could be addressed by identifying:
- the sequence of land occupation or cumulative impact. This could be done cartographically for different periods of time.
- the environmental effects related to the critical parameters, of the actual and potential modalities of land use. This also should be synthesized on matrix format;
-the perception of the environmental quality of the studied area by the users. This should be done based on questionnaires.
The planning team should have the highlights of sensitivity on maps at working scale less that 1:10.000. Thus the planner can overlap the selected area for the project on the cartography and search for spatial restrictions and vital interaction that have to be protected. Finally, the planning team must have a training course to use the handbook and it must be updated.

Acknowledgement
The authors are grateful to Ernesto Diaz from the Ministry of Environment and Natural Resources for the comments on the paper, to Victor Vásquez for editing the material and to Nicola Cerone for the drawing.

References
Amir,S.(1982). "Tools of Coastal Resource Management. Framework with Multiple Objectives". *Journal of Environmental Management*,17:121-132.
Clark,J.(1977). *Coastal Ecosystem Management*. Wiley- Interscience Publication, 928 pp.
Gomez,E. and V. Giménez. *Areas protegidas de Venezuela*. Cuadernos de Divulgación de Lagoven, S.A.
Ministerio del Ambiente y de los Recursos Naturales Renovables (MARNR) (1979). Dictamen 1602. Consultoría Jurídica, November 11.
Ministerio del Ambiente y de los Recursos Naturales Renovables (MARNR) (1985). "Plan del Sistema Nacional de Areas Protegidas" Primera Etapa. Marco Conceptual. Dirección Sectorial de Planificación y Ordenación del Ambiente.
Mitsch,W. and J. Gosselink (1987). *Wetlands*. Van Nostrand Reinhold Company. New York 539pp.
República de Venezuela (1966). "Ley Forestal de Suelos y Aguas". *Gaceta Oficial* #1004 E, January 26.
República de Venezuela (1974). "Decreto 110 de Protección de Manglares y sus Areas de Vecindad". *Gaceta Oficial* #30413, June 6.
República de Venezuela (1976). "Ley Orgánica del Ambiente". *Gaceta Oficial* #31004, June 16.
República de Venezuela (1976). "Ley Orgánica de la Administración Central". *Gaceta Oficial* #1932E, December 28.
República de Venezuela (1977). "Decreto de Reforma Parcial del Reglamento de la Ley de Suelos y Aguas". *Gaceta Oficial* #2022E, April 28.
República de Venezuela (1978). "Reglamento Parcial No.4 de la Ley Orgánica del Ambiente sobre Clasificación de las Aguas". *Gaceta Oficial* #2323E, October 20.
República de Venezuela (1983). "Ley Orgánica para la Ordenación del Territorio". *Gaceta Oficial* #31363, November 17.
República de Venezuela (1989). "Decreto 623. *Gaceta Oficial* #4158E, January 25.
República de Venezuela (1989). "Decreto 276 Reglamento Parcial de la Ley Orgánica para la Ordenación del Territorio sobre Administración y Manejo de Parques Nacionales y Monumentos Naturales". *Gaceta Oficial* , September 6.
Sebastiani,M.(1989). "Metodología para Identificar Areas Sensibles con fines de Ordenamiento Territorial". Universidad Simón Bolívar (Trabajo de Ascenso, Mimeograph).
Sorensen,J. and A. Brandani (1987). "An Overview of Coastal Management Efforts in Latin America". *Coastal Management*,15:1-25.

SUBJECT INDEX
Page number refers to first page of paper.

Accretion, 193
Aerial photography, 281, 347, 362
Alaska, 1
Aquatic plants, 16, 247, 259, 281, 347, 439, 454, 477

Bays, 75, 125, 193, 424, 463
Beach erosion, 193, 215

California, 16, 418, 477
Cartography, 268
Chemicals, 84
Clean Water Act, 75, 301
Coastal environment, 31, 61, 75
Coastal management, 1, 100, 142, 207, 215, 235, 247, 332, 384, 405, 454, 490, 503
Compliance, 127
Contaminants, 84
Contamination, 164

Deltas, 85, 157, 418
Demographic projections, 46
Design, 181
Development, 31, 490
Discharge, 75
Diseases, 84
Diversion, 157
Dredging, 164, 424, 454, 463

Ecology, 31, 164, 226
Economics, 164
Ecosystems, 85, 317
Environmental impacts, 207
Environmental issues, 490
Environmental planning, 142
Environmental Protection Agency, 301
Erosion, 454
Estuaries, 1, 46, 61, 75, 85, 100, 125, 142, 439, 463
Eutrophication, 61, 125
Evaluation, 31

Federal-state relationships, 207
Fish habitats, 46, 235, 259, 418
Fish management, 259
Fish protection, 84
Fisheries, 16, 46, 110, 259, 317, 418
Florida, 347
Fresh water, 85, 215

Geographic information systems, 247, 332, 376
Geography, 268
Geomorphology, 110
Global positioning, 332
Grasses, 16, 281, 439, 454, 477
Gulf of Mexico, 215, 376
Gulfs, 1

Harbors, 477
Heavy metals, 127
Hydraulic design, 181
Hydrology, 176, 226, 405

Inventories, 376
Islands, 418

Land usage planning, 31, 503
Land usage regulations, 301
Losses, 100
Louisiana, 215, 235, 317, 405

Management, 176
Mapping, 247, 268, 281, 332, 347, 362, 376
Marshes, 110, 176, 193, 226, 235, 317, 405, 424
Methodology, 362, 490
Mississippi River, 157
Monitoring, 75, 281, 405, 463, 477

National Oceanic and Atmospheric Administration, 259
Natural resources, 317
Nitrogen, 61
Nonpoint pollution, 142
Numerical models, 181
Nutrient loading, 439
Nutrients, 61, 125

Ocean environments, 247
Oceans, 376, 490
Offshore drilling, 384

Permits, 226
Petroleum, 207, 384
Planning, 384
Ports, 384
Productivity, 235, 405
Programs, 268
Public participation, 384

Regional planning, 362
Regulation, 75, 207, 463
Regulations, 127
Remote sensing, 259, 268, 332, 347
Resource management, 332, 347, 454
Restoration, 164, 176, 418, 439
Rivers, 85
Runoff, 85

San Francisco, 125, 424
Sea level, 110
Sediment, 84, 164
Sediment deposits, 157
Sedimentation, 424
Sewage, 61
Shellfish, 110
Shore protection, 503
Shoreline changes, 193
Simulation models, 226

Texas, 193
Tidal marshes, 215

Tidal waters, 181
Toxic wastes, 84
Turbidity, 193

United States, 100, 125
U.S. Army Corps of Engineers, 301

Vegetation, 16, 226, 424, 463
Venezuela, 503

Wastewater disposal, 127
Wastewater treatment, 127
Water pollution, 46, 85, 301, 439
Water quality, 61, 142, 176
Wetlands, 1, 16, 31, 46, 100, 164, 176, 181, 207, 215, 226, 247, 259, 268, 301, 317, 347, 362, 376, 384, 405, 418, 463, 503
Wildlife habitats, 1, 418
Wildlife management, 157, 235

Zoning, 181